Mycoplasmas in swine

Edited by

Dominiek Maes, Marina Sibila & Maria Pieters

Mycoplasmas in swine

CABI

Published exclusively in the Benelux region (Belgium, The Netherlands and Luxembourg)
ACCO, with ISBN 9789463797962

Academische Coöperatieve Vennootschap cvba
Uitgeverij Acco
Sluisstraat 10
3000 Leuven
Belgium
E-mail: uitgeverij@acco.be
Website: www.acco.be

Published exclusively throughout the world excluding the Benelux region (Belgium, The Netherlands and Luxembourg) by *CABI*, with:
ISBN 9781789249941 (Hbk)
ISBN 9781789249958 (epdf)
ISBN 9781789249965 (epub)

CABI
Nosworthy Way
Wallingford
Oxfordshire OX10 8DE
UK
Tel: +44 (0)1491 832111
Fax: +44 (0)1491 833508
E-mail: info@cabi.org
Website: www.cabi.org

CABI
WeWork
One Lincoln St
24th Floor
Boston, MA 02111
USA
Tel: +1 (617)682-9015
E-mail: cabi-nao@cabi.org

SILVER PARTNER

ECO

BRONZE PARTNERS

HUVEPHARMA®

BioChek

BIOCHEK, SMART VETERINARY DIAGNOSTICS

HIPRA

The Reference
in Prevention
for Animal Health

Contents

CONTRIBUTORS

Editors

Dominiek Maes
Faculty of Veterinary Medicine
Ghent University
Ghent, Belgium
Dominiek.Maes@UGent.be

Marina Sibila
Centre de Recerca en Sanitat Animal (CReSA)
Institut de Recerca i Tecnologia Agroalimentàries (IRTA)
Campus de la Universitat Autònoma de Barcelona
Bellaterra, Spain
marina.sibila@irta.cat

Maria Pieters
College of Veterinary Medicine
University of Minnesota
St. Paul, USA
piet0094@umn.edu

Authors

Alyssa Betlach
College of Veterinary Medicine
University of Minnesota
St. Paul, USA

Swine Vet Center, P.A.
St. Peter, USA

Anne Gautier-Bouchardon
Ploufragan-Plouzané-Niort Laboratory
French Agency for Food, Environmental and Occupational Health and Safety (Anses)
Ploufragan, France
Anne.bouchardon@anses.fr

Filip Boyen
Faculty of Veterinary Medicine
Ghent University
Ghent, Belgium
Filip.Boyen@UGent.be

John Carr
College of Public Health, Medical and Veterinary Sciences
James Cook University
Queensland, Australia
swineunit1@yahoo.com

Chanhee Chae
College of Veterinary Medicine
Seoul National University
Seoul, Republic of Korea
swine@snu.ac.kr

Céline Deblanc
Swine Virology and Immunology Unit
French Agency for Food, Environmental and Occupational Health and Safety (Anses)
Ploufragan, France
Celine.deblanc@anses.fr

Odir Antonio Dellagostin
Unit of Biotechnology
Federal University of Pelotas
Pelotas, Brazil
odir@ufpel.edu.br

Steven Djordjevic
The ithree institute
University of Technology Sydney
Sydney, Australia
Steven.Djordjevic@uts.edu.au

Christelle Fablet
Ploufragan-Plouzané-Niort Laboratory
French Agency for Food, Environmental and Occupational Health and Safety (Anses)
Ploufragan, France
Christelle.Fablet@anses.fr

João Carlos Gomes Neto
Nebraska Innovation Campus
University of Nebraska-Lincoln
Lincoln, USA
jgomesneto2@unl.edu

Freddy Haesebrouck
Faculty of Veterinary Medicine
Ghent University
Ghent, Belgium
Freddy.Haesebrouck@UGent.be

Katharina Hoelzle
Department Behavioral physiology of livestock
University of Hohenheim
Stuttgart, Germany
Katharina.Hoelzle@uni-hohenheim.de

Ludwig Hoelzle
Department Livestock infectiology and environmental hygiene
University of Hohenheim
Stuttgart, Germany
ludwig.hoelzle@uni-hohenheim.de

Sam Holst
Swine Vet Center, P.A.
St. Peter, USA
sholst@swinevetcenter.com

Derald Holtkamp
College of Veterinary Medicine
Iowa State University
Ames, USA
holtkamp@iastate.edu

Veronica Jarocki
The ithree institute
University of Technology Sydney
Sydney, Australia
Veronica.Jarocki@uts.edu.au

Jörg Jores
Vetsuisse Faculty
University of Bern
Bern, Switzerland
joerg.jores@vetsuisse.unibe.ch

Peter Kuhnert
Vetsuisse Faculty
University of Bern
Bern, Switzerland
peter.kuhnert@vetsuisse.unibe.ch

Enrique Marco
Marco VetGrup SLP
Barcelona, Spain
emarco@marcovetgrup.com

Corinne Marois
Ploufragan-Plouzané-Niort Laboratory
French Agency for Food, Environmental and Occupational Health & Safety (Anses)
Ploufragan, France
Corinne.Marois@anses.fr

Heiko Nathues
Vetsuisse Faculty
University of Bern
Bern, Switzerland
heiko.nathues@vetsuisse.unibe.ch

Tanja Opriessnig
College of Veterinary Medicine
Iowa State University, USA

The Roslin Institute and The Royal (Dick) School of Veterinary Studies
University of Edinburgh, UK
Tanja.Opriessnig@roslin.ed.ac.uk

Andreas Palzer
Veterinary Pig Practice Scheidegg
Scheidegg, Germany
Andreas.Palzer@med.vetmed.uni-muenchen.de

Mathias Ritzmann
Faculty of Veterinary Medicine
Ludwig-Maximilians-Universität München
München, Germany
Ritzmann@med.vetmed.uni-muenchen.de

Andrew Rycroft
Department of Pathobiology & Population Sciences
Royal Veterinary College
London, UK
ARycroft@rvc.ac.uk

Joaquim Segalés
Facultat de Veterinària (Universitat Autònoma de Barcelona)
Centre de Recerca en Sanitat Animal (CReSA)-Institut de Recerca i Tecnologia Agroalimentàries (IRTA)
Campus de la Universitat Autonoma de Barcelona
Bellaterra, Spain
joaquim.segales@irta.cat

Guoqing Shao
Institute of veterinary science
Jiangsu Academy of Agriculture Sciences
Nanjing, Jiangsu, China
gqshaonj@163.com

Joachim Spergser
University of Veterinary Medicine Vienna
Vienna, Austria
Joachim.Spergser@vetmeduni.ac.at

Artur Summerfield
Institute of Virology and Immunology
Faculty of Veterinary Medicine
University of Bern
Bern, Switzerland
artur.summerfield@vetsuisse.unibe.ch

Paul Yeske
Swine Vet Center, P.A.
St. Peter, USA
pyeske@swinevetcenter.com

Reviewers

Rachel Derscheid
Veterinary Diagnostic Laboratory
Iowa State University
Ames, USA
rdersch@iastate.edu

Mathias Devreese
Faculty of Veterinary Medicine
Ghent University
Ghent, Belgium
Mathias.Devreese@UGent.be

Bert Devriendt
Faculty of Veterinary Medicine
Ghent University
Ghent, Belgium
B.Devriendt@UGent.be

Jeroen Dewulf
Faculty of Veterinary Medicine
Ghent University
Ghent, Belgium
Jeroen.Dewulf@UGent.be

Marcelo Gottschalk
Faculty of Veterinary Medicine
University of Montreal
Québec, Canada
marcelo.gottschalk@umontreal.ca

Roberto Maurício Carvalho Guedes
Veterinary School
Universidade Federal de Minas Gerais
Belo Horizonte, Brazil
guedesufmg@gmail.com

Luís Guilherme de Oliveira
School of Agricultural and Veterinarian Sciences
São Paulo State University (Unesp)
Jaboticabal, Brazil
luis.guilherme@unesp.br

Isabel Hennig-Pauka
Field Station for Epidemiology
University of Veterinary Medicine Hannover
Bakum, Germany
Isabel.Hennig-Pauka@tiho-hannover.de

Paolo Martelli
Department of Veterinary Science
University of Parma
Parma, Italy
paolo.martelli@unipr.it

Guy-Pierre Martineau
National Veterinary School of Toulouse
Toulouse, France
g.martineau@envt.fr

Chris Minion
Veterinary Medicine
Iowa State University
Ames, USA
fcminion@iastate.edu

Jens Peter Nielsen
Det Sundhedsvidenskabelige Fakultet
Københavns Universitet
Copenhagen, Denmark
jpni@sund.ku.dk

Katharina Stärk
Royal Veterinary College
London, UK
kstaerk@rvc.ac.uk

Karine Ludwig Takeuti
Federal University of Rio Grande do Sul, Brazil
Porto Alegre, Brazil
karine.takeuti@ufrgs.br

Pablo Tamiozzo
Facultad de Agronomía y Veterinaria
Universidad Nacional de Río Cuarto
Río Cuarto, Argentina
topo.vet@gmail.com

Tijs Tobias
Faculty of Veterinary Medicine
Utrecht University
Utrecht, The Netherlands
t.j.tobias@uu.nl

Per Wallgren
Dept of Animal Health and Antimicrobial Strategies
National Veterinary Institute
Uppsala, Sweden
per.wallgren@sva.se

PROLOGUE

Glenn F. Browning

Mycoplasmas are among the most important bacterial pathogens of pigs. They establish chronic infections that are difficult to eliminate, from the pig and the farm. They result in significant economic loss, predispose pigs to disease caused by more acute pathogens, and have an impact on the welfare of the pigs.

Control of mycoplasmosis in the intensive chicken industry has reached a point where antimicrobial therapy for mycoplasmas is rarely required in many countries. This has been achieved by establishing mycoplasma-free breeding stock, by development of accurate diagnostic tests, implementation of regular testing, development of vaccines that limit infection, eliminate its effects and reduce transmission, and by implementing rigorous biosecurity protocols. Control of mycoplasmoses is an important goal for the global pig industry, and recent eradication efforts attest to the level of determination in some countries to achieve this goal. However, the limitations we still face in effectively controlling porcine mycoplasmoses is one of the leading reasons for using antimicrobials in pig farms. Total disease control should be achievable with sufficient ongoing research and development to enhance our tools for control and their application in the field.

Better management of the mycoplasmoses in pigs requires a comprehensive approach. We need better understanding of many fundamental aspects of the biology of these pathogens. While there are several pathogenic mycoplasmas of pigs, *Mycoplasma hyopneumoniae,* is undoubtedly the most important. *M. hyopneumoniae* was first identified in the 1960s (Goodwin et al., 1967; Goodwin and Whittlestone, 1963; 1966; Mare and Switzer, 1966), although the disease it causes, enzootic pneumonia, was recognized as a distinct entity in the early 1950s (and named viral pneumonia of pigs at the time) (Gulrajani and Beveridge, 1951).

ABBREVIATIONS

ADG	average daily gain
ADWG	average daily weight gain
AFLP	amplified fragment length polymorphism
AMP	antimicrobial proteins
BALF	broncho-alveolar lavage fluid
BALT	bronchus-associated lymphoid tissue
BIP	broncho-interstitial pneumonia
BM-DC	bone-marrow-derived dendritic cell
CA-SFM	Comité de l'Antibiogramme de la Société Française de Microbiologie
CBPP	contagious bovine pleuropneumonia
CCU	color changing unit
CD	coding sequence
CDCD	caesarian derived colostrum deprived
CDS	cytosolic DNA sensor
CF	continuous flow
CFU	colony forming unit
CLR	C-type lectin like receptor
CLSI	Clinical and Laboratory Standards Institute
CRAMP	cathelin-related antimicrobial peptide
CT	cholera toxin
Ct	cycle threshold
CVPC	cranioventral pulmonary consolidation
DC	dendritic cell
DIC	disseminated intravascular coagulation
DNA	deoxyribonucleic acid
DON	deoxynivalenol
DPI	days post inoculation
DUF	domains of unknown function
ECOFF	epidemiological cut-off
EF-Tu	elongation factor thermo unstable
ELISA	enzyme linked immunosorbent assay
EP	enzootic pneumonia
EUCAST	European Committee on Antimicrobial Susceptibility Testing
Fc	fragment crystallizable
FCR	feed conversion ratio
FIGE	field-inversion gel electrophoresis
GAG	glycosaminoglycans
GAPN	glyceraldehyde-3-phosphate dehydrogenase
GDU	gilt developer unit

GLpO	glycerolphosphate-oxidase
HI	hemagglutination inhibition
HSP70	heat shock protein 70
IAP	infectious anemia of pigs
IAV-S	influenza A virus of swine
ICE	integrative conjugative element
ID	intradermal
IFA	immuno-fluorescence assay
IFN-γ	interferon gamma
Ig	immunoglobulin
IHC	immunohistochemistry
Il	interleukin
IM	intramuscular
IN	intranasal
IRPCM	International Research Programme on Comparative Mycoplasmology
IS	insertion sequence
ISH	*in situ* hybridization
LAMP	lipid associated membrane proteins
LDH	lactate dehydrogenase
LPS	lipopolysaccharide
LTB	heat-labile enterotoxin
MALDI-ToF	matrix-assisted laser desorption-ionization – time of flight
MAMPS	microbial-associated molecular patterns
MBC	minimum bactericidal concentration
MDA	maternally derived antibodies
MGE	mobile genetic elements
MHC	major histocompatibility complex
MIB	mycoplasma Ig binding protein
MIC	minimum inhibitory concentration
MIP	mycoplasma Ig protease
MLST	multi-locus sequencing typing
MLV	modified live virus
MLVA	multiple-locus variable number tandem repeats analysis
NET	neutrophil extracellular traps
NLR	nucleotide-binding oligomerization domain-like receptors (NLR)
NOD	nucleotide-binding oligomerization domain (NOD)-like receptors (NLR)
NrdF	ribonucleoside-diphosphate reductase subunit beta
OR	odds ratio
ORF	open reading frame

PCMV	porcine cytomegalovirus
PCR	polymerase chain reaction
PCV-2	porcine circovirus type 2
PCVD	porcine circovirus disease
PdhB	pyruvate dehydrogenase E1 beta subunit
PFGE	pulsed-field gel electrophoresis
pH	power of hydrogen
PLGA	polylactic-co-glycolic acid
PM	porcine mycoplasmosis
PPLO	pleuropneumonia-like organism
PRDC	porcine respiratory disease complex
PRR	pathogen recognition receptor
PRRSV	porcine reproductive and respiratory syndrome virus
PRV	pseudorabies virus
qPCR	quantitative PCR
QRDR	quinolone resistance-determining regions
RAPD	random amplified polymorphic DNA
RFLP	random fragment length polymorphism
RIG-I	retinoic acid-inducible gene-I
RLR	retinoic acid-inducible gene-I-like receptor
Rn	reproduction ratio
RNA	ribonucleic acid
Rt-PCR	real time PCR
ROS	reactive oxygen species
SC	subcutaneous
SCID	severe combined immunodeficiency
SNP	single nucleotide polymorphism
SP-A	surfactant protein
SPF	specific-pathogen-free
ST	sequence type
TBM	tracheobronchial mucus
Th	helper T cells
TLR	toll-like receptor
TNF	tumor necrosis factor
T-reg	regulatory T cells
VLP	variable surface lipoprotein
VNTRs	variable number of tandem repeats
WGS	whole genome sequencing
WHO	World Health Organization

CHAPTER 1

OVERVIEW OF THE GENERAL CHARACTERISTICS AND CLASSIFICATION OF PORCINE *MYCOPLASMA* SPECIES

Andrew Rycroft[1]

1 Department of Pathobiology & Population Sciences
Royal Veterinary College, London, UK

1.1 Introduction

1.1.1 Characteristics of *Mollicutes* and mycoplasmas

The bacteria which permanently lack cell wall peptidoglycan are placed in the class *Mollicutes* (Latin for soft skin). This is a distinct class of prokaryotes which includes genera such as *Mycoplasma* and *Ureaplasma*. They are free-living prokaryotes with small cells (0.2 – 0.4 μm) and are pleomorphic because they have no shape-defining cell wall. They also tend to produce very small colonies in culture. These bacteria carry a small genome (usually between 700 and 1000 kbp). This presumably confers only a relatively limited anabolic and catabolic ability compared to many bacteria. This may reflect the failure to acquire or evolve more functions or, alternatively, the loss of genetic capacity as the organism has degenerated to rely on its close association with the host.

The first mollicute to be successfully cultured was *Mycoplasma mycoides*, the organism we now recognize as the causative agent of bovine pleuropneumonia. Later isolations of mycoplasmas or other *Mollicutes* resembling *M. mycoides* were referred to as pleuropneumonia-like organisms or PPLO. This name was widely used but was gradually replaced by the term *Mycoplasma* for all bacteria of this type causing disease in humans and animals. The term *Mollicutes* was agreed for the entire class of organisms in 1967 but the word "*Mycoplasma*" that in the correct sense only refers to the genus, still persists among biologists as the general or trivial term for these wall-less bacteria in many situations today.

1.1.2 Phylogenetic relationships of mycoplasmas

Until the advent of molecular (DNA-based) investigations, the classification of *Mollicutes* was reliant on culture-based characteristics and serology. Often growth requirements in culture were difficult to determine because successful culture usually requires animal serum. Since serum contains a wide variety of undefined chemicals and nutrients, it was usually not possible to recognize which components were essential requirements and which were not. The growth-inhibition test became widely established as the method of choice for recognition and classification of *Mollicutes*. This used antibody in hyperimmune serum, usually applied to an absorbent paper disc, to inhibit the growth of a mollicute on a culture plate. This activity was specific and only caused the inhibitory effect on strains of the same species or those very closely related to the strain used as antigen to raise the hyperimmune serum. Unfortunately, the method depended on the sharing of sera and was not always reproducible between laboratories. It was also dependent upon the key antigen or antigens for inhibition being present on all members of the species but not on any other species. This worked well within the boundaries of a laboratory dealing with a relatively restricted range of organisms and with known anti-sera. But as the number of possibilities increased, the number of sera needed also increased, as did the possibility of growth-restricting antigens being shared with unrelated organisms.

With the availability of DNA sequence analysis, particularly that encoding ribosomal RNA, estimates could be made of evolutionary distance between species. A number of attempts to do this have been made, perhaps with increasing accuracy, and these have greatly helped our understanding of the relationships between *Mollicutes* (Toth et al., 1994).

Mollicutes are sometimes described in the literature as Gram-negative, but the reason for this is unclear, and it is mistaken, since they possess neither outer membrane nor lipopolysaccharide, the key elements of a Gram-negative cell envelope. Almost certainly, *Mollicutes* are distantly related to the Gram-positive bacteria (*Firmicutes*, Latin: firm skin). Evidence for this was first published by Woese et al. (1980) using 16S rRNA sequence analysis. They concluded that the *Mollicutes* (*Mycoplasma*, *Spiroplasma* and *Acholeplasma*) arose by "degenerative" or "regressive" evolution. This apparently happened in a branch of *Prokaryotes* related to ancestors of *Clostridium* that led to the genera *Bacillus* and *Lactobacillus*. There remained considerable uncertainty from this study and the authors concluded that the *Mollicutes* were "not a phylogenetically coherent group" but they were all related to the family of Gram-positive bacteria, the *Bacillaceae*.

Revision and refinement of this analysis of 16S rRNA sequence by Weisburg et al. (1989) and Manilov (1992) from 48 different species of *Mollicutes* led to the recognition of 5 distinct phylogenetic groups or clades. These were the *Spiroplasma*, *Hominis* and *Pneumoniae* groupings together with the *Anaeroplasma* group (which included *Acholeplasma*) and a separate group containing the single species *Asteroleplasma anaeroboium*.

Phylogenetic relatedness based on the analysis of 16S rRNA is not reflected in the characteristics most easily seen by a bacteriologist. Thus, the site of disease, the ease and speed of growth in culture, nutritional requirements and the colonial appearance have all been thought of as meaningful in showing the similarity or differences between species of *Mollicutes*. Nevertheless, the DNA evidence must carry considerable weight and the superficial phenotypic features that were previously considered important in classifying these organisms are now seen to be of little significance in the phylogenetic relationships of these organisms (Figure 1.1).

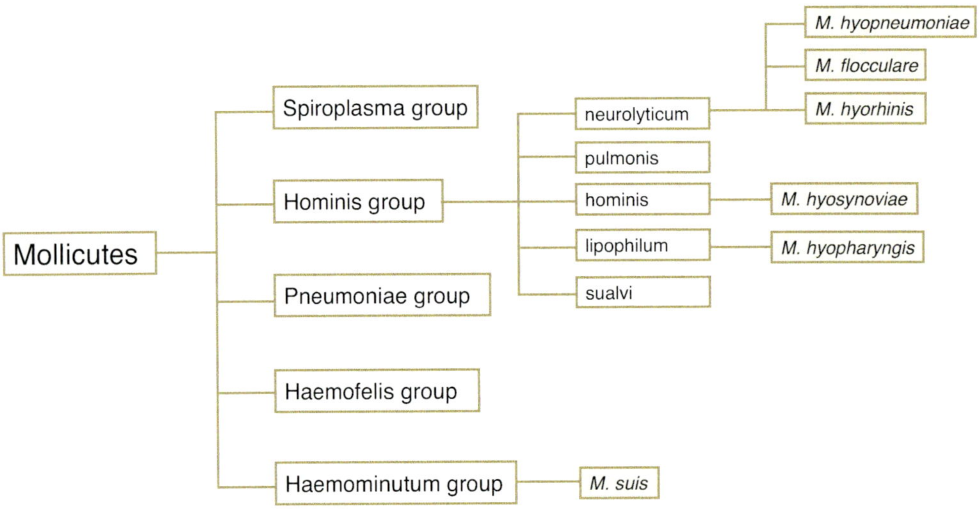

Figure 1.1. Simplified groupings of porcine *Mycoplasma* species based on the 16S rRNA derived groups and clusters of Weisburg et al. (1989) but modified to include results of the hemotropic mycoplasmas from Peters et al. (2008) and Siqueira et al. (2013).

The *Spiroplasma* group consists of four clusters, one of which is the agent of contagious bovine pleuropneumonia (CBPP) *M. mycoides,* and another is *M. capricolum*. The majority of members known to be in this group are the helical spiroplasmas: usually found on plants, insects and arachnids.
The *Pneumoniae* group includes the human pathogen *M. pneumoniae* and the avian respiratory pathogen *M. gallisepticum* among others.
The *Hominis* group includes many of the mollicute animal pathogens including the porcine mycoplasmas.

It has been suggested that the evolutionary distance from Gram-positive bacteria must be very great. One characteristic of mycoplasmas that speaks volumes is the differential use of the TGA codon. Bacteria of almost all types use TGA (UGA) as a stop codon while all mycoplasmas (excluding acholeplasmas and phytoplasmas) use this to code for the amino acid tryptophan. Such a fundamental difference that is seen so widely among the

Mollicutes indicates a long isolation of these organisms from other bacteria. However, Weisburg et al. (1989) proposed that from the molecular evidence of DNA sequence, *Mollicutes* appear to be normal bacteria and they were described as having an "unspectactular" phylogenetic position. Despite the very different phenotypic appearance arising from the lack of peptidoglycan, the evidence suggests that this is not because *Mollicutes* are phylogenetically distant from other eubacteria.

Another characteristic of *Mollicutes* that has come to light with genome sequencing and annotation, is the apparent lack of pseudogenes or intermittent junk DNA. The presence of "rusting hulks of working genes", as Steve Jones refers to them (Jones, 1993), might be expected if a proportion of the genetic makeup of Gram-positive bacteria was no longer required and had become redundant in *Mollicutes*. Perhaps, and we can only speculate, genome sized reductions have allowed removal of unnecessary, non-functional DNA more efficiently than appears to happen in eukaryotes particularly.

Yet another characteristic of *Mollicutes* is their inclusion of cholesterol in their cytoplasmic membrane. This is thought to give some stability to the membrane. Most *Mollicutes* acquire the cholesterol from an exogenous source. Acholeplasmas (which also lack peptidoglycan) do not require exogenous cholesterol because they have the capacity to synthesize it and incorporate this in their membrane for stability (Khan et al., 1981).

Finally, a characteristic of mycoplasmas is their very close association, in many cases, with other organisms (Razin et al., 1998). Their niche is to have an intimate contact with the mucosal surfaces of animals and plants. It is this very close association with host cells, of which we understand relatively little, that is the key to understanding mycoplasmas. These organisms have evolved alongside animals and plants to be "quiet" intruders: living, like many commensal organisms, without prompting a response from the host nor causing it perceptible damage. Some, of course, do prompt an inflammatory response and these we see as the pathogens. The reason for those mycoplasmas causing a response may be to assist in their spread to other animals or it may be accidental as environments change or the genetics and physiology of the host alters in some way. It is the subtleties of this close association between host and *Mycoplasma*, and perturbation of the balance in that close relationship, that will provide the solution to understanding mycoplasmas and *Mycoplasma* pathogenicity in the future.

1.1.3 Pathogenicity in mycoplasmas

Specific aspects of pathogenicity of porcine mycoplasmas will be dealt with in later chapters. In general, however, the mechanisms by which mycoplasmas cause disease often remain either obscure or speculative. Because mycoplasmas are so different from many other pathogens studied there is limited opportunity to recognize homologues of genes known in other pathogens. It is not surprising that biosynthetic genes for amino acids and growth factors are missing; but global regulatory genes such as the sigma factor *rpoS*, transcriptional repressor *crp* and the leucine-responsive regulatory protein *lrp*, are also absent (Salyers and Whitt, 2002). It is therefore likely that mycoplasmas have independently developed their own regulatory systems. Similarly, they could have as yet unrecognized functions acting to engage with host cells, enable their survival *in vivo* and promote transmission between hosts.

It appears clear that adhesion to mucosal surfaces is an important aspect of colonization of host tissues. Mycoplasmas are extracellular surface parasites although some have been reported to get inside host cells (Yavlovich et al., 2004; McGowin et al., 2009; Burki et al., 2015). Because mycoplasmas lack the peptidoglycan of a cell wall or the lipopolysaccharide of the outer membrane, they present a very different exposed surface to that of most bacteria. The antigenic composition and topography of the single plasma membrane must be critical in the avoidance of recognition and evasion of damage from the innate immune response of the host. Some important progress has been made in understanding the role of this membrane surface (as will be detailed in later sections) but investigations of the host-pathogen relationship of mycoplasmas and their animal hosts is difficult and much remains to be understood at the cellular and molecular level.

Early suggestions of one or more toxins produced by *M. hyopneumoniae* (Debey and Ross, 1994) have not yet been corroborated. It was suggested that cytopathological effects observed from the human pathogen *M. pneumoniae*, including loss of cilia, was likely to be related to elaboration of a toxic substance or enzyme by Gabridge et al. (1974). However, the existence of any mycoplasma-derived protein exotoxins is not yet established. In contrast, the involvement of small molecules such as H_2O_2 in cellular damage has gained some support, at least in specific (bovine) mycoplasmas but also in porcine mycoplasmas (Galvao Ferrarini et al., 2018). In contrast, mutants of *M. gallisepticum* unable to produce H_2O_2 from glycerol were consistently virulent in the respiratory tracts of experimental chickens implying that

H_2O_2 is simply not required for this organism to cause disease, at least in some circumstances (Szczepanek et al., 2014). The role of membrane phospholipases, perhaps acting on the host cell membrane, is still unclear (Shibata et al., 1995; Rottem and Naot, 1998).

What is clear is that mycoplasmas are stealthy. They are able to persist in the body of the host, either on the mucosal surface or even, it appears, in major organs, without alerting the innate immune system to their presence. Or perhaps the immune system is alerted, but then suppressed or silenced. This is seen in the poor immunological responses to infection and in the relatively limited inflammatory response. Related to this is the antigenic variation displayed by many *Mycoplasma* organisms such as *M. hyorhinis* (Wise et al., 1992). The importance of altering surface antigens over time and the role it might play in evasion of the host immune defenses remains to be seen.

For many years it had been recognized that mycoplasmas bind immunoglobulin onto their surface. This was considered likely to be a passive means of coating the bacteria with "self" antigen and thereby evading recognition by the innate immune system. More recently, a two-protein immunoglobulin (Ig) binding system was described by Arfi et al. (2016). This was originally found in the ruminant mycoplasma *M. mycoides* subsp. *capri*. One protein, the mycoplasma immunoglobulin binding protein (MIB), captures antibody molecules with high affinity while the second, a serine protease known as mycoplasma Ig protease (MIP), cleaves the heavy chain of the Ig molecule. Analysis of genomes suggests that this system is widespread among the *Mycoplasma* pathogens and may contribute to avoidance of recognition and consequent immune evasion by mycoplasmas invading the body.

One important aspect of *Mycoplasma* pathogenicity that has become apparent is the lack of ability of phagocytic cells (neutrophils and macrophages) to eliminate these pathogens in the body. The reason for this is not yet known but mycoplasmas appear to be hiding effectively from the cells that are intended to destroy them. In some cases there is a clear cytokine response to *Mycoplasma* infection (Rottem and Noat, 1998). Lymphocytes are activated but sometimes in an inappropriate manner such that B cells are non-specifically stimulated and generate ineffective antibodies. Perhaps this is a "deliberate" attempt to subvert the immune response as can be seen in other bacterial pathogens. In *Staphylococcus aureus* and *Streptococcus equi* superantigens such as TSST-1 are produced. These directly activate T cells without the requirement for an antigen to be presented in the context of

Class II MHC. To date, only the rodent pathogen *M. arthritidis* is recognized as producing such a superantigen (*M. arthritidis* mitogen) but the role of this component in pathogenesis of disease in mice is apparently minimal (Luo et al., 2008). Alternative mechanisms of actively diverting the immune response may be operating and these may await discovery in the future.

1.2 The mycoplasmas of the pig

In the pig, the *Mollicutes* are important pathogens. There are six species of *Mycoplasma* that are uniquely or most commonly found in the pig. These are *M. hyopneumoniae*, *M. hyorhinis*, *M. flocculare*, *M. hyosynoviae*, *M. hypopharyngis* and *M. suis*. In addition, some species of *Acholeplasma* (such as *A. laidlawii*, *A. granularum* and *A. axanthum*) may be present in the pig but also occur elsewhere and are of no known significance in porcine disease. Based on comparison of the 16S rRNA sequences, five of these are phylogenetically related within the *Hominis* group (Figure 1.1).

Most of these species have been studied over the last 60 years or so. That mode of study has changed as understanding has grown and new techniques have become available. Some porcine mycoplasmas are relatively easy to culture *in vitro* but others are really quite difficult and the hemoplasma *M. suis* is so far uncultivable. Add to this the problem of identification, particularly between different laboratories using different reagents, and the chances of confusion increase. This has led, unfortunately, to some misunderstanding and misrepresentation of *Mycoplasma* species with isolates being misidentified and properties of one having been attributed to another. With the advent of PCR-based identification (Stemke et al., 1994; Stemke, 1997) rather than the traditional method of growth inhibition by hyperimmune serum, the community is now in a far clearer position. However, there are certainly reports in the literature and isolates in collections that are incorrect and there is a need for research laboratories to be aware of this when faced with contradictory results or incomplete understanding.

1.3 Phylogenetic relationship between porcine mycoplasmas

As noted above, five of the porcine mycoplasmas fall into the same phylogenetic group, the *Hominis* group. Analysis of the 16S rRNA sequence alignments suggested a close relationship between *M. hyopneumoniae, M. hyorhinis* and *M. flocculare* which caused them to be placed together in the *Neurolyticum* cluster of the *Hominis* group as described by Weisburg (Stemke et al., 1992). *M. hyosynoviae* is phylogenetically distinct and falls into the *Hominis* cluster; *M. hyopharyngis* is associated with the *Lipophilum* cluster. The *Hominis* group was further investigated using 16S rRNA sequence (Pettersson et al., 2000). This added further species to the group and reinforced the notion of the *Hominis* cluster as a distinct clade of *Mycoplasma* species within the *Hominis* group (Figure 1.1).

Recent, very detailed analysis of the porcine mycoplasmas by Siqueira et al. (2013) used a genome-wide analysis of the species of *Mycoplasma*. This placed the porcine species that were included in the study into a small *hyopneumoniae* clade consisting of closely-related *M. hyorhinis*, *M. conjunctivae*, *M. flocculare* and *M. hyopneumoniae*. It was shown that *M. hyorhinis* was basal (ancestral) to the other three species in all of the analyses. This *hyopneumoniae* clade was still within the same major grouping as *M. hominis*, but it was concluded that the *Hominis* grouping was separate. Later genomic analysis of the entire *Mollicutes* by Gupta et al. (2018) has proposed substantial changes to the overall taxonomy of the *Mollicutes*. This has confirmed the close relationship of the 4 species in the *hyopneumoniae* group but again puts *M. hyosynoviae* into a separate grouping more closely associated with *M. hominis*.

M. suis, as might be anticipated, falls into a different family, of hemoplasmas, taxonomically in the *Pneumoniae* group of mycoplasmas (Neimark et al., 2001). This was revised by Peters et al. (2008), again using 16S rRNA but also with *rnpB* sequences. From this work, the *Hemoplasma* group was considered distinct from the *Pneumoniae* group and was subdivided into *Hemominutum* and *Hemofelis* groups; *M. suis* was found to belong within the *Hemominutum* group.

1.4 *Mycoplasma hyopneumoniae*

M. hyopneumoniae is the agent of porcine enzootic pneumonia (EP). This is a respiratory disease of worldwide distribution and substantial economic impact. Primarily, the disease is seen as coughing with reduced growth performance. Lesions are seen primarily in the apical, cardiac and accessory lobes (Figure 1.2). These are regions of consolidation in which there is a lymphocytic hyperplasia. The lesions are not usually seen in adults probably because they are protected by past exposure, neither in very young animals because the disease takes time to develop. The most common stage is when animals are growing quickly, at about 8 to 16 weeks of age. In an outbreak of disease, a high proportion of the animals will become diseased but mortality is unusual unless secondary infection with virus or other bacteria such as *Pasteurella* superinfects the animals. Those animals infected with EP are very much more susceptible to secondary infections than those free of the infection.

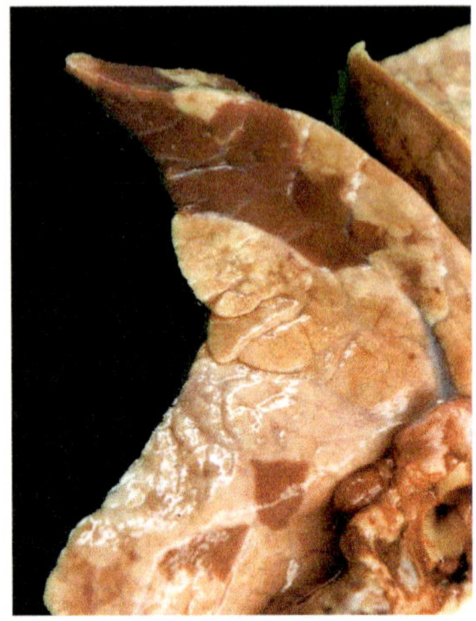

Figure 1.2. Lesions of enzootic pneumonia showing lesions of consolidation in the cardiac lobe 35 days post intra-tracheal inoculation. Experimental infection in pigs free of other respiratory pathogens. (Source: A. N. Rycroft)

1.4.1 Cultivation of *Mycoplasma hyopneumoniae*

M. hyopneumoniae has always been considered difficult and tedious to grow. It is a slow-growing mollicute, particularly when cultured directly from clinical material (lung lesions). This is exacerbated because overgrowth with a few contaminant bacteria, including *M. hyorhinis*, quickly overwhelms cultures and because primary cultures of *M. hyopneumoniae* can take as long as 10 – 15 days to become visible (see Chapter 8). This has limited our progress in understanding the virulence of the organism and the pathogenesis of EP because it has impeded experimental work and perhaps deterred competent researchers from effective investigations.

A reliable method of culture, primarily in liquid medium, was established by Nils Friis in Copenhagen in 1975. This medium that became widely known as Friis medium was prepared with great attention to detail and the procedures used for diagnosis were meticulous. In this way, culture could be used for routine isolation of *M. hyopneumoniae* from clinical cases. One problem was the regular co-culture of *M. hyorhinis* from lesions of EP. *M. hyorhinis* outgrows *M. hyopneumoniae* in the liquid medium and it becomes almost impossible to retrieve the latter. Friis showed that addition of 5% hyperimmune anti-*M. hyorhinis* serum together with 500 µg/ml cycloserine effectively inhibited *M. hyorhinis* allowing study of *M. hyopneumoniae* in isolation (Kobisch and Friis, 1996). There are also commercially available media used both in solid and liquid form. The constituents of such media are not publicly available but some of them have been widely used for experimental and clinical applications for many years. *M. hyorhinis* rapidly outgrows *M. hyopneumoniae* producing much larger colonies on the solid medium (Figure 1.3).

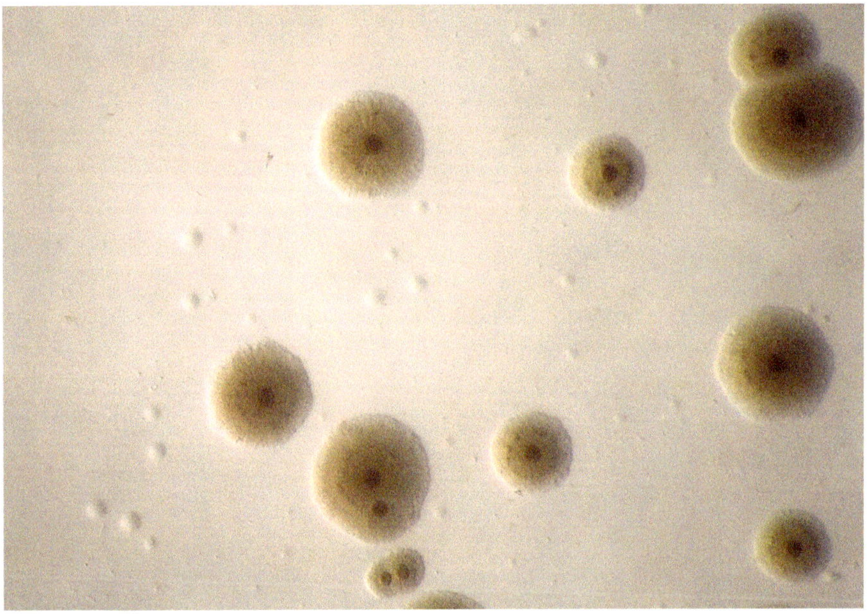

Figure 1.3. Colonies of *M. hyorhinis* (large granular colonies with central core spot) alongside *M. hyopneumoniae* (small translucent colonies) illustrating the size difference on *Mycoplasma* solid medium after 7 days incubation at 37°C. (Source: A. N. Rycroft)

A more recent development in the culture of *M. hyopneumoniae* and its selective separation from *M. hyorhinis* was discovered by Cook et al. (2016). Culture of lung tissues in solidified Friis medium containing a low concentration of the antibiotic kanamycin allowed suppression of *M. hyorhinis*. In this way, *M. hyopneumoniae* could be more readily retrieved in isolation from lung lesions carrying both organisms.

1.4.2 The discovery of *Mycoplasma hyopneumoniae*

EP is a major respiratory disease of the pig. While it rarely causes life-threatening disease, it is known to impede growth performance in pig herds and, together with a number of viral and bacterial agents (particularly *P. multocida*), causes more severe pneumonia, the porcine respiratory disease complex. EP is now recognized to be caused by *M. hyopneumoniae* (see Chapter 5).

The name *M. hyopneumoniae* was first used in 1965 when it was proposed as a new species by Maré and Switzer (1965). However, the disease it causes in pigs was known for many years before and there was considerable rivalry and competition to clarify and identify the causative agent of what we now know as EP.

Early descriptions of what was probably porcine EP began in 1933 when Köbe referred to Ferkelgrippe and in 1939 when Glässer described Schweinegrippe (Glasser, 1939). In 1941, Blakemore and Gledhill wrote, "In England, infectious pneumonia of pigs has been recognized by practitioners for many years - in severe cases there was a chronic phase which lasted several weeks. The pigs then became unthrifty and suffered from a chronic cough but they eventually recovered." This observation was later corroborated by Goodwin et al. (1967) who further stated the condition to be associated with a reduction in growth rate and that "the cumulative economic loss is so great, both nationally and internationally, that EP has been described as the most important disease of pigs today." This foresighted statement predicted the significance of a disease that we now recognize as a major cause of lost production in the global pig industry.

The disease was also termed infectious pneumonia and parainfluenza and in 1951 the pneumonia was recognized as a separate disease syndrome; in particular it was distinct from swine influenza. Gulrajani and Beveridge (1951) described the "chronic infectious pneumonia" as "differentiated from swine influenza." They used different criteria. First, they were unable to isolate influenza virus from outbreaks of pig pneumonia in East Anglia and Northern Ireland. Second, they could not demonstrate antibodies against influenza virus in the sera of recovered animals. Third, bacteria-free filtrates of pneumonic lung were infective, but, compared with swine influenza, the onset of the disease was slower, the course more protracted and the infective agent persisted longer in the lungs. From then, it was referred to variously as "virus pneumonia", "influenza-like disease", "enzootic virus pneumonia" and "transmissible pneumonia".

A remarkably accurate account of the EP was given by Alan Betts in 1952. While he referred to the condition as virus pneumonia of pigs, it is apparent from the distribution of lesions, the association with secondary bacterial pneumonia and the retarded growth rate, that the disease transmitted was EP. Efforts were made to characterize the transmissible agent, and filtration was a key tool in early microbiological investigations. The agent of this disease could be transmitted to pigs in lung homogenate and yet could pass through a bacteriological filter. This indicated it was a virus. However, like certain other filter-passing infectious agents known at the time (*chlamydiae, rickettsiae*), it was also found to be sensitive to tetracycline.

1.4.3 What is the evidence that *Mycoplasma hyopneumoniae* is responsible for EP?

Work conducted at Cambridge in the 1950s, and published in 1963, showed that the agent in lung homogenate could be passaged in cell culture and then transmitted 4 times in pigs, inducing the typical disease. Pleomorphic organisms were visible in the lung lesions: the microorganism associated with the lung lesions was 0.2 – 0.45 μm in size and it could be seen following staining with *Giemsa* stain. These were described in detail by Peter Whittlestone in 1957 (Figure 1.4). For this work, the Cambridge group used the J strain of EP (Goodwin and Whittlestone, 1963) originally derived from an outbreak of respiratory disease in Northern Ireland and transmitted by inoculation of pigs designated H and then J at Cambridge University (Gulrajani and Beveridge, 1951). They chose this material because it was unable to make colonies resembling the pleuropneumonia-like organism (PPLO) on solid medium. This was deliberate because PPLO, isolated from pig lung lesions, had been found to be non-pathogenic. Almost certainly, the PPLO colonies that were cultured on solid medium at that time were *M. hyorhinis*. The transmissible agent of EP could not so readily be cultured but it seems that the lung material, named the J strain and used for the experiments, was uncontaminated with *M. hyorhinis*.

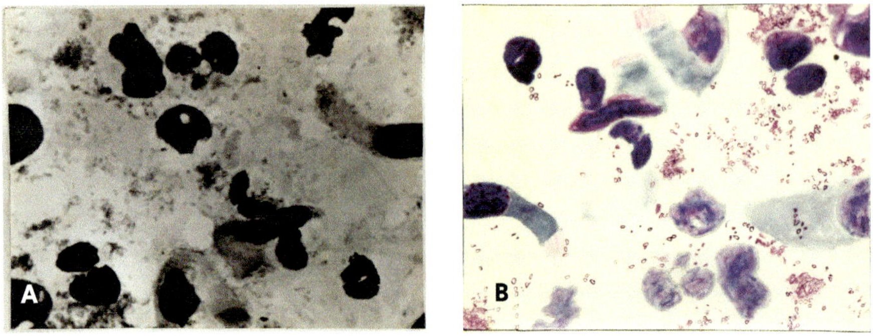

Figure 1.4. **(A)** Original photograph of pleomorphic organisms in a Giemsa-stained impression smear from lesions in the "pneumonic lung of a pig infected with J strain showing large numbers of the various forms of the pleomorphic organism". **(B)** Colored drawing of the image in A made by P. Whittlestone (From Whittlestone, P. 1957).

Researchers in the USA were also working to try and isolate the causal agent of EP with some success. L'Ecuyer and Switzer in 1963 attempted to propagate the causative agent of 'virus pneumonia of pigs' (EP) using infected lung tissue from field cases of 'virus pneumonia' which were found to be free of *M. hyorhinis*, yet still able to transmit the disease experimentally to pigs.

They inoculated 12 different types of cell culture. Only 2 of the 12 cell culture types, primary swine kidney cells and human cervical carcinoma cells (HeLa cells), allowed the infectious agent to propagate as shown by the induction of microscopic and macroscopic lung lesions in pigs 3 – 4 weeks post infection. However, following serial passages in HeLa cells the field strains lost infectivity. They also observed that there was no cytopathic effect on the infected cell cultures (L'Ecuyer and Switzer, 1963). The inability of L'Ecuyer and Switzer to show the causative agent growing in culture, by methods other than inoculating pigs to produce the disease, made further investigations into the causative agent of 'Virus pneumonia' almost impossible at that time.

By 1965, Maré and Switzer were able to isolate a microorganism from a case of EP in the USA and were able to grow it in broth modified from that described by Goodwin and Whittlestone (1964). Furthermore, they were able to induce lesions of EP with it. Maré and Switzer were able to grow this organism on their solid medium containing 1% agar, both from broth cultures and pneumonic lung suspension, obtaining colonies which they believed were visually different from *M. hyorhinis*. They named these new colonies *M. hyopneumoniae*.

The Cambridge researchers called their J strain organism *M. suipneumoniae*. Although much of the work was conducted before 1957, the first mention of this name as a new species in the public literature was in October 1965 (Goodwin et al., 1965). Two months earlier, in Iowa, Maré and Switzer (1965) published the name *M. hyopneumoniae*. This name took precedence for the new species.

There was nevertheless a degree of uncertainty whether *M. hyopneumoniae* and *M. suipneumoniae* were the same pathogen. Both groups had published pictures of the colonies on solid medium, both also showed the typical histopathology of lesions and both showed the microscopic appearance of the organism. Controversially at the time, the American group had reproduced the disease not from colonies grown on solid medium but from pneumonia-inducing fluid. The organism they named *M. hyopneumoniae* was then recovered onto solid medium but the inoculating fluid could, conceivably, have carried virus or another pathogen. The British group, in contrast, had used colonies to reproduce experimental disease and so there was some certainty that the organism was indeed the causative agent and free of virus. The name "*M. suipneumoniae*" persisted for some time; certainly until 1978. Nevertheless, once serological identity between *M. hyopneumoniae* and "*M. suipneumoniae*" had been demonstrated by growth-inhibition and metabolic-inhibition tests (Goodwin et al., 1967), the former was confirmed as the accepted name.

1.5 *Mycoplasma flocculare*

M. flocculare was described by Anders Meyling and Neils Friis in 1972 as a new species based upon the Growth Inhibition and Metabolic Inhibition tests. It closely resembles *M. hyopneumoniae* in characteristics and appearance, with no central core to the colony, although strains are usually seen to grow more slowly than *M. hyopneumoniae*. Friis also found it produced a more intense acid reaction *in vitro* and was able to grow at 30°C: a feature he considered may be related to the natural niche of the upper respiratory tract (Kobisch and Friis, 1996).

M. flocculare is found in the nasopharynx of the pig and it appears to be non-pathogenic. It is distributed throughout the global pig population as a normal inhabitant of the upper respiratory tract. Only in germ-free piglets was any pathological change detected. This was described as a lympho-histiocytic cell proliferation in the lamina propria with epithelial damage in the nasal cavity and lung (Friis, 1973). More recent experimental challenge of 5 week-old, caesarean-derived, colostrum-deprived piglets with *M. flocculare* yielded a similar result with no detectable disease (Gomes-Neto et al., 2014). This result was all the more meaningful because the mycoplasma was detected in the piglets at 7 days post-infection and evidence of infection was present from positive serological responses.

While there is no evidence that it causes disease alone, it is sometimes isolated alongside *M. hyopneumoniae* from EP. A recent investigation by Fourour et al. (2019) corroborated the lack of disease from *M. flocculare* inoculation of specific-pathogen-free (SPF) pigs. However, a small additive effect on the acute phase protein response, and on weight gain, was detected when the *M. flocculare* infection was co-infected with *M. hyopneumoniae*.

The 16S rRNA sequence of the two species also shows them to be very closely related. This led to the suggestion that *M. flocculare* was an avirulent form of *M. hyopneumoniae*. Indeed, strains considered to be *M. hyopneumoniae* in mycoplasma collections have later been confirmed to be *M. flocculare* (Rycroft, unpublished findings). With whole genome analysis, these organisms have been definitively shown to be distinct. Using PCR-based identification (Stakenborg et al., 2006a) or whole genome sequencing, it is now straightforward to distinguish between the species.

Arising from the close relationship with the pathogen *M. hyopneumoniae* is the issue of shared antigens. Meyling and Friis (1972) detected cross-reaction between *M. flocculare* and *M. hyopneumoniae* by agar gel immunodiffusion, and this translated into the concern that antibody generated by pigs against *M. flocculare* might be mistaken for that due to infection with *M. hyopneumoniae*, causing a false positive test result in ELISA tests. In reality, this cross reactivity was shown to be small (Bereiter et al., 1990). Nevertheless, the problem of cross-reactive antibody was overcome by enhancing the specificity of serological tests (ELISA) with specific monoclonal antibodies such as was used in the commercial competition ELISA (blocking ELISA) devised by Feld et al. (1992).

Analysis of the genome of *M. flocculare* shows that it shares 105 coding sequences with *M. hyopneumoniae* and the two species were recognized as having a high degree of similarity. However, there were clearly differences in genomic structure and organization (Sequeira et al., 2013). Some of the genes for adhesins were different. *M. flocculare* carries genes for P97 copy 2 adhesin and for P97-like adhesin but does not have an orthologue for the primary cilium adhesin, copy 1 of the P97, found in *M. hyopneumoniae*. The conclusion from scrutiny of the gene clustering was that P97 copy 2 and P97-like genes were present in the organism ancestral to both species but that the duplication event causing copy 1 of P97 gene must have occurred since the separation of *M. hyopneumoniae* and *M. flocculare*.

A further difference was noted by Calcutt et al. (2015). The gene encoding the glycerol 3-phosphate oxidase, *glpD*, is missing in *M. flocculare*. This enzyme and orthologues in other *Mycoplasma* species, is needed for production of H_2O_2 from glycerol. This activity is considered to be important in *M. mycoides* subsp. *mycoides* SC in the pathogenesis of contagious bovine pleuropneumonia (CBPP) through cytotoxicity of H_2O_2 for host cells (Pilo et al., 2005). If H_2O_2 production in vivo is crucial to the ability of *M. hyopneumoniae* to induce lesions, this difference in *M. flocculare* may be decisive in the inability of *M. flocculare* to induce disease.

1.6 *Mycoplasma hyopharyngis*

This organism was first described by Eriksen et al. (1986). Blank et al. (1986) suggested it belonged to the *Fermentans* group of *Mollicutes*, but *M. fermentans* itself is within the *Hominis* group in the *Lipophilum* cluster (Weisburg et al., 1989). Pettersson et al. (2001) also placed it in the *Lipophilum* cluster. Kobisch and Friis (1996) gave little information about this species. They stated it to be an arginine-metabolising mycoplasma from nasal and pharyngeal samples from the pig. Growth-Inhibition tests showed *M. hyopharyngis* to differ antigenically from other mycoplasmas, thereby conforming it as a separate species.

M. hyopharyngis was not considered to be pathogenic or associated with disease in pigs. However, isolates were later reported by Bradbury et al. (1994) to have been recovered from diseased joints in pigs, while Friis recovered the organism from pharyngeal scrapings in Denmark but reported that it was difficult to maintain culture in broth (Kobisch and Friis, 1996). Friis et al. (2003) considered the organism to be rare and since that time, there has been almost no further interest or analysis of this *Mycoplasma* which appears to be a commensal of the respiratory tract, able, on occasion, to become involved in joint disease of pigs.

1.7 *Mycoplasma hyosynoviae*

M. hyosynoviae was described by Richard Ross and Judith Karmon in 1970. Taxonomically, *M. hyosynoviae* was shown to fall into the *Hominis* group of *Mollicutes* (Figure 1.1). It was found in respiratory tracts and in the joints of pigs with arthritis. The joint disease associated with infection is mostly seen in pigs older than 10 weeks (3 – 5 months) and can be acute with swelling particularly visible in the hock joints, difficulty in moving, stiffness, lameness and an arched back (Blowey, 1993) (see Chapter 13). The possibility that this age restriction was due to inherent resistance of young pigs to this infection was disproven when young pigs were shown clearly to be just as susceptible to experimental infection as older pigs (Lauritsen et al., 2008). It is therefore most likely to be the decline of protective maternal immunity that enables the pathogen to cause disease in pigs as they reach 8 - 10 weeks of age and this has now been corroborated by Tølbøll Lauritsen et al. (2017).

The outcome of infection with *M. hyosynoviae* can vary considerably. It probably spreads between pigs as a respiratory organism. The natural habitat appears to be the palatine tonsil lymphoid tissue from where it can be isolated in carrier animals (Figure 1.5). In some cases, the organism progresses no further and animals can become carriers (Hagedorn-Olsen et al., 1999b). Under the right conditions, the pathogen invades from the lymphoid tissue and, by hematogenous spread, localizes in the joints, for which it has great affinity, and sometimes other sites in the pig such as the pericardium. The means of hematogenous spread is still not known: is it within or on the surface of macrophages or other cells, or is it by a cell-free mechanism? As yet, this is not known.

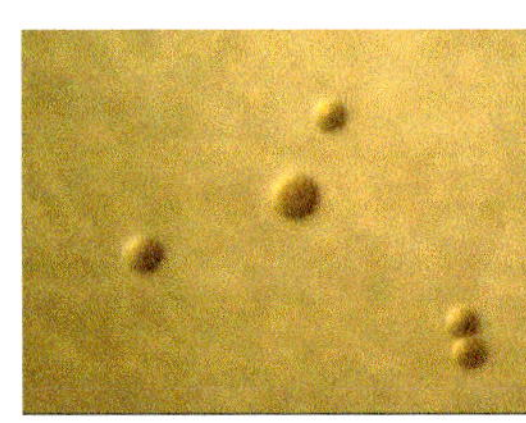

Figure 1.5. Colonies of *M. hyosynoviae* showing the "fried egg" appearance typical of many species of *Mycoplasma*. (Source: A. N. Rycroft)

Progression of infection to cause joint disease is associated with risk factors such as the level of immunity in individuals or the herd, management practices, breed of pig and aspects of the environment such as temperature. However, the factor most associated with development of active joint disease is transport of animals and the stress this brings to the animals (Kobisch and Friis, 1996). The contribution of strain virulence, if this varies between strains, is not known.

1.8 *Mycoplasma hyorhinis*

M. hyorhinis, to a bacteriologist, is the weed of the porcine mycoplasmas. It grows more quickly (one to two days) and forms larger colonies than its close relative, *M. hyopneumoniae* (Figure 1.3). Moreover, it is almost ubiquitous in the respiratory tract of pigs including the apparently normal lung tissue. It is also a pathogen responsible for polyserositis (serofibrinous lesions in the pericardial, pleural and peritoneal cavities) and arthritis in the pig: usually post weaning at 3 - 7 weeks of age (Friis and Feenstra, 1994) (See Chapter 13). As such, it is thought to migrate to those sites via a hematogenous route, probably from the pharyngeal and tonsillar surface where it naturally resides.

Phylogenetically, *M. hyorhinis* is close to *M. hyopneumoniae* and *M. flocculare* in the *Neurolyticum* cluster of the *Hominis* group of mycoplasmas. Considerable research on *M. hyorhinis* has been focused on the variation in surface antigen expression (the VLP system) as a means of immune evasion (Wise et al., 1992).

1.8.1 What is the evidence that *Mycoplasma hyorhinis* can also act as the causative agent of EP?

Although *M. hyopneumoniae* is believed to be the primary agent of EP in swine, and numerous experiments with purified cultures of *M. hyopneumoniae* have confirmed this, there have also been studies showing that *M. hyorhinis* alone is found in pneumonic lesions (Kobisch and Friis, 1996) and that *M. hyorhinis* may be capable of inducing pneumonia in the pig, at least in gnotobiotic piglets (Poland et al., 1971; Gois et al., 1971; Lin et al., 2006). This has led to divided opinions on the possible role of *M. hyorhinis* in pig pneumonia and questions over the primary agent of EP. Uncertainty has been compounded by the recognition that EP disease apparently occurs in animals vaccinated against EP in the field, from which only *M. hyorhinis* has been isolated.

Several investigators have suggested that *M. hyorhinis* plays a secondary role in the disease to *M. hyopneumoniae* (L'Ecuyer and Switzer, 1963; Fourour et al., 2019). It has been shown by a number of researchers that *M. hyorhinis* can be isolated from the nasal cavities and fresh lungs of clinically 'normal' pigs. In 1963 Ross, Switzer and Maré isolated *M. hyorhinis* from 30% of clinically normal pigs and Hartwich and Niggeschulze (1966) were able to isolate *M. hyorhinis* from 75% of clinically normal lungs from pigs (Poland et al., 1971). Goodwin et al. (1968) concluded that *M. hyorhinis* is commonly present in the lung lesions of EP and that this mycoplasma may often prevent the isolation of "*M. suipneumoniae*". However, there have been few studies looking into the production of pneumonia in pigs using pure strains of *M. hyorhinis*.

One such study was carried out by Jenny Poland in 1971. This was an attempt to show that a strain of *M. hyorhinis* was able to induce pneumonia when inoculated into germ-free piglets. The piglets were mono-infected with a strain of *M. hyorhinis*, TR32. Prior to infection the strain, isolated from pig lungs in Czechoslovakia, was cloned three times then passaged through a pig before being inoculated by aerosol into 9 pigs at 8 - 9 days of age. From this study, it was concluded that at least one strain of *M. hyorhinis* could produce pneumonia in pigs and therefore *M. hyopneumoniae* was not the only mycoplasma able to cause pneumonia in pigs (Poland et al., 1971). A similar study by Gois et al. (1971) also showed that intranasal infection of gnotobiotic pigs with *M. hyorhinis* was capable of causing pneumonia with macroscopic pneumonic lesions alone or with polyserositis and arthritis. Histological findings further showed peribronchiolar and perivascular infiltration 'cuffing' in the lungs at 14 days post infection (Gois et al., 1971).

Work carried out by Poland further demonstrated that following intranasal, intravenous or aerosol infection of gnotobiotic pigs, classical lesions of polyserositis and/or arthritis could be produced with four field strains of *M. hyorhinis*. Two strains produced pleuropneumonia in a total of three out of 39 infected pigs, and all four strains produced bronchopneumonia in a total of five out of 39 infected pigs. It was also noted that, macroscopically, lung lesions resembled those caused by *M. hyopneumoniae*. However, the histological appearance differed in this case: peribronchiolar and perivascular infiltration of lymphocytes 'cuffing' was rarely a feature of the lesions reportedly caused by *M. hyorhinis*. Therefore, considering the discrepancy in the findings of Gois et al. (1971) and Poland et al. (1971), differentiation of the causal agent of the lesions based on microscopic appearance would not be possible.

Different studies have also observed differences in virulence between strains of *M. hyorhinis* following experimental infection. Five field strains were used to infect SPF-pigs at 2 weeks of age, and clinical signs were monitored. Infection with four out of five strains, clinical signs were observed from one week post infection such as elevated temperature, retarded growth, joint swelling, lameness, polyserositis and arthritis. The fifth strain only produced pleural adhesion in one pig out of 5; the other 4 strains caused acute and chronic stages of the disease including pericarditis, pleurisy and peritonitis. However, no pneumonic lesions were noted at necropsy (Kobisch and Friis, 1996). Similar results were reported by Gomes-Neto et al. (2014).

A recent study in Taiwan has also suggested that *M. hyorhinis* alone is capable of causing pneumonia (Lin et al., 2006). Using a combination of three wild-type *M. hyorhinis* isolates or a single isolate, typical mycoplasma pneumonic lung lesions were seen in two out of the six 6-weeks-old *M. hyopneumoniae* and *M. hyorhinis*-free pigs challenged with the combined wild-type isolates. No lesions were observed in the six pigs challenged with the single isolate or in the control group. The authors concluded that *M. hyorhinis* may play a significant role in swine respiratory disease and that *M. hyopneumoniae* is not the only mycoplasma involved in EP (Lin et al., 2006).

It seems that, taking the evidence together, it is possible that *M. hyorhinis* may be capable of causing lesions in the lungs of highly susceptible pigs such as those without normal microbiota or any maternal or active immunity. The relevance of such vulnerable animals to farm practice is obviously limited. While *M. hyorhinis* is frequently present in healthy lung tissues and may exacerbate lesions of EP (Fourour et al., 2019), the established causal agent of EP is undoubtedly *M. hyopneumoniae*.

1.9 *Mycoplasma suis*

M. suis (formally *M. hemosuis*) is now recognized as the name for the agent of porcine eperythrozoonosis or porcine hemoplasmosis. This is seen worldwide in domestic and wild pigs and is considered an economically significant infection in swine production (see Chapter 14). It is an infectious form of anemia first described by Kinsley (1932) who considered it as a protozoal infection and by Doyle (1932) describing it as a Rickettsial disease.

The hemoplasmas, as the hemotrophic mycoplasmas became known, affect a wide range of different animal species. They were only recognized as *Mollicutes* in 2001 using analysis of the 16S rRNA (Neimark et al., 2001). Until then, they were thought of as Rickettsia-like organisms in the genera *Hemobartonella* and *Eperythrozoon* species. They are considered to be most closely related to the *Pneumoniae* group of mycoplasmas (Tasker et al., 2003).

The pathogen attaches to and parasitizes erythrocytes of the host leading to damage and causing hemoglobin release with consequent anemia and jaundice. As yet, these organisms are not cultivable *in vitro* (Hoelzle, 2008). This alone has retarded progress in understanding the host-pathogen interaction and there is relatively little knowledge of the immunology, genetics or physiology of the pathogen. There is evidence that *M. suis* may be transmitted to humans and thereby considered as a zoonotic pathogen (Yuan et al., 2009).

CHAPTER 2

DIVERSITY OF *MYCOPLASMA HYOPNEUMONIAE* STRAINS

Veronica M. Jarocki[1]

Steven P. Djordjevic[1]

1 The ithree institute, University of Technology Sydney, Australia

2.1 Introduction

Mycoplasma hyopneumoniae is a geographically dispersed and economically significant pathogen affecting global pork production. Infections caused by *M. hyopneumoniae* are insidious because the pathogen persists in herds despite widespread vaccination and medication and predisposes the respiratory tract to secondary bacterial pathogens (Chae, 2016). Diagnostic tests based on serology are challenged by the need to differentiate immune responses induced as a result of vaccination and infection and the time it takes for the pathogen to mount a measurable response. Bacterin formulations persist despite multiple efforts to develop subunit vaccines (Djordjevic et al., 1997; Jorge et al., 2014; Marchioro et al., 2014b; Woolley et al., 2014). Molecular assays are applied according to the prevailing view that the organism is restricted to the upper respiratory tract, whereas the organism may reside in organ sites distant to the lung such as spleen, liver and kidney (Marchioro et al., 2014b; Marois et al., 2007; Raymond et al., 2018c). Only a handful of genome sequences have been completed (Minion et al., 2004; Vasconcelos et al., 2005; Liu et al., 2011, 2013; Siqueira et al., 2013; Han et al., 2017) thus the knowledge of the *M. hyopneumoniae* pangenome is severely limited. Furthermore, approximately 40% of the predicted open reading frames (ORFs) have an unassigned function, significantly impeding knowledge of the mechanisms that underpin virulence.

Both low and high virulence *M. hyopneumoniae* strains circulate globally in swine herds, and strains have demonstrated diversity at genomic, antigenic, and proteomic levels leading several comparative studies to seek links with these variations to pathogenesis. A broader understanding of these differences could aid in establishing more efficacious control measures. No standardized methods for characterising strains currently exist for this organism, which hinders direct comparison of published results, although an effort to standardize molecular comparisons was recently proposed (Betlach et al., 2019). This chapter provides a summary of the literature regarding diversity in *M. hyopneumoniae* strains, pointing out discrepancies, and highlighting that there is still much to be gleaned from this genome-reduced, yet highly complex pathogen.

2.2 Genomic diversity

Genomic diversity allows bacterial species to adapt to their surrounding environment, enabling them to optimize colonisation of their host and avoid both immune and pharmaceutical driven clearance. Genomes of the same bacterial species possess a set of common genes known as the core genome. However, genetic variations can be introduced by means of point mutations, homologous recombination, and horizontal gene transfer. These changes lead to sequence variations, gene insertions and deletions, and chromosomal rearrangements (Li et al., 2009a).

Elucidating genomic diversity in *M. hyopneumoniae* is important for both epidemiological, and pathogenicity studies, as well as for the optimization of control measures (Charlebois et al., 2014). Discriminating bacterial strains based on their genetic content is referred to as genotyping. Currently, the most thorough method of determining genetic variability within a bacterial species is via whole-genome sequencing (WGS) followed by genome comparisons. While WGS technology is becoming increasingly available and affordable (Wyrsch et al., 2015), this has not always been true, especially for routine diagnostics, and WGS may still not present the best option for large scale field isolate studies. As such, researchers have employed other genotyping methods to ascertain which *M. hyopneumoniae* strains reside in various populations, including field-inversion gel electrophoresis (FIGE), pulsed-field gel electrophoresis (PFGE), amplified fragment length polymorphism (AFLP), random amplified polymorphic DNA (RAPD), microarrays, PCR-random fragment length polymorphism (PCR-RFLP), multi-locus sequencing typing (MLST), multiple-locus variable number tandem repeats analysis (MLVA), and sequencing of variable number of tandem repeats (VNTRs).

2.2.1 *Mycoplasma hyopneumoniae* WGS comparisons

To date, seven *M. hyopneumoniae* genome sequences have been published: *M. hyopneumoniae* strain 232 (232; Minion et al., 2004), *M. hyopneumoniae* strains J and 7448 (J, 7448; Vasconcelos et al., 2005), *M. hyopneumoniae* strain 168 and the attenuated version 168-L (168, 168-L; Liu et al., 2011, 2013), *M. hyopneumoniae* strain 7422 (7422; Siqueira et al., 2013) and *M. hyopneumoniae* strain KM014 (KM014; Han et al., 2017). The general characteristics of each genome are summarized in Table 2.1. What is immediately evident is that strain KM014, isolated from a diseased pig in South Korea, is considerably larger at 964503 bp, than the other strains. Other notable differences are the greater number of pseudogenes in strains 168 (44), 168L (32) and 7422 (41) versus 0-1 in other strains, and the lower number of repeat regions in strains 7422 (46) and 232 (54), while others range from 73 to 94. All sequenced *M. hyopneumoniae* genomes have a high abundance of genes encoding hypothetical proteins (40.5 - 43% of all genes).

Table 2.1. General characteristics of seven *M. hyopneumoniae* genomes (strains 232, J, 7448, 168, 168L, 7422 and KM014).

Characteristic	232	J	7448	168	168L	7422	KM014
Length (bp)	892758	897405	920079	925576	921093	898495	964503
GC content (%)	28.6	28.5	28.5	28.46	28.5	28.5	28.37
Total CDS	744	753	769	764	766	762	979
Total known proteins	441	448	447	449	446	447	558
Hypothetical proteins	303	305	322	315	320	315	421
Pseudogenes	0	0	1	44	32	41	0
Repeat regions	54	77	89	85	73	46	94
tRNA	30	30	30	30	30	30	32
rRNA	2	2	2	3	3	3	2
Virulence Factors (VFDB)	9	7	7	6	6	6	15
Insertion sequences (IS)	1	8	8	13	9	2	7
Protein synthesis genes	147	147	147	148	148	149	173
Protein Fate genes	8	8	8	8	8	8	11
Cofactor/vitamin metabolism genes	19	19	19	19	19	20	25
Carbohydrate metabolism genes	14	14	14	14	14	14	17
Nucleotide metabolism genes	6	6	6	6	6	6	6
Amino acid metabolism genes	4	4	4	4	4	4	6
Fat metabolism genes	2	2	2	2	2	2	4
Metabolite repair/mitigation genes	1	1	1	1	1	1	0
Energy genes	34	34	34	34	34	35	45
DNA processing genes	27	25	25	29	29	26	33
Stress response, defence, virulence genes	25	25	25	31	31	25	49
RNA processing genes	22	22	22	26	26	22	32
Cellular processes genes	6	6	6	6	6	6	7
Membrane transport genes	3	3	3	3	3	3	3
Not assigned to PATRIC subsystems (% genes)	21.1	22.05	21.59	32.72	30.68	32.41	31.77
Hypothetical proteins (% genes)	40.73	40.5	41.88	41.23	41.78	41.34	43
Country of origin	USA	UK	Brazil	China	China	Brazil	South Korea
MLST	ST27	ST28	ST15	ST61	ST61	ST41	ST28

*Values obtained using PATRIC (Wattam et al., 2017) which may differ slightly from original RefSeq/published values. Genomes are annotated using the submitting researcher's choice of methods, PATRIC reannotates genomes using RASTtk to provide consistency. Cells are presented in a colour range from dark green (indicating the lowest value or an absence of variation) to yellow (indicating comparatively higher values). (Source: V. Jarocki)

In terms of gene products, sequenced *M. hyopneumoniae* genomes do not vary extensively, with most differences attributed to the absence or presence of hypothetical proteins, transposases and insertion sequences (ISs), ABC transporter ATP-binding proteins, subtilisin-like serine proteases, and type III restriction system endonucleases (Minion et al., 2004; Vasconcelos et al., 2005; Liu et al., 2011, 2013; Siqueira et al., 2013; Han et al., 2017; Figure 2.1). *M. hyopneumoniae* strains J and 168-L are both attenuated, made avirulent after continuous culture passaging (Zielinski and Ross, 1990; Liu et al., 2013). The most extensively studied virulence factors of *M. hyopneumoniae* are adhesins (such as P97, P102, P146, P159, P216), none of which are missing in these two strains. Interestingly, a P102 paralogue appears to be missing or truncated in strain KM014 (Figure 2.1), which does not evidently undermine its ability to cause disease. This suggests complex yet subtle mechanisms behind pathogenesis.

Indeed, a comprehensive comparative analysis of strains 168 and 168-L (attenuated 168), evidenced that the two genomes were remarkably similar. 168L was only 4483 bp smaller, and this size discrepancy was mainly due to fewer IS elements. The researchers did, however, identify 330 genetic variations in 168-L (Liu et al., 2013). While most of these variations were single nucleotide polymorphisms (SNPs), often leading to either no change in amino acids or isofunctional amino acid substitutions, there were some sequence insertions, deletions and truncations noted. Some genetic variations occurred in adhesins and other virulence-associated genes, as well as in metabolic genes. However, many occurred in hypothetical proteins (40.71%). This highlights the need for future studies to characterize *M. hyopneumoniae* hypothetical proteins, especially when taking into account that many have been confirmed through proteomic studies (Pendarvis et al., 2014; Tacchi et al., 2016; Berry et al., 2017; Paes et al., 2018), have been found to be antigenic (Galli et al., 2013), and possess long A/T runs thought to cause adaptive mutations to facilitate immune system circumvention (Mrázek, 2006).

Horizontal gene transfer allows for rapid bacterial genomic diversification (Cury et al., 2017). While mobile genetic elements (MGEs) such as plasmids and phages have been identified in other mycoplasma species (Breton et al., 2012; Djordjevic et al., 2001; Dybvig et al., 2005), they have not been described in *M. hyopneumoniae*. Nonetheless, in addition to IS elements, two putative integrative conjugative elements (ICEs) have been discovered (Liu et al., 2013; Pinto et al., 2007b; Siqueira et al., 2013; Vasconcelos et al., 2005). These elements are shown in detail in Figure 2.2, are marked 1 and 2 in Figure 2.1 and are shown in pink and yellow in Figure 2.3. As some ICEs are associated with virulence (Clawson et al., 2016; Obi et al., 2018; Roy Chowdhury et al., 2016), their initial discovery in pathogenic strains 232 and 7448, but not in strain J, prompted researchers to suggest their involvement in *M. hyopneumoniae* virulence (Pinto et al., 2007a). Support for this hypothesis has waned since the discovery of both ICE elements in 168-L (Liu et al., 2013). Thus, their role in some *M. hyopneumoniae* strains is currently unknown (Maes et al., 2018).

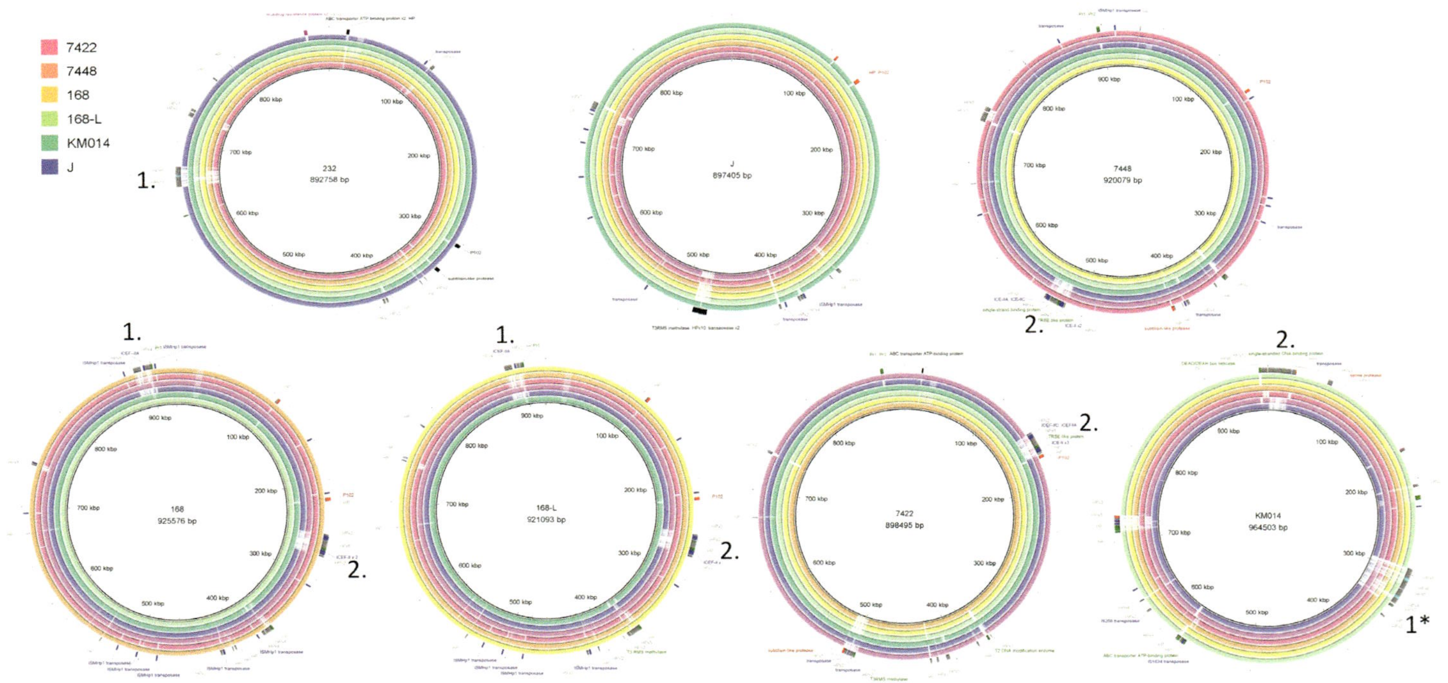

*in KMO14, the ICE elements are adjacent to additional variable regions.

Figure 2.1. Comparisons of *M. hyopneumoniae* genomes. BRIG (Alikhan et al., 2011) genome comparisons are illustrating regions of genetic variations for each sequenced strain, marked above their corresponding outer rings. Hypothetical proteins are shown in grey, Mobile genetic elements (MGEs) are in blue, proteases in red, antimicrobial resistance proteins in pink, domains of unknown functions (DUFs) in aqua, others (i.e. ABC transporters and restriction endonucleases) in green and various combinations in black. The putative Integrative and Conjugative Elements (ICEs) are marked 1 and 2, and these relate to 1 and 2 in Figure 2.2 below. (Source: V. Jarocki; Schematic generated using Geneious, Kearse et al., 2012)

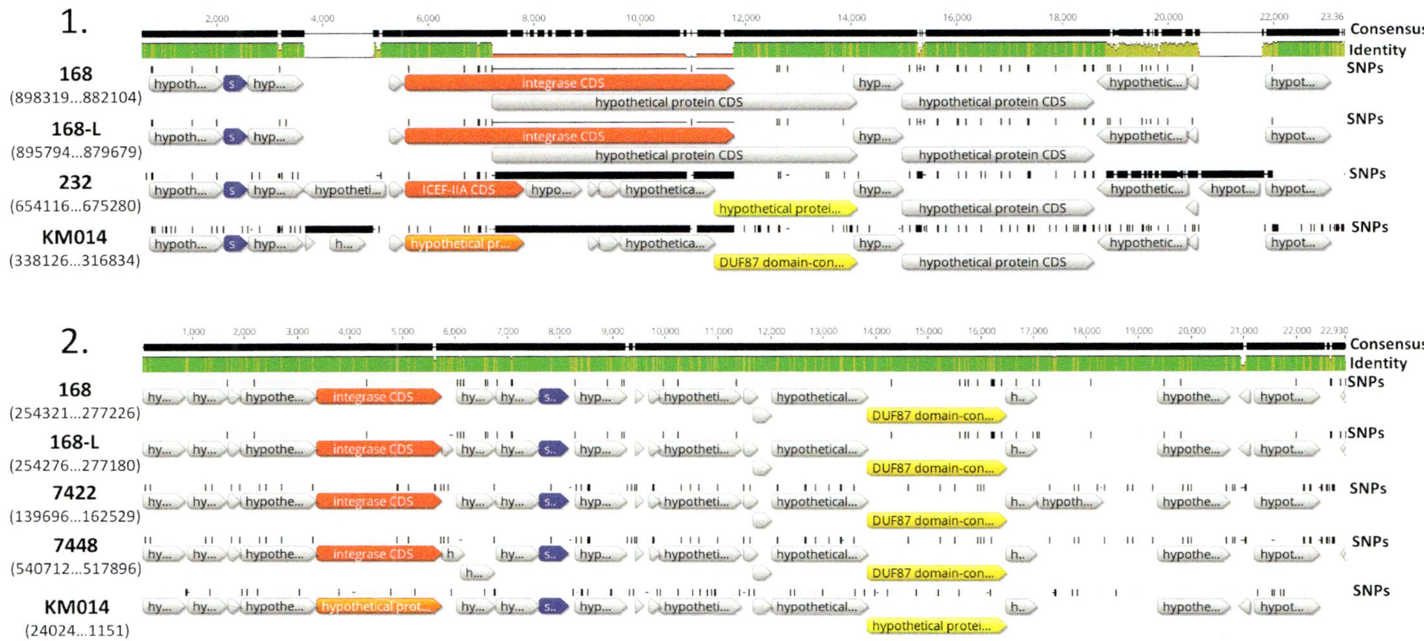

Schematic generated using Geneious (Kearse et al., 2012).

Figure 2.2. *M. hyopneumoniae* putative Integrative and Conjugative Elements (ICEs). 1. Sequence alignment illustrating an ICE in 168, 168-L, 232 and KM014. The intergrase coding sequence (CDS) (red) is significantly different in 168 and 168-L than in 232 and KM014 (shown in orange as it is annotated as a hypothetical protein). 2. Sequencing alignment illustrating an ICE in 168, 168-L, 7422, 7448 and KM014. Apart from some deviations in hypothetical proteins (grey), this region is generally homologous across all strains. Single-stranded DNA binding proteins are shown in blue, DUF87 domain protein with homology to *traE* conjugative protein shown in yellow. Location of each region is in brackets under each strain name. (Source: V. Jarocki)

Bacterial genome arrangement has profound effects on phenotype and is thought to impact gene expression (Darling et al., 2008). Figure 2.3 illustrates the arrangement of *M. hyopneumoniae* chromosomes. While most of the strains possess a similar gene organisation, four regions in strain 232 spanning 100 kbp to 350 kbp are inverted compared to other genomes. The ICEs are dispersed throughout the chromosomes, consistent with their mobile characteristics. Most strikingly however is that strain KM014 does not possess a single positionally orthologous region. This strain also contains an additional region, adjacent to an ICE, around ~300 kbp. This region consists solely of hypothetical proteins.

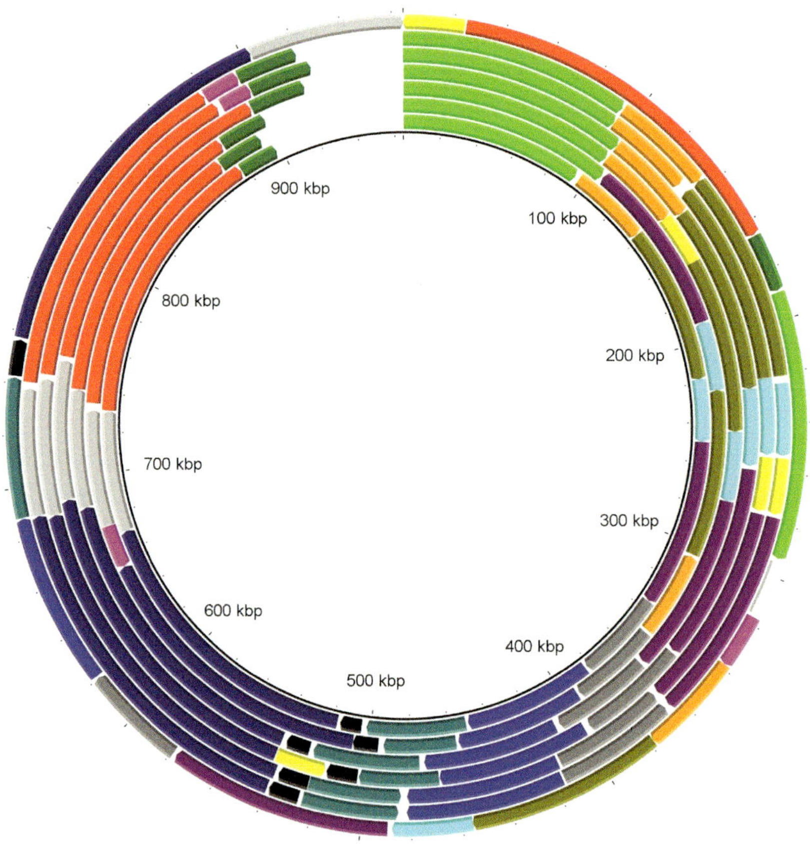

Figure 2.3. *M. hyopneumoniae* chromosomal arrangements. Shown is a mauve (Darling et al., 2004) alignment incorporated onto a ring. From outer to inner ring: KM014, 168- L, 168, 7422, 7448, 232, J. Homologous regions have matching colours. ICEs are shown in pink and yellow. Unique KM014 region in white. (Source: V. Jarocki)

Though substantial variations in genome arrangements within species are known to occur (Darling et al., 2008), it is still perhaps surprising that *M. hyopneumoniae* strain KM014 and J are the same sequence type (ST), ST28, implying that they share a common ancestor. A phylogeny tree places the two strains within a clade (Figure 2.4), however, with only seven strains to compare, little can be inferred by their proximity. Strain J was originally isolated in the late 1950s (Goodwin and Whittlestone, 1963) and deposited at the Mycoplasma Reference Laboratory (MRL) in 1967 (Whittlestone, 1973). KM014 on the other hand, was isolated in 2014 (Han et al., 2017). If these two strains do share a direct history, delving into the evolutionary processes that resulted in extensive chromosomal rearrangement and the loss or acquisition of the unique KM014 region would be fascinating.

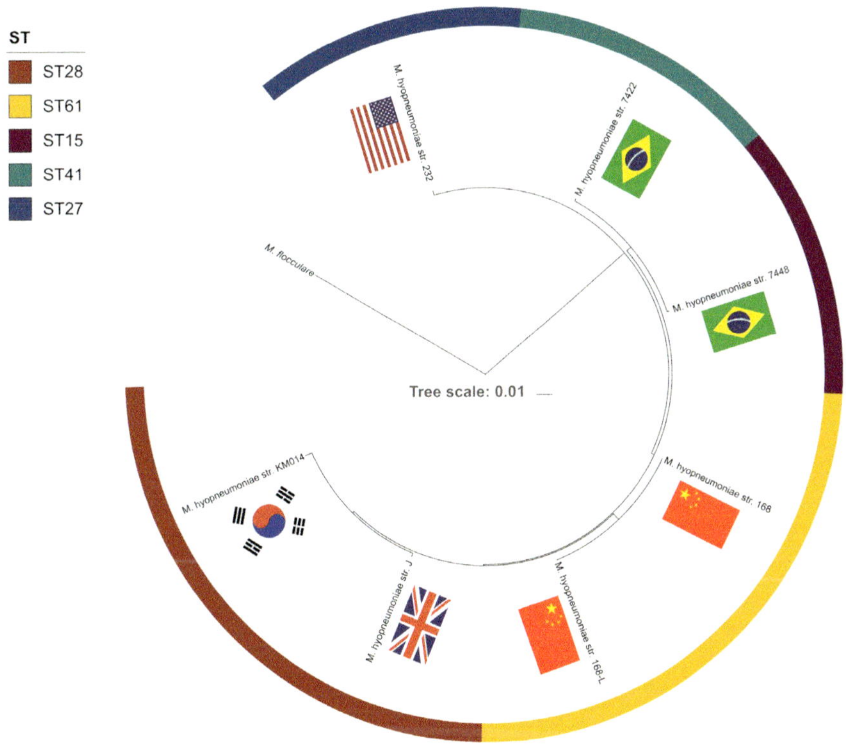

Figure 2.4. Phylogenetic relationship of *M. hyopneumoniae* sequenced strains. STs are illustrated by the outer ring; flags represent the country of origin. (Source, V. Jarocki; Phylogenetic tree generated by Phylosift (Darling et al., 2014) and FastTree (Price et al., 2010), and visualised using iTOL (Letunic and Bork, 2019).)

2.2.2 Genetic diversity in *Mycoplasma hyopneumoniae* field isolates

Regardless of the genotyping method used, a consensus across the literature is that *M. hyopneumoniae* strains demonstrate high diversity at the international level (Frey et al., 1992; Stakenborg et al., 2005b, 2006b; Mayor et al., 2008; Dos Santos et al., 2015a). Studies have also shown that vaccine strains differ significantly from field strains (Tamiozzo et al., 2015; Michiels et al., 2017, Garza-Moreno et al., 2019b), that the field strains themselves differed between vaccinated farms and non-vaccinated farms (Garza-Moreno et al., 2019b), and that vaccination does not necessarily reduce diversity within farms (Michiels et al., 2017; Rebaque et al., 2018).

The literature regarding genetic diversity in *M. hyopneumoniae* field isolates between and within herds is not as unambiguous (details provided in Table 2.2). In a single closed herd, Rebaque et al. (2018) demonstrated that the same *M. hyopneumoniae* strain could persist for at least two years. Similarly, Mayor et al. (2007; 2008) showed that within herds, isolates were identical, as were isolates sourced from herds in geographical proximity. Furthermore, based on *p146* polyserine repeat motif genotyping, Tamiozzo et al. (2013) demonstrated that the same *M. hyopneumoniae* strains can persist post depopulation programs, and Overesch and Kuhnert (2017) reported a persistence of *M. hyopneumoniae* STs post strict control measures. Nevertheless, other studies observed homology within a single herd and heterogeneity between different herds (Pantoja et al., 2016; Stakenborg et al., 2006b, 2005b; Vranckx et al., 2012a), and both inter- and intra- herd isolate diversity has also been reported (Nathues et al., 2011; Pulgarón et al., 2015; Tamiozzo et al., 2015; Michiels et al., 2017; Felde et al., 2018b; Tadee et al., 2018). Moreover, multiple strains have been identified in individual pigs in some (Charlebois et al., 2014; Dos Santos et al., 2015a; Michiels et al., 2017; Nathues et al., 2011), but not all (Garza-Moreno et al., 2019b) studies. Batches of slaughter pigs where several *M. hyopneumoniae* strains were identified using MLVA showed more severe lung lesions (Michiels et al., 2017). However, at the individual level, no correlation could be established between the presence of specific isolates in slaughter pigs and the severity of lung lesions (Charlebois et al., 2014). The variations between studies could be attributed to many factors, such as differences in genotyping technique and targets used, sample numbers, and types of clinical samples used, as well as housing conditions and animal management practices upheld in the isolate's country of origin.

Table 2.2. Literature reporting genetic diversity in *M. hyopneumoniae* clinical samples and field isolates.

	Isolate origin and numbers	Findings	Reference
AFLP	Total = 9 Denmark (n=8, field isolates) Reference strain (n=1; J)	• No indistinguishable banding observed in *M. hyopneumoniae* isolates, as opposed to *M. genitalium* and *M. hyorhinis* isolates	Kokotovic et al., 1999
FIGE	Total = 10 Canada (n=1, field isolate) France (n=1, field isolate) Switzerland (n=5, field isolates) Reference strains (n=3; NCTC 10110, ATCC 25934, and JF 184a – all originating from strain J)	• Clonal homogeneity in Swiss isolates • Heterogeneity in geographically distant isolates	Frey et al., 1992
MLVA	Total = 66 Belgium (n=66, MLVA positive clinical samples from 4 herds)	• 3 loci (P97R1, P146R3, H4) • Each herd had a single distinct strain with 2-3 clonal variants	Vranckx et al., 2012a
	Total = 355 Brazil (n=95, clinical samples, 76 farms) Mexico (n=26, clinical samples, 16 farms) Spain (n=25, field isolates, 24 farms) USA (n=209, clinical samples, 208 farms) Reference strain (n=1; ATCC-25095)	• 2 loci (P97R1, P146R3) • 87 MLVA types from the USA, 12 MLVA types in Mexican isolates, 39 MLVA types in Brazil and 22 MLVA types found in Spanish isolates. • Isolates from Mexico related to 2 production systems, 1 production system had only one MLVA type, the other had 11. • Higher variation in P146 than P97 VNTRs. • More than 30 repeats in P146 were only observed in isolates from Spain and Brazil • Evidence of multiple variants in single samples	Dos Santos et al., 2015a
	Total = 64 Cuba (n=63 clinical samples from 5 farms from 1 slaughterhouse) Reference strain (n=1; J)	• 4 loci (P97R1, P146R3, H2R1, H4) • High genetic variability in isolates detected in healthy and sick animals, from different and same herd. • 22% of isolates from pneumonic lungs had only 3 repeat units in P97R1, and isolates from 40 and 43 VNTR copy numbers in P97R1 were only found in healthy lungs.	Pulgarón et al., 2015
	Total = 33 Argentina (n=28 clinical samples from 3 herds) 5 vaccine strains (2 corresponding to strain J, 3 unspecified)	• 5 loci (P97R1, P146R3, H4, H5, P95) • Different genotypes among and within herds and among bacterin strains	Tamiozzo et al., 2015

	Isolate origin and numbers	Findings	Reference
MLVA	Total = 100 USA (n=100 clinical samples used for MLVA, from 3 farms, same production operation)	• 2 loci (P97R1, P146R3) • 4 MLVA types • 1 variant found in all 3 farms	Pantoja et al., 2016
	Total = 495 Belgium (n= 495 clinical samples from 10 herds from one slaughterhouse)	• 4 loci (P97R1, P146R3, H5R2, H1) • 135 different variants found • Average number of genotypes per batch was 7 • Three different genotypes found in two lungs from one herd • Prevalence and severity of lesions significantly higher in batches with >1 genotype • High variation in genotype at batch and animal level • Vaccination does not lead to decreased diversity	Michiels et al., 2017
	Total = 35 Argentina (n=35 clinical samples from one closed farm, 27 samples taken in 2013, 8 samples taken in 2015)	• 3 loci (P97R1, P146R3, H4) • Persistence of same strain in closed farm over a two-year period, with all 2015 samples demonstrated the same 3 loci type as most 2013 samples	Rebaque et al., 2018
	Total = 53 Argentina (n=48 clinical samples from 5 herds) Reference strains (n=5; 7448, 7422, J, PMS, 232)	• 2 loci (P97R1/P97R1A, P146R3) • Two herds had only 1 genotype present, while the other three herds exhibited more diversity. • In general, local specimens were more closely related than to reference strains.	Sosa et al., 2019
MLST	Total=54 Switzerland (n=46, field isolates) Australia (n=1, field isolate) France (n=2, field isolates) Canada (n=1, field isolate) Reference strains (n=4; 232, J, JF184a, 7448)	• High genetic diversity overall • Isolates from a single farm were identical • Farms in close geographical or operational contact showed identical clones	Mayor et al., 2008
	Total = 63 Switzerland (n=63, 16 clinical samples from 10 farms, 47 clinical samples from wild boar)	• Some identical genotypes found in commercial pigs and wild boars • Different genotypes found in wild boars before outbreaks, then same genotypes post-outbreak suggesting direct transmission	Kuhnert and Overesch, 2014

MLST	Total = 16 Switzerland (n=16, clinical samples; 8 from 2014, 5 from 2015, 3 from 2016)	• Persistence of genotypes represent a possible partial failure of elimination measures • Two prevalent STs noted– ST26 and ST37	Overesch and Kuhnert, 2017
	Total = 16 Thailand (n=16 field isolates from 9 farms)	• Two ST48s, 14 unique STs when compared to 229 strains in the MLST database (https://pubmlst.org/mhyopneumoniae/) • Diverse genotypes found on 2 farms	Tadee et al., 2018
Multiple methods	Total = 45 Belgium (n=31 field isolates, 21 farms) Denmark (n=3 isolates) Lithuania (n=4 field isolates, 2 farms) The Netherlands (n=2 field isolates, 2 farms) United Kingdom (n=3 isolates) Reference strains (n=2; 232, J)	• Used AFLP, PCR-RFLP and RAPD • Multiple isolates from 3 farms had indistinguishable AFLP banding • Two isolates from two different Belgian farms had AFLP indistinguishable banding • AFLP and PCR-RFLP of p146 more discriminatory than RAPD, and RFLP of P97 • Extensive variation of the p146 gene in different isolates, however, 60% of isolates originating from the same herd had identical profiles. • RAPD results not reproducible between different runs • Most isolates originating from the same herd and different farms had identical RAPD patterns • No clear relationship between country of origin and patterns obtained • More resemblance between strains isolated from the same herd	Stakenborg et al., 2006b
	Total = 53 Germany (n=52, field isolates from 45 pigs from 21 herds) Reference strain (n=1; J)	• Used MLVA and RAPD • 3 loci (P97R1, P146R3, H4) • High genetic heterogeneity of *M. hyopneumoniae* in single pigs, within the same herds and between different herds. • Did not identify any 5000bp RAPD fragments	Nathues et al., 2011

	Isolate origin and numbers	Findings	Reference
Multiple methods	Total = 173 Canada (n=155, 145 clinical samples from 48 farms from 2 slaughterhouses, 10 field isolates) France (n=10, field isolates) 6 vaccine strains Reference strains (n=3; 232, ATCC 25095, ATCC 25934)	• Used PCR-RFLP of *p146* and MLVA • 4 loci used (P97R1, P97R2, Locus 1, Locus 2) • A total of 87 MLVA types, all samples typable • Genotypes lacking locus 1 associated with lower virulence • Some pigs were infected by multiple genotypes • Lungs harbouring more than one genotype were not associated with higher lesion scores • Lungs harbouring co-infections not associated with higher lesion scores • High diversity between farms using both typing methods • Field isolates and vaccine strains had <55% homology • PCR-RFLP of p146 was unable to type all samples • Identified 83 different PCR-RFLP patterns	Charlebois et al., 2014
	Total = 44 Hungary (n=40, field isolates) Slovenia (n=3, field isolates) Czech Republic (n=1, field isolate)	• Used MLST and MLVA • 23 STs from Hungary, 3 STs from Slovenia and 1 ST from the Czech Republic, no common ST across the three countries • 38 MLVA types • Some isolates with common herd origin had same ST and MLVA, but mostly high diversity observed • Most STs were further differentiated by MLVA, representing multiple variants	Felde et al., 2018b
PFGE	Total = 39 Belgium (n=31 isolates from 21 farms) Denmark (n=2, field isolates) Lithuania (n=4 isolates from 2 farms) Reference strains (n=2; 232, J)	• High heterogeneity between isolates originating from different herds, including proximate farms • More homogeneity in isolates originating from a single herd	Stakenborg et al., 2005b
RAPD	Total = 24 USA (n=23, field isolates) Reference strain (n=1; J)	• Significant heterogeneity in field isolates, even in isolates from same geographical locations	Artiushin and Minion, 1996

RAPD	Total = 7 Belgium (n=6, 3 isolates from pigs with EP, 3 isolates from pigs with subclinical EP, all isolates from different herds) Denmark (n=1 isolate)	• Isolates classified as highly, moderately or low virulent after challenge infections based on the severity of clinical signs and lung lesions • 5000 bp RAPD fragment only found in highly and moderately virulent isolates	Vicca et al., 2003
Sequencing of VNTRs	Total = 115 Switzerland (n=95, clinical samples from 34 farms, 3 isolates) France (n=10 isolates) Canada (n=1 isolate) Australia (n=1 isolate) Reference Strains (n=5; 232, NCTC 10110, JF184a, J, 7448)	• Heterogeneity in *M. hyopneumoniae* field isolates with no strong geographic association, • Clones were responsible for local outbreaks	Mayor et al., 2007
	Total = 51 Spain (n=46, 25 clinical samples from 10 vaccinated farms, 21 clinical samples from 10 non-vaccinated farms) 5 vaccine strains	• Using 4 loci, 72.1% of isolates could be typed, identified 12 genotypes in vaccinated farms and 12 genotypes in non-vaccinated farms • Using 2 loci, 88.4% of isolates could be typed, identified 9 genotypes in vaccinated farms, genotypes in non-vaccinated farms and one genotype common to a vaccinated and non-vaccinated farm • 1-3 types found per farm • In vaccinated farms, the isolates differed to strains used in vaccines • Using 2 loci, the same type was found in pigs with same breeding origins, using 4 loci this link was not found • No multiple genotypes found in individual animals	Garza-Moreno et al., 2019b

expression of surface proteins to be turned on or off (Chopra-Dewasthaly et al., 2012, 2008; Citti et al., 2010), and reciprocal recombination, in which genetic material is mutually exchanged between two genes (McGowin and Totten, 2017). While these processes have not been described in *M. hyopneumoniae,* their surface antigens do vary, namely through size variation by both genetic and proteomic mechanisms, as described below.

2.4.1 Size variation through VNTRs

Variable number tandem repeats (VNTRs) arise due to homologues recombination and slipped-strand mispairing during DNA replication (Vergnaud and Pourcel, 2009).VNTRs in intergenic regions can promote genome rearrangements (Rocha and Blanchard, 2002). However, when they occur within CDSs, they produce protein variants with a distinct number of amino acid repeats, known as variable number tandem amino acid repeats (VNTARs).

Variations in the size of surface proteins not only assist masking *M. hyopneumoniae* from the porcine immune system, but can also affect binding ability. Near the P97 C-terminal, in a region designated repeat region 1 (R1 or RR1), a minimum of eight repeats was needed to mediate attachment to swine cilia (Hsu and Minion, 1998; Minion et al., 2000; Wilton et al., 1998). Deutscher et al. (2010) also observed that the VNTRs in R1 as well as another region, R2, are involved in binding extracellular matrix proteins in Mhp271, a P97 paralogue. Nevertheless, Pulgaron et al. (2015) observed that a portion of *M. hyopneumoniae* strains isolated from diseased lungs only had three repeats in P97R1, and strains with forty and forty-three repeats were only detected in apparently healthy lungs. Castro et al. (2006) reported that VNTRs sizes differed considerably after examining twelve VNTR-containing CDSs from four pathogenic *M. hyopneumoniae* strains (232, 7448, 7442 and PMS), and one avirulent strain (J). Strain J had the least amount of VNTRs in three CDSs (P97R1, P146R3, P216R1), the most in three (H2R1, H2R2, H5R2), and was mid-range in the other fourteen regions assessed. Wilton et al. (1998) also demonstrated that the numbers of repeats in P97R1, as well as in P97R2, differed across a selection of international strains. While VNTRs provide *M. hyopneumoniae* with a repertoire of surface variants, their role in virulence, if any, warrants further investigation.

Variations in P97 have also been identified, with Zhang et al. (1995) reporting it missing in one pathogenic field isolate and one avirulent reference strain, though size variations were reported. Likewise, Assunção et al. (2005b), reported 10/18 isolates with full length P97, seven that had undergone size variation, and one isolate had P97 missing altogether. Petersen et al. (2019) showed that sera generated from pigs naturally and experimentally infected reacted poorly to P97, as well as P102, though they reacted strongly to P97 and P102 paralogues.

Polymorphisms in VNTRs form the basis of genotyping techniques such as MLVA, yet it is prudent to note that mycoplasmas have a high mutation rate (Citti et al., 2010). While Assunção et al. (2005a,b) and Dos Santos et al. (2015a) did not detect variations in VNTRs within individual colonies after various serial passages, after 380 passages Lui et al. (2013) detected several mutations within strain 168-L VNTRs, including SNPs in P97 and an insertion in P146 (the two prime targets for MLVA). It is unclear how environmental changes and stressors could alter VNTRs within a strain and whether these changes could affect genotyping.

2.4.2 Size variation through proteolytic processing

M. hyopneumoniae can further diversify its cell surface antigens by proteolytic processing (Tacchi et al., 2016). This unique post-translational modification is irreversible and by either trimming a protein at its termini (exoproteolysis) or cleaving a protein into shorter fragments (endoproteolysis), proteolytic processing can alter a protein's size, structure, and activity (Lange and Overall, 2013).

All proteins ultimately undergo regulated or processive proteolysis in which unrequired, damaged or misfolded proteins are tagged and degraded into their amino acid constituents (Taylor, 1993). However, some proteins undergo limited or non-processive proteolysis in which a protein is not fully degraded, and cleavage gives rise to either the functional form of the protein, or to biologically active cleaved fragments (Konovalova et al., 2014).

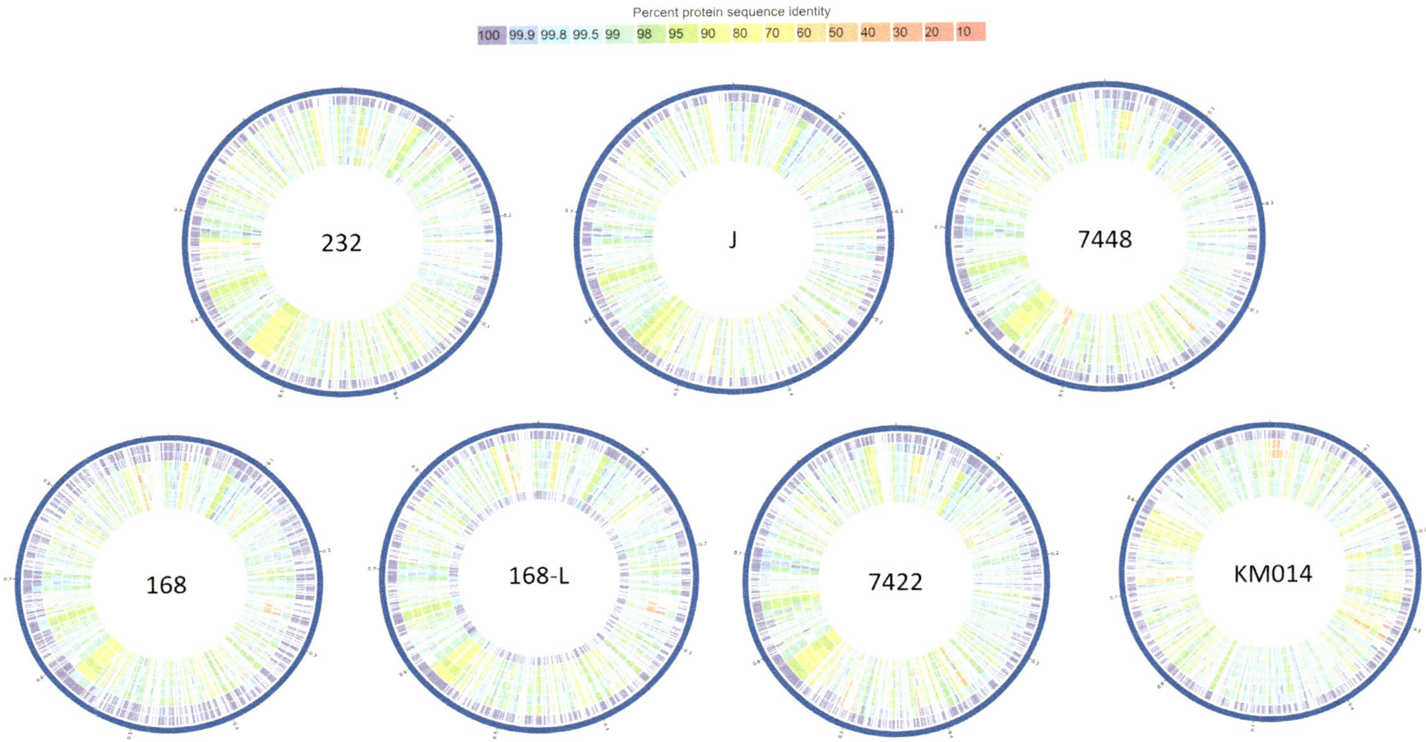

Figure 2.5. Proteomic comparison of sequenced *M. hyopneumoniae* strains. The blue outer ring signifies the strain being compared (strain name in the centre). Made by the comprehensive proteomic comparison tool in PATRIC (Wattam et al., 2017). (Source: V. Jarocki)

2.6 Concluding remarks

2

Despite a small genome, *M. hyopneumoniae* can diversify its surface proteins through size variation, causing antigenic variations and frustrating efforts to control and effectively eliminate it from commercial production facilities. *M. hyopneumoniae* also presents enough genetic diversity to enable classification into various strains. However, the lack of standardized terminology and characterization approaches currently impedes cross literature comparisons and epidemiological studies. Comparative proteomic studies focusing on expression profiles have identified potential virulence factors, however many suffer the limitations of all *in vitro* experiments. Much remains in comprehending the diverse biology of this perfidious pathogen.

CHAPTER 3

MYCOPLASMA HYOPNEUMONIAE PATHOGENICITY: THE KNOWN AND THE UNKNOWN

Peter Kuhnert[1]

Jörg Jores[1]

1 Vetsuisse Faculty, University of Bern, Switzerland

3.1 Introduction

Pathogenicity by definition is the ability of a microorganism to cause disease. This ability depends on virulence factors that drive host-pathogen interactions, enabling bacteria to replicate and spread in the host by both, sabotaging and avoiding the immune system. During infection, *Mycoplasma hyopneumoniae* (*M. hyopneumoniae*) must adhere to, and colonize the ciliated lung epithelium, and then fight the hosts' continuous efforts to expel the pathogen through mucociliary clearance and various immune responses. While single cytopathic factors have been described (Geary and Walczak, 1983; Paes et al., 2017b), classical toxins have not been reported for *M. hyopneumoniae* and it can be assumed that several virulence traits drive the host-pathogen interactions. It is known that mycoplasmas depend on scavenging nutrients, and that they release molecules that induce chemokines, which in turn attract immune cells that drive inflammation. Recent investigations uncovered moonlighting functions of several *M. hyopneumoniae* molecules and extensive proteolytic processing of *M. hyopneumoniae* surface proteins. Tacchi et al. (2016) identified 35 proteins that are endoproteolytically cleaved in *M. hyopneumoniae*. These include not only adhesins, but also lipoproteins and multifunctional cytosolic proteins moonlighting at the cell surface. This massive processing leads to a very dynamic surface topography of *M. hyopneumoniae* that could well be involved in host evasion and modulation of the immune response. Currently, our understanding of *M. hyopneumoniae* pathogenicity as a sequence of events is incomplete. Given the paucity of porcine *ex vivo* and *in vivo* data generated in combination with well-defined *Mycoplasma* mutants, we are unable to delineate how a myriad of candidate virulence traits fits into a sound pathogenicity model. Therefore, it is desirable to rigorously demonstrate the role of candidate virulence factors in target tissues, *ex vivo* systems, and finally in the native host (Figure 3.1). In that respect, it should be noted that robust *ex vivo* models and a universal challenge model would foster research on host-pathogen interactions. A challenge model that resembles the natural route of infection employing an aerosol chamber (Czaja et al., 2002) would be most appropriate and in line with the principles of Replacement, Reduction and Refinement (the 3R's) governing humane experimental expectations.

Below we summarize the current knowledge of *M. hyopneumoniae* virulence and provide a simple model of pathogenicity.

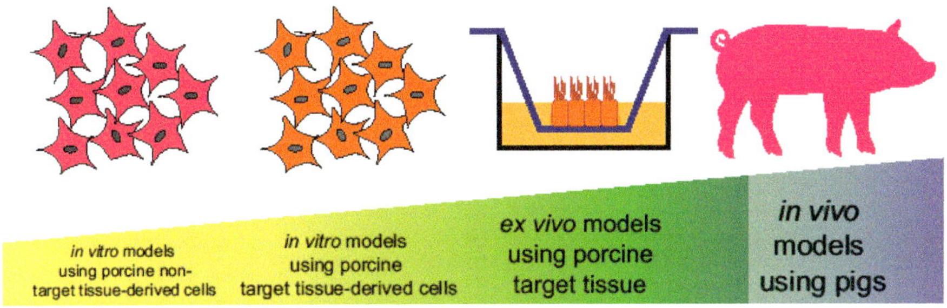

3

Figure 3.1. Likelihood of biological relevance of models used to decipher host-pathogen interactions and pin point virulence traits in *M. hyopneumoniae*. (Source: P. Kuhnert and J. Jores)

3.2 Sequence of pathogenesis

M. hyopneumoniae is an exclusive porcine pathogen that causes enzootic pneumonia (EP). The pathogenesis is not yet fully understood and therefore likely to be complex. The inspiratory flow of droplets containing the infectious agent transports the *M. hyopneumoniae* cells to their target tissue where they have to establish themselves. This flow tends to be turbulent in the larger airways including the trachea while laminar in the lower airways (Gamage et al., 2018). In order to colonize the respiratory tissue, mycoplasmas have to withstand cough (a forceful airflow exceeding 100 km/h), anatomical barriers, aerodynamic changes and the primary innate immune defense mechanism of the lung, namely the mucociliary clearance (Sjaastad et al., 2016). Healthy airways up to the bronchioles are lined by a surface layer that consists of the sticky mucus secreted by goblet cells and the low viscosity periciliary layer underneath, which covers the ciliated epithelial cells and enables their metachronal beating to expel the mucus containing microorganisms and diverse particles at a speed of several millimeters per minute (Bustamante-Marin and Ostrowski, 2017). This is the primary defense mycoplasmas have to overcome to initiate colonization of the target tissues. Highly specific adhesins are required to initiate the tissue and host specific attachment. After inspiration, *M. hyopneumoniae* organisms adhere to the ciliated epithelial cells of the trachea, bronchi and bronchioles underneath the mucous layer (Jenkins et al., 2006; Wilton et al., 2009), while it is likely that an initial adhesion step to the mucus prepones the adhesion to the ciliated cells. Already one day after infection, *M. hyopneumoniae* organisms can be detected in close association with the cilia. With transmission electron microscopy, they can also be observed between the cilia and on the tips of the microvilli of the epithelial cells (Tajima and Yagihashi, 1982). Afterwards, a cascade of events driven by the innate immune system is taking place. The recruitment of granulocytes is likely to contribute to bystander tissue damage. It was shown that alveolar macrophages, even in the presence of convalescent serum, poorly engulf *M. hyopneumoniae* (Deeney et al., 2019). Serum killing in the presence of opsonizing antibodies seems to be poor too (Deeney et al., 2019). This underlines the immune modulating capacity of *M. hyopneumoniae*, which contributes to pathogenesis. Anyway, in most cases the infection clears after several weeks to months, likely as a result of cooperation of the adaptive and innate immune system.

3.3 Adhesion

A prerequisite of *M. hyopneumoniae* infection is adhesion to the ciliated epithelium of the respiratory tract (trachea, bronchi, and bronchioles) of pigs. Adhesion is followed by the induction of ciliostasis, loss of cilia, and eventually epithelial cell death (Debey and Ross, 1994). The primary adhesin of *M. hyopneumoniae* is P97 (Hsu et al., 1997; King et al., 1997; Hsu and Minion, 1998). P97 contains two repeat regions (R1 and R2) with R1 being the cilium binding region of the protein (Hsu and Minion, 1998). For cilium binding, several pentapeptide repeats (AAKPV/E) in R1 are necessary (Minion et al., 2000). On the other side, both R1 and R2 are required to bind heparin (Jenkins et al., 2006). P97 was first described in strain 232 and has a variable size in other strains, as *e.g.* P94 in type strain J, due to variable numbers of pentapeptide repeats in the R1 region (Wilton et al., 1998; Djordjevic et al., 2004). The adhesin P97 is expressed as pre-protein P125 in strain 232 (P123 in strain J) and processed by cleavage, and mature P97 is further extensively cleaved, resulting in a functionally diverse array of fragments that bind multiple host molecules and alter the *Mycoplasma* surface (Djordjevic et al., 2004; Raymond et al., 2015). The other major adhesin, located on the same operon as P97, is P102 (Minion et al., 2004; Vasconcelos et al., 2005). P102 can also be proteolytically cleaved resulting in surface located P60 and P42 proteins involved in binding to PK15 cells (Seymour et al., 2012). Both proteins bind plasminogen, while P42 additionally binds fibronectin.

There are six paralogs for P97 adhesins identified in the *M. hyopneumoniae* genome (Table 3.1) as defined by >30% identity over 70% of the amino acid sequence (Minion et al., 2004). These paralogs are highly expressed *in vivo* (Adams et al., 2005). P146, also called Mhp684 or LppS, is a paralog of P97 containing a serine-rich region. This region provides the basis for using its sequence in genotyping (Mayor et al., 2007). A homolog identified in *M. conjunctivae* was hypothesized to be involved in the pathogenesis of infectious keratoconjunctivitis (Belloy et al., 2003). In *M. hyopneumoniae*, the protein P146 is proteolytically processed and cleaved into three major fragments P50, P40 and P85 (Bogema et al., 2012). The fact that *M. hyopneumoniae* alters the number of consecutive serine repeats in P146 [so far observed to vary from 7 up to 48, personal observation and (Garza-Moreno et al., 2019b)] might indicate that it is involved in antigen variation and immune evasion (Betlach et al., 2019). Mhp271 is another paralog of P97 and also contains repeat regions R1 and R2 (Deutscher et al., 2010). Mhp271 binds porcine cilia but this binding is less restricted to R1. In addition to heparin, it also binds fibronectin and for both binding partners an interaction with R2 is crucial. Like P97, Mhp271 is proteolytically processed resulting in a similar surface-located protein. Mhp385 is yet another example of a P97 paralog that binds to porcine cilia, heparin and fibronectin (Deutscher et al., 2012). It also contains a short R1 region and is proteolytically cleaved resulting in three fragments (P115, P88 and P27), which are located at the surface. While the P97 paralog Mhp107 also binds heparin, fibronectin and plasminogen, different protein regions are responsible for specific binding to these three compounds (Seymour et al., 2011). Each of the three portions of Mhp107 were able to bind porcine cilia but none contained an R1 motif. Proteolytic cleavage has been reported by Tacchi et al. (2016). The P97 paralog P216 (Mhp493) is cleaved into surface accessible heparin-binding proteins P120 and P85 (Wilton et al., 2009). It also binds porcine cilia and lacks the R1 motif. No functional data is available for the P97 paralog Mhp280 (Adams et al., 2005).

Besides P97 paralogs described above, *M. hyopneumoniae* has six highly expressed paralogs of P102, which are described below. P116 (Mhp108) is one of them (Seymour et al., 2010) and adheres to PK15 cells as well as porcine cilia, and binds fibronectin and plasminogen. It is located at the cell surface and can be extensively cleaved to multiple fragments. The P102 paralogs Mhp272, Mhp274 and Mhp275 have not been functionally characterized yet (Adams et al., 2005). The P102 paralog Mhp384 is located at the cell surface and cleaved into P60 and P50 (Deutscher et al., 2012). It binds to cilia and heparin. Finally, the P102 paralog P135 (Mhp683) is cleaved into the three fragments P45, P48 and P50 residing on the cell surface (Bogema et al., 2011). It binds to heparin and porcine cilia.

Table 3.1. Adhesins reported for *Mycoplasma hyopneumoniae*

Adhesin name	Adhesion group	Comment	Reference
P97 (Mhp183)	P97	Membrane protein, binds cilia, heparin and actin	Hsu et al. (1997)
P146 (Mhp684)	P97 (paralog)	Binds cilia, heparin and plasminogen	Bogema et al. (2012)
Mhp271	P97 (paralog)	Binds cilia, heparin and fibronectin	Deutscher et al. (2010)
Mhp385	P97 (paralog)	Binds cilia, heparin and fibronectin	Deutscher et al. (2012)
Mhp107	P97 (paralog)	Binds cilia, heparin, plasminogen and fibronectin	Seymour et al. (2011)
P216 (Mhp493)	P97 (paralog)	Binds cilia and heparin	Wilton et al. (2009)
Mhp280	P97 (paralog)	n.a.	Adams et al. (2005)
P102 (Mhp182)	P102	Binds fibronectin and plasminogen	Seymour et al. (2012)
P116 (Mhp108)	P102 (paralog)	Binds cilia, fibronectin and plasminogen	Seymour et al. (2010)
Mhp272	P102 (paralog)	n.a.	Adams et al. (2005)
Mhp274	P102 (paralog)	n.a	Adams et al. (2005)
Mhp275	P102 (paralog)	n.a.	Adams et al. (2005)
Mhp384	P102 (paralog)	Binds cilia and heparin	Deutscher et al. (2012)
P135 (Mhp683)	P102 (paralog)	Binds cilia and heparin	Bogema et al. (2011)
P159 (Mhp494)	n.a.	Binds cilia and heparin	Burnett et al. (2006)
P216 (MHJ_0493)	n.a.	Binds cilia and heparin	Tacchi et al. (2014)
P68 (Mhp390)	n.a.	Membrane lipoprotein, binds cilia, induces apoptosis in alveolar macrophages	Liu et al. (2019)
MHJ_0125 (glutamyl aminopeptidase)	Cytosolic protein	Moonlighting at membrane, binds heparin and plasminogen	Robinson et al. (2013)
MHJ_0461 (leucine aminopeptidase)	Cytosolic protein	Moonlighting at membrane, binds heparin, plasminogen and foreign DNA	Jarocki et al. (2015)
L-lactate dehydrogenase (LDH)	Cytosolic protein	Moonlighting at membrane, binding heparin, actin, fibrinogen and epithelial surface proteins	Tacchi et al. (2016)
Elongation factor Tu (Ef-Tu)	Cytosolic protein	Moonlighting at membrane, binding plasminogen, heparin, fibronectin and actin	Widjaja et al. (2017) Yu et al., (2018b)
Glyceraldehyde 3-phosphate dehydrogenase (GAPDH)	Cytosolic protein	Moonlighting at membrane, binding plasminogen, heparin, fibronectin and actin	Berry et al. (2017)
Fructose-1,6-bisphosphate aldolase	Cytosolic protein	Moonlighting at membrane, binds fibronectin and tracheal epithelial cells	Yu et al. (2018a)

In addition to the P97 and P102 paralogs and their parental proteins accounting for 14 adhesins, *M. hyopneumoniae* offers more candidate adhesins. P159, an adhesin unrelated to the P97/P102 families (Burnett et al., 2006; Raymond et al., 2013) adheres to cilia, PK15 cells and binds heparin. It is surface located and is also proteolytically processed to dominant fragments P27, P52 and P110 and further fragments P35 and P76 (Raymond et al., 2013). P216 is another cilium adhesin that is surface located, where it undergoes cleavage to fragments P120 and P85 which can be further cleaved and bind heparin, as well as other proteins from host epithelial cells (Tacchi et al., 2014). P68 is a recently described membrane-associated cilia-binding lipoprotein (Liu et al., 2019). It was shown to induce apoptosis in porcine peripheral blood mononuclear cells (PBMC) as well as porcine alveolar macrophages and contributes to the induction of pro-inflammatory cytokines TNF-α and IL-1ß. Surprisingly, several cytosolic and metabolic proteins have been reported to moonlight at the cell surface and act as adhesion molecules (Table 3.1). This includes the glutamyl aminopeptidase MHJ_0125 binding to heparin and plasminogen (Robinson et al., 2013) and the leucine aminopeptidase MHJ_0461 that binds heparin, plasminogen, and foreign DNA (Jarocki et al., 2015). Both aminopeptidases were shown to facilitate conversion of plasminogen to plasmin.

A study by Tacchi et al. (2016) indicated that several metabolic proteins including L-lactate dehydrogenase (LDH) undergo post-translational processing and cell surface location. They reported cleavage of LDH at seven sites and observed binding of LDH to heparin, fibrinogen, actin, and other porcine epithelial surface proteins indicating that LDH might also act as an adhesin. Similarly, elongation factor Tu (Ef-Tu) was shown to bind plasminogen, heparin, fibronectin, actin and is also extensively processed by cleavage at 26 sites (Berry et al., 2017; Widjaja et al., 2017; Yu et al., 2018b). The glyceraldehyde 3-phosphate dehydrogenase (GAPDH) is another cytosolic protein of *M. hyopneumoniae* moonlighting at the cell surface, being cleaved at four specific sites and binding heparin, fibronectin, plasminogen, and actin (Berry et al., 2017). Recently, Yu et al. (2018a) identified seven possible virulence-associated factors in *M. hyopneumoniae* strain 168. One of them was the cytosolic fructose-1,6-bisphosphate aldolase moonlighting as an adhesin at the cell surface and binding fibronectin.

As shown, adhesin receptors on ciliated epithelial cells are mainly glycosaminoglycans (GAG; in particular heparin), fibronectin, and plasminogen. Recently, cell surface-associated actin has been identified as an additional receptor for P97 that could allow *M. hyopneumoniae* to stabilize colonization even once cilia are destroyed (Raymond et al., 2018b). Besides P97, a number of other putative actin-binding proteins were detected on the surface of *M. hyopneumoniae* in this study. The ability of using ubiquitous extracellular actin as a receptor would also explain the ability of *M. hyopneumoniae* to colonize extrapulmonary sites.

The high redundancy of adhesion molecules and the fact that *M. hyopneumoniae* dedicates much of its reduced genome for that purpose is surprising. It indicates that adherence is of utmost importance for a bacterium without cell wall and it also reflects the multifunctional character of these adhesins and their cleaved fragments. Several comparative studies indicate that adhesins play a major role in virulence and pathogenesis of *M. hyopneumoniae* (Liu et al., 2013; Siqueira et al., 2013). Nevertheless, candidate adhesins must be rigorously compared and tested in porcine systems that model the disease, in order to establish a meaningful hierarchy of adhesins that are targets for disease intervention.

3.4 Candidate virulence factors

Adhesion enables colonization as the starting point of infection, which is then assisted by other virulence factors allowing *e.g.* invasion, dissemination, host cell and tissue destruction. However, classical virulence factors like toxins are generally lacking in *Mycoplasma*. Therefore, for a long time it has remained unclear how these small self-replicating organisms could elicit tissue destruction. For many years the production of toxic metabolites like H_2O_2 were suggested as a virulence mechanism of certain *Mycoplasma* species. The bovine pathogen *M. mycoides* subsp. *mycoides* is able to take up glycerol efficiently, which is then metabolized by glycerolphosphate-oxidase (GlpO) leading to the production of H_2O_2. Similar metabolic pathways were described for virulence in the human pathogen *M. pneumoniae* (Hames et al., 2009). For *M. hyopneumoniae,* Ferrarini et al. (2016) reconstructed a metabolic model based on genome data. They postulated an ability of *M. hyopneumoniae* to use glycerol as a carbon source, thereby enabling the production of H_2O_2. A homologous gene (*glpD*) is present in *M. hyopneumoniae* and its protein has been suggested to act as glycerol-3-phosphatase (GlpO), rather than a dehydrogenase (Ferrarini et al., 2018). In fact, the authors could detect production of H_2O_2 in the presence of glycerol. However, this was strain dependent and the attenuated type strain J did not produce detectable amounts of H_2O_2. Indeed, functional assays for testing production of H_2O_2 in our lab were to date also negative for various strains of *M. hyopneumoniae*. Thus, whether production of hydrogen peroxide in the presence of glycerol is a possible *in vivo* virulence mechanism in *M. hyopneumoniae,* given the common presence of catalase in the host, remains to be investigated. This could explain variation in virulence reported for different strains of *M. hyopneumoniae* (Vicca et al., 2003; Villarreal et al., 2009; Woolley et al., 2012).

3

Apart from a glycerol conversion pathway described above, Ferrarini et al. (2016) used a genome-scale metabolic modeling approach and identified the myo-inositol uptake and catabolism pathway in *M. hyopneumoniae,* which is absent from the phylogenetically closely related commensal *M. flocculare*. Based on genome comparisons, *M. hyopneumoniae* is the only *Mycoplasma* species harboring the genes for the catabolism of myo-inositol. In fact, *M. hyopneumoniae* was shown to be able to take up myo-inositol and use it as an alternative energy source in the absence of glucose (Ferrarini et al., 2018). The catabolic pathway is composed of at least 10 genes located on a transcriptional unit and among others leads to the production of acetyl coenzyme-A, an essential co-factor in many metabolic reactions. Furthermore, the genes *iolA, iolC* and *iolB* from the pathway locus have the highest number of transcript reads in the *M. hyopneumoniae* genome (Siqueira et al., 2014). Since myo-inositol is freely available in the serum of pigs, it might be a suitable alternative energy source for *M. hyopneumoniae* living in the highly vascularized lungs, thereby adapting to this niche. Infection trials with a knock-out mutant of one of the pathway genes could provide further evidence for its role during infection and whether *M. hyopneumoniae* is still able to colonize the porcine lung and cause disease. In a recent study, approximately 100 non-essential and species-specific genes were identified in the highly virulent strain F7.2C using Tn-sequencing of a transposon mutant library (Trueeb et al., 2019). This set of genes contains putative virulence determinants and genes of the myo-inositol pathway, besides others, were part of that pool. Corresponding transposon mutants can now be tested in experimental infections and compared to the wild type.

Lipid associated membrane proteins (LAMP) have also been discussed as being implicated in pathogenicity of mycoplasmas by interacting with the host immune system mainly through Toll-like receptors (TLRs) (You et al., 2006; Zuo et al., 2009). In *M. hyopneumoniae*, whole membrane lipoprotein fractions were shown to induce apoptosis via caspases 3 and 8 activation *in vitro* in various cell types, including porcine PBMC (Bai et al., 2013;

Bai et al., 2015; Ni et al., 2015). Furthermore, LAMPs activate production of nitric oxide (NO) and reactive oxygen species (ROS) in the host cell. Cholesterol exacerbates apoptosis induced by *M. hyopneumoniae* and leads to increased caspase 3 activity, TNF-α expression and enhanced H_2O_2 and NO production *in vitro,* compared to *M. hyopneumoniae* alone (Liu et al., 2017).

As parasitic organisms, mycoplasmas need to scavenge nutrients including nucleotides from their environment, *i.e.* the host (Razin et al., 1964). *Mycoplasmas* are well known for their potent membrane nucleases that often make genetic experiments difficult (Minion and Goguen, 1986; Minion et al., 1993). A well-recognized membrane nuclease is MnuA, first described in *M. pulmonis* (Jarvill-Taylor et al., 1999). In *M. bovis*, the membrane located nuclease MnuA was studied in detail and shown to degrade DNA-based neutrophil extracellular traps (NETs), thereby enabling *M. bovis* to escape these traps (Gondaira et al., 2017; Mitiku et al., 2018) and not being exposed to the toxic arsenal of granulocytes. A similar mechanism was described for *M. hyopneumoniae,* since the nuclease-encoding *mnuA* gene is also present in its genome (Henthorn et al., 2018). Therefore, MmuA could be a nuclease that allows innate immune evasion of *M. hyopneumoniae* by degrading NETs and using at the same time the nucleotides for proliferation. Infection trials with a knock-out mutant of *mnuA* will provide evidence about its role as a virulence trait and whether its absence will enable clearance of the pathogen or not.

M. hyopneumoniae releases extracellular DNA that allows the organism to form biofilms on abiotic surfaces as well as *in vivo* (Raymond et al., 2018b). Biofilms are recognized as a way to resist antimicrobials, as well as the host immune responses. A capsule-like structure has been reported based on electron microscopy (Tajima and Yagihashi, 1982). It is known from *M. pulmonis* that a capsule prevents phagocytosis (Bolland et al., 2012), but interestingly, *in vitro* data from other species such as *M. mycoides* and *M. agalactiae* (Gaurivaud et al., 2016; Schieck et al., 2016) has shown that capsulated cells are less protected from serum complement than non-capsulated mutants. Recently, at least for *M. mycoides,* a capsule has been shown to be a virulence trait in the native host (Jores et al., 2018). However, the presence of a capsule in *M. hyopneumoniae* was not further evidenced or confirmed by other researchers. The annotated genome sequences of *M. hyopneumoniae* do not appear to encode any capsule biosynthesis pathway. Theoretically a novel set of proteins encoded by 'hypothetical proteins' or wrongly annotated genes do not exclude the presence of a capsule.

Factors allowing immune evasion are subtle or discrete virulence traits that do support a parasitic lifestyle. Recently, *M. genitalium* was shown to possess an immunoglobulin (Ig) G binding protein called protein M. Protein M not only fixes IgG very efficiently but also prevents subsequent antigen-antibody binding and subsequent signaling pathways of the bound antibodies (Grover et al., 2014). *M. mycoides* has been shown to possess a protein that has a high structural similarity to protein M. *In silico* experiments have shown that these genes encode a *Mycoplasma* Ig binding protein (MIB) and a *Mycoplasma* Ig protease (MIP). The complex of MIB and Ig is necessary for the proteolytic activity of MIP. Moreover, it has been shown that members of the MIB-MIP system are membrane localized *in vitro* (Krasteva et al., 2014), translated *in vivo* (Weldearegay et al., 2016) and active *in vivo* (Jores et al., 2019). The two proteins are encoded by two genes organized in tandem and often detected in multiple copies in various mycoplasmas, including *M. hyopneumoniae* (Arfi et al., 2016). Preliminary studies in our laboratory indicate that *M. hyopneumoniae* shows proteolytic activity towards Ig. However, further studies including knock-out mutants of this redundant system are needed to investigate the role of the MIB-MIP system in virulence and immune evasion of *M. hyopneumoniae*. The same holds true for the recent finding of Jarocki et al. (2019) reporting absence of formylated N-terminal methionines in *M. hyopneumoniae*. Since formylated peptide receptor covers white blood cells, the lack of formylated peptides could aid *M. hyopneumoniae* in immune evasion.

Several large-scale comparisons on the genome, transcriptome, proteome, metabolome and secretome level were performed in order to define virulence and pathogenesis of *M. hyopneumoniae* (Pinto et al., 2009; Liu et al., 2013; Siqueira et al., 2013; Siqueira et al., 2014; Ferrarini et al., 2016; Paes et al., 2017a). Comparing the pathogenic strain 168 with its attenuated derivative, Liu et al. (2013) reported that besides the known virulence-associated proteins (mainly adhesins), mutations were also found in genes involved in metabolism and growth. In a genome-reduced organism like *M. hyopneumoniae* that is characterized by a limited set of biosynthetic pathways, it is not surprising that further loss of enzymatic functions has a dramatic influence on survival and growth of the microorganism. This also holds true for lipoproteins involved in nutrient acquisition (Browning et al., 2011). A genome comparison of *M. hyopneumoniae* and *M. flocculare* revealed differences in genome structure and organization (Siqueira et al., 2013). In *M. flocculare,* some genes of the P97 adhesin family were absent, showed sequence differences, or were missing domains involved in adhesion to host cells. However, there were no specific factors identified that could explain the pathogenic nature of *M. hyopneumoniae* compared to the commensal *M. flocculare*.

Integrative and conjugative elements (ICE) are self-transmissible chromosomal mobile elements involved in horizontal gene transfer, thereby providing new virulence and/or antibiotic resistance traits (Johnson and Grossman, 2015). Such ICE have been identified in mycoplasmas as well (Citti et al., 2018). For *M. hyopneumoniae* it was shown that ICE are part of the genomes of pathogenic strains 7448 and 232, but not of the attenuated strain J (Pinto et al., 2007a). However, an ICE was identified in the pathogenic strain 168 and likewise in its attenuated variant (Liu et al., 2013). Thus, their role in encoding virulence traits of *M. hyopneumoniae* remains unclear.

Variation in expression levels of virulence-associated genes like adhesins is likely to contribute to pathogenicity. A proteome analysis of *in vitro* cultures comparing the non-pathogenic strain J to virulent strains 7448 and 7422 supports this hypothesis (Pinto et al., 2009). The authors identified 64 proteins, including noticeably P97, being expressed at higher level in the pathogenic strains compared to the non-pathogenic strain.

Finally, the secretomes of *M. hyopneumoniae* and *M. flocculare* grown in a serum reduced medium were compared recently, looking more precisely at factors directly involved in *Mycoplasma*-host interaction (Paes et al., 2017b). A higher number of secreted proteins was identified in *M. hyopneumoniae* compared to *M. flocculare* (62 vs. 26). While *M. hyopneumoniae* secreted adhesins, methylases, nucleases and lipoproteins, *M. flocculare* secreted only two adhesins that were also identified in the *M. hyopneumoniae* secretome. However, there were no clear-cut factors that differed between pathogenic and non-pathogenic strains or differed between the pathogen *M. hyopneumoniae* and the commensal *M. flocculare*.

3.5 Immune modulation and *Mycoplasma* host interaction

The host immune response is considered to be the main driver of pulmonary lesions. At the same time, *M. hyopneumoniae* modulates the immune response in order to persist in the host. Inflammatory cytokines such as IL-1 and TNF-α are induced during infection and plasmin is recognized to be central for the regulation of inflammatory responses (Choi et al., 2006; Woolley et al., 2013). A number of the P97/P102 adhesin family members interact with plasminogen of the porcine host and enhance its activation to plasmin (Meyns et al., 2007; Raymond and Djordjevic, 2015), a serine protease which in turn stimulates macrophage signaling resulting in production of ROS and cytokine release, thereby contributing to inflammation (Syrovets et al., 2012). The moonlighting leucine aminopeptidase was shown to act as a multifunctional protein, including binding of plasminogen and enhancing its conversion to plasmin on the *M. hyopneumoniae* cell surface (Jarocki et al., 2015). Plasminogen is readily available in the ciliated airways affected by *M. hyopneumoniae* (Seymour et al., 2012; Woolley et al., 2013).

The chronic nature of the disease caused by *M. hyopneumoniae* suggests difficulties of the host immune system to effectively clear the bacterium, indicating a certain degree of immune evasion. It was shown that *M. hyopneumoniae* can evade phagocytic uptake by porcine alveolar macrophages (Deeney et al., 2019). On the other hand, interaction with surface accessible actin on the epithelial cells and causing cytoskeletal rearrangements could allow the organism to be endocytosed (Raymond et al., 2018b). It was shown that a subpopulation of *M. hyopneumoniae* that binds to ß1 integrin can invade porcine epithelial cells. These mycoplasma can survive within the endosomes, escape them, and reside within the cytoplasm (Raymond et al., 2018). Thereby, this mechanism might not only contribute to evasion from the immune system but could also support the dissemination of the pathogen to internal organs and persistence within the pig without causing disease. In fact, experimental studies have demonstrated that *M. hyopneumoniae* can persist in lungs for a very long period, at least 214 days, and that infected pigs also remain contagious during at least this period (Pieters et al., 2009). Isolation of *M. hyopneumoniae* from tissue other than the lung has also been reported, suggesting that spread may occur via the lymphatic or blood circulation. On that note, the agent was re-isolated from the liver, spleen and kidneys of experimentally infected and in contact-infected pigs (Le Carrou et al., 2006; Marois et al., 2007; Marchioro et al., 2013) and *M. hyopneumoniae* DNA has been detected in these same tissues (Woolley et al., 2012), as well as in other body fluids, such as piglet processing fluids (Vilalta et al., 2019). However, this spread within the body appears to be transient and no association has been shown between the presence of *M. hyopneumoniae* in these organs and lesions. Nevertheless, such evasion strategies of *M. hyopneumoniae* strains could be the reason for recurrent outbreaks of EP in an otherwise EP-free country like Switzerland (Overesch and Kuhnert, 2017).

3.6 Pathogenicity model

As stated above, a sound pathogenicity model is currently missing. What we know for certain is that binding to epithelial cells enables colonization and results in ciliostasis and cilial destruction (Figure 3.2). This is followed by an influx of cells of the innate immune system such as neutrophil granulocytes and macrophages (Figure 3.3.A), that are triggered by chemotaxis resulting from *M. hyopneumoniae* molecules. The destruction of the mucociliary apparatus, together with down-modulating the immune response at later stages, can also enhance the susceptibility of *M. hyopneumoniae* infected pigs to secondary pathogens (Thacker et al., 1999; Muneta et al., 2008; Shen et al., 2017). Which mechanisms described above are contributing to what extent, and in which sequence, is currently unknown and deserves more research into that interesting subject of innate immunity. With time, the number of granulocytes diminishes and lymphoid cells increase (Figure 3.3.B). As to how much cells such as granulocytes contribute to pathomorphological changes, like edema, needs to be studied in future efforts.

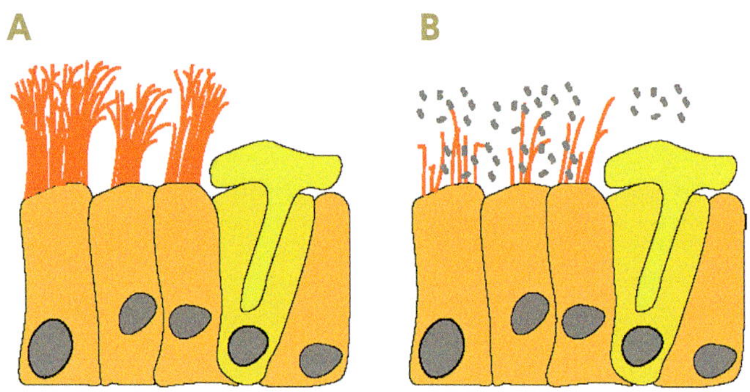

Figure 3.2. Loss of microvilli of the trachea, bronchus and bronchiole after infection with *M. hyopneumoniae*. **(A)** Physiological state. **(B)** State after infection. The non-ciliated mucus-producing goblet cells are displayed in green. (Source: P. Kuhnert and J. Jores)

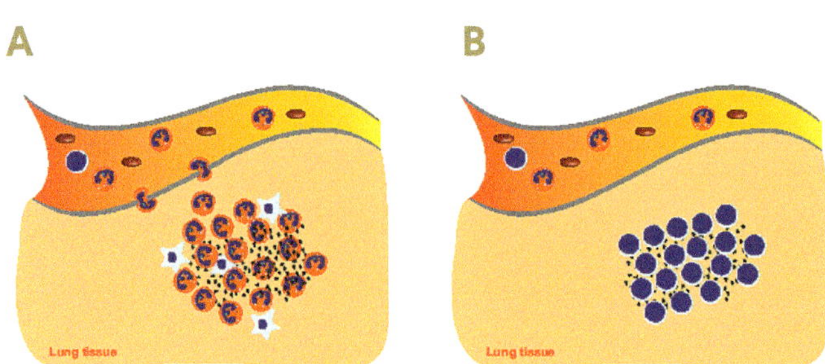

Figure 3.3. Sequence of involvement of immune cells in disease. **(A)** *Mycoplasma* cells trigger the recruitment of innate immune cells such as neutrophil granulocytes and macrophages to the infected tissue. **(B)** The number of neutrophil granulocytes diminishes and lymphoid cells increase. (Source: P. Kuhnert and J. Jores)

CHAPTER 4

EPIDEMIOLOGY OF *MYCOPLASMA HYOPNEUMONIAE* INFECTIONS

Maria Pieters[1]

Alyssa Betlach[1,2]

Heiko Nathues[3]

Dominiek Maes[4]

1 College of Veterinary Medicine, University of Minnesota, St. Paul, USA
2 Swine Vet Center, P.A., St. Peter, USA
3 Vetsuisse Faculty, University of Bern, Switzerland
4 Faculty of Veterinary Medicine, Ghent University, Ghent, Belgium

4.1 Introduction

A clear understanding of the epidemiology of *Mycoplasma hyopneumoniae* (*M. hyopneumoniae*) is key for disease control and prevention, especially in times when pathogen elimination is highly pursued. Thus, in this chapter we focus on aspects related to prevalence, infection dynamics, transmission, risk factors, and distribution of *M. hyopneumoniae*. Information in this chapter has been compiled from published literature and practical experiences in the field, and is based on the most recent molecular characterization techniques used for this bacterium.

4.2 Prevalence

Being non-reportable, infections caused by *M. hyopneumoniae* are usually not officially registered. However, infections with this pathogen are noticed in the field. In general, it is thought that more than 70% of swine herds around the world are infected with *M. hyopneumoniae* (Pieters and Maes, 2019). While most data on infections is associated with the use of vaccine and antimicrobials for disease control, there is no data or certification to account for pathogen presence or absence in most countries, other than data on the SPF-program in Denmark, which includes approximately 2500 herds of serological monitoring. Following eradication programs at the national level, Switzerland (Stark et al., 2007), Norway (Gulliksen et al., 2019), and Finland (Rautiainen et al., 2001) claim to be *M. hyopneumoniae* free, although eventual detection of *M. hyopneumoniae* in wild boars in Switzerland has been reported (Kuhnert and Overesch, 2014; Kuhnert et al., 2011).

A common practice (in certain countries) for detection of issues suggestive of *M. hyopneumoniae* infections is lung lesion scoring at abattoirs. Nevertheless, the lack of pathognomonic lesions and the potential for lesions to heal prior to market age (Noyes et al., 1990), make it difficult for lung lesion evaluation at abattoir to show a full picture of pathogen prevalence in the field.

At the farm level, most infections with *M. hyopneumoniae* are diagnosed in growing pigs, with finishers being more commonly detected positive for the bacterium than nursery pigs. However, it is important to note that pigs of all ages are susceptible to *M. hyopneumoniae* infections, and in fact, circulation of this bacterium in breed-to-wean herds is very relevant for disease transmission and has become a topic of significant discussion in recent years.

While mortality induced by *M. hyopneumoniae* is generally low, in uncomplicated cases, the morbidity of this bacterium is high in farms. Once established in a herd, the infection remains endemic. Most likely, the slow reproduction ratio (R_n) of *M. hyopneumoniae* (Meyns et al., 2004; Villarreal et al., 2011a) is responsible for this behavior. All pigs in a herd will not become infected with the bacterium at once, on the contrary, a small proportion of pigs gets infected, leaving a significant number of naïve pigs ready to encounter shedding pigs at any moment, thus perpetuating disease transmission in the population. Recent studies have indicated that approximately 20% of gilts are still negative to *M. hyopneumoniae* up to first farrowing, regardless of being introduced into endemically infected farms (Takeuti et al., 2017c).

4.3 Infection dynamics

A slow pace characterizes *M. hyopneumoniae* in many senses. For example, bacterial duplication in the laboratory takes several hours (Meyns et al., 2007), transmission rates have been estimated to be low (Meyns et al., 2004), clinical signs become apparent only after weeks of infection and many infected pigs do not show overt clinical signs, and recovery from infection is lengthy (Pieters et al., 2009). In general, the dynamics of *M. hyopneumoniae* infections are characterized by non-clinical periods in which the actual presence of the pathogen in the host may be suspected, but confirmation is difficult to obtain.

Infection with *M. hyopneumoniae* occurs via host contact with respiratory secretions contaminated with the bacterium. The actual route of entry of this pathogen into a susceptible host is considered respiratory only. Whether it is by nose-to-nose contact, barn air space sharing, or transmission in the air over longer distances, infection will only result as the pathogen gains access to and colonizes the respiratory tract. Little information is available about particle size of the infectious matrix and the capability of *M. hyopneumoniae* to colonize pigs. However, it can be hypothesized that a specific matrix particle size may be needed for infection to take place, based on respiratory physiology and the capability of *M. hyopneumoniae* to enter deep into the airways.

The minimal dose to induce infection in pigs has been determined under experimental conditions. Marois et al. (2010) compared various infectious

4.6 Molecular epidemiology

The circulation of different strains has suggested potential variability of *M. hyopneumoniae*. Several techniques focused on nucleic acid amplification, genomic sequencing, and the identification of Variable Number Tandem Repeat (VNTR) differences in surface-adhesin loci have been developed to molecularly characterize this microorganism. Partial (Mayor et al., 2007; Bogema et al., 2012) or complete genome sequencing (Minion et al., 2004; Vasconcelos et al., 2005) and typing methods, such as Multiple-locus Variable number tandem Repeat Analysis (MLVA; de Castro et al., 2006; Vranckx et al., 2011; Dos Santos et al., 2015a), have been commonly employed due to their high reproducibility, feasibility, and discriminatory power. Due to the fastidious nature of *M. hyopneumoniae,* these techniques have been validated using clinical specimens, as a bacterial isolate is difficult to obtain. Thus far, the utilization of these molecular methods has been primarily employed in research settings. However, they have been used to aid in surveillance and outbreak investigations (Anderson et al., 2017).

Through the employment of MLVA, a wide circulation of *M. hyopneumoniae* VNTR types has been identified within and across multiple countries and individual states (Vranckx et al., 2011; Dos Santos et al., 2015a; Takeuti et al., 2017b). Furthermore, the presence of several *M. hyopneumoniae* VNTR types within infected, closed herds and even individual pigs has been shown (Nathues et al., 2011; Michiels et al., 2017). Although genomic heterogeneity has been identified across different population dynamics, the potential sources and drivers for this molecular variability has been poorly explored. Michiels et al. (2017) observed that groups of pigs in which more than one *M. hyopneumoniae* VNTR type was detected, exhibited increased lung lesion severity. Other studies did not observe an association between genetic clusters of *M. hyopneumoniae* and lung lesion severity (Charlebois et al., 2014). In the field, differences in clinical performance and disease severity has been suggested to vary among *M. hyopneumoniae* strains. Therefore, further insight in the molecular epidemiology of *M. hyopneumoniae* infections may support the control of this microorganism.

In general, the epidemiology of *M. hyopneumoniae* has not changed drastically in the most recent years, but the understanding of it has. Newer and more accurate techniques and models have allowed for the development of a different picture of the disease dynamics. However, to date no diagnostic techniques or methodologies allow forecasting virulence or the degree of vaccine protection.

CHAPTER 5

MYCOPLASMA HYOPNEUMONIAE CLINICAL SIGNS AND GROSS LUNG LESIONS, INCLUDING MONITORING

John Carr[1]

Marina Sibila[2]

Joaquim Segalés[3]

1 College of Public Health, Medical and Veterinary Sciences, James Cook University, Queensland, Australia
2 Centre de Recerca en Sanitat Animal (CReSA), Institut de Recerca i Tecnologia Agroalimentàries (IRTA), Campus de la Universitat Autònoma de Barcelona, Bellaterra, Spain
3 Facultat de Veterinària (Universitat Autònoma de Barcelona), Centre de Recerca en Sanitat Animal (CReSA)-Institut de Recerca i Tecnologia Agroalimentàries (IRTA), Campus de la Universitat Autonoma de Barcelona, Bellaterra, Spain

5.1 Introduction

Mycoplasma hyopneumoniae (*M. hyopneumoniae*) is a respiratory swine pathogen, and therefore, the relationship between the bacteria, the pig, the microbiota of the nasopharynx/bronchi and the general environment of the animal is key to understand the clinical outcome. Importantly, within a population of infected farmed pigs, there is a group of individuals, each with their own immune stimulation state and different stages and timings of infection. Examination of a few pigs is only going to provide a snap shot of the overall clinical-pathological situation at best. In addition, understanding this complex environment, as the clinical expression associated with *M. hyopneumoniae* is always going to be complicated by the other inhabitants of the nasopharynx and upper respiratory tract (Table 5.1).

M. hyopneumoniae infection may follow a clinical or subclinical course. In fact, in the majority of pig farms, most of the adult pigs are infected by this pathogen without displaying *M. hyopneumoniae* compatible clinical signs (Maes et al., 2018).

M. hyopneumoniae is associated with three major clinical-pathological conditions, depending mainly on the co-infecting pathogens.The respiratory disease originated from the sole infection by *M. hyopneumoniae* is known as Porcine Mycoplasmosis (PM). In contrast, when other secondary bacterial agents are co-infecting the animal, the disease is named Enzootic Pneumonia (EP). The bacterial agents most frequently found in this multifactorial condition are *Pasteurella multocida, Actinobacilus pleuropneumoniae, Glaesserella (Haemophilus) parasuis* and *Streptococus suis,* among others (Sibila et al., 2009). Parasites such as *Ascaris suum* and possibly *Metastrongylus* spp. may also contribute to the overall respiratory picture. Finally, when viral pathogens (such as *Porcine circovirus 2, Porcine reproductive and respiratory syndrome virus, Aujeszky's disease virus* and others) are also contributing to the respiratory outbreak, the term Porcine Respiratory Disease Complex (PRDC) is used. However, the term PRDC can also be used when describing a multimicrobial respiratory outbreak in which *M. hyopneumoniae* is not participating (Opriessnig et al., 2011a; Maes et al., 2018). *M. hyopneumoniae* interactions with other respiratory and/or systemic pathogens are revised in Chapter 7.

Table 5.1. Pathogens and commensals that may be identified inhabiting the nasopharynx of the pig.

Bacteria	Virus	Metazoan
Actinobacillus pleuropneumoniae	African swine fever virus	*Ascaris suum*
Actinobacillus suis	Aujeszky's disease virus	*Metastrongylus apri*
Bordetella bronchiseptica	Classical swine fever virus	
Glaesserella (Haemophilus) parasuis	Porcine circovirus 1,2,3	
Mycoplasma flocculare	Porcine cytomegalovirus	
Mycoplasma hyopneumoniae	Porcine reproductive and respiratory syndrome virus	
Mycoplasma hyorhinis	Porcine respiratory coronavirus	
Pasteurella multocida	Swine influenza A virus	
Streptococcus spp.		
Trueperella pyogenes		

5

5.2 Clinical signs

The severity of clinical signs associated with *M. hyopneumoniae* infection depends on the above mentioned clinical-pathological conditions.

Porcine mycoplasmosis is characterized by a mild and non-productive cough, with minimal effect on average daily weight gain (ADWG) and other productive parameters. This scenario would be equivalent to the outcome of experimental inoculations with *M. hyopneumoniae*. Under experimental conditions, where seronegative animals are infected solely with a high dose of *M. hyopneumoniae* (10^7 - 10^8 color changing units/pig) at the same time, the appearance of a mild dry cough is expected at 1 – 2 weeks post infection (Marois et al., 2007; Vicca et al., 2007; Wolley et al., 2012). The coughing may continue for several weeks, reaching its maximum at 4 - 5 weeks after infection (Kobisch et al., 1993; Villarreal et al., 2009; García-Morante et al., 2016b, 2017b). However, the sole infection by *M. hyopneumoniae* is rarely observed under commercial conditions, since the participation of other concomitant pathogens is very frequent.

Enzootic pneumonia affects mainly growing and finishing pigs and is characterized by a non-productive cough of variable severity and duration,

Under field conditions, where secondary bacteria are frequently co-infecting the animals (EP), a suppurative bronchopneumonia is observed (Figure 5.5.A). In this case, a catarrhal-purulent exudate can be extruded from the bronchial tree under pressure (Figure 5.5.B). The catarrhal-purulent exudate results in full consolidation of the lung tissue and, therefore, sinks in water. In a PRDC scenario, besides CVPC, affected animals show interstitial pneumonia characterized by enlarged and non-collapsed lungs, rubbery in consistency, sometimes with interstitial edema (Figure 5.6). Frequently, enlargement of tracheobronchial lymph nodes does occur as well (Pieters and Maes, 2019).

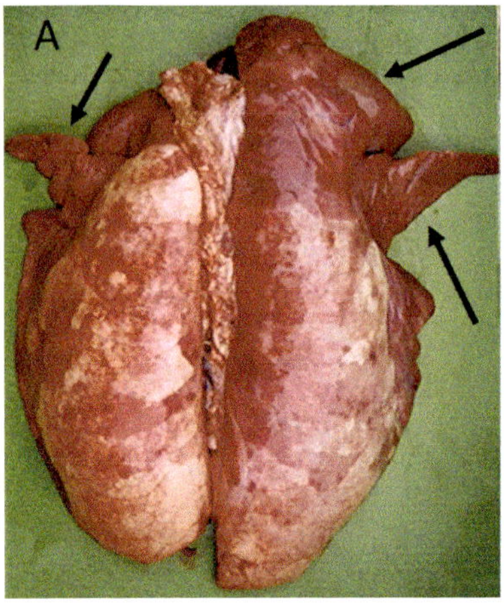

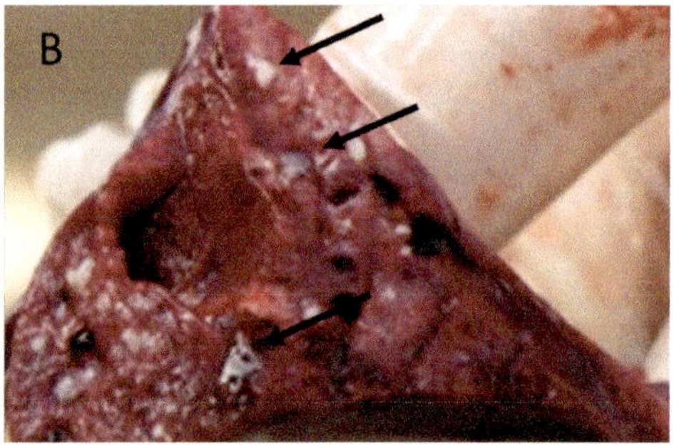

Figure 5.5. (A) Lungs of a finishing pig showing pulmonary cranioventral consolidation (arrowed) and suppurative bronchopneumonia-like lesions. (B) Catarrhal-purulent material exudates from the cut of the bronchial airways (arrowed). (Source: J. Carr)

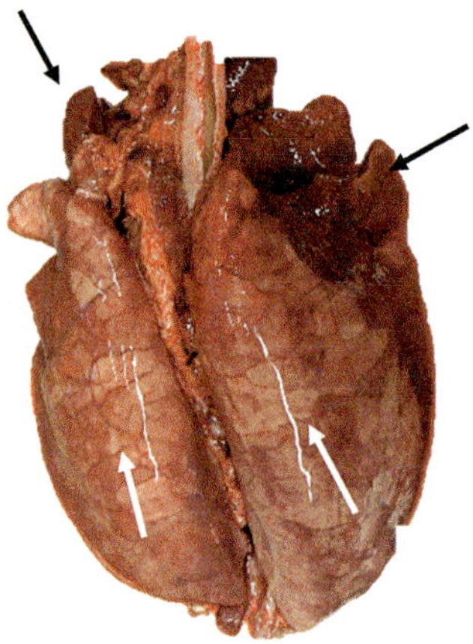

Figure 5.6. Lungs showing cranioventral consolidated areas (black arrows) together with interstitial pneumonia-like lesions (lack of pulmonary collapse, interstitial edema and tan-mottled areas, white arrows). (Source: CReSA-IRTA)

5.4 Clinical-pathological monitoring

Monitoring of potential infections by *M. hyopneumoniae* should include clinical as well as pathological assessments.

Presence of respiratory clinical signs can be visually evaluated daily by the farmer, coughing being the most prominent and indicative clinical sign. This evaluation may be assisted by sonic recordings of possible coughing events. However, the diagnostic value of the clinical assessment is limited as it can be produced by many infectious and non-infectious factors (Michiels et al., 2015) and subclinically infected pigs go unnoticed. Therefore, confirmation of the association of these signs with *M. hyopneumoniae* infection, especially in previously naïve herds, requires laboratory confirmation (Nathues et al., 2012) (see Chapter 8).

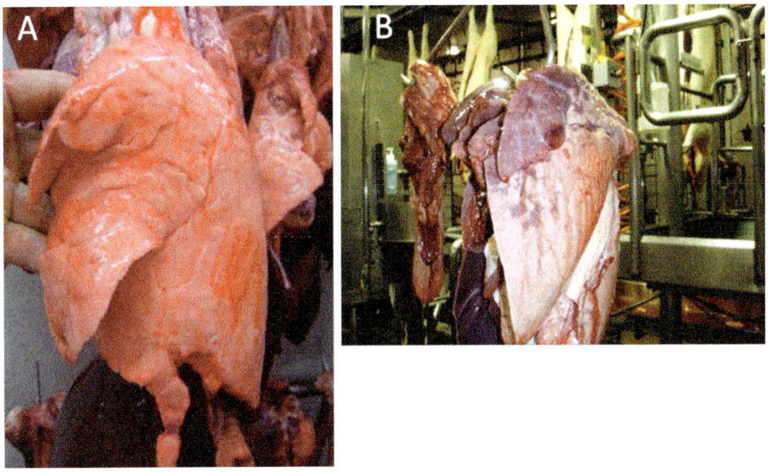

Figure 5.7. Examination, at slaughterhouse chain, of a lung without (A) and with (B) cranioventral pulmonary consolidation. (Source: J. Carr)

Reference	Ph. Eur	Morrison et al. (1985)	Straw et al. (1986)	Christensen et al. (1999)
Scoring system	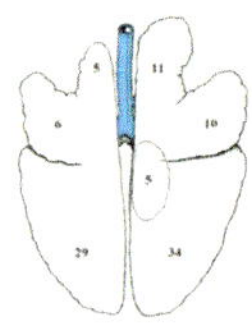	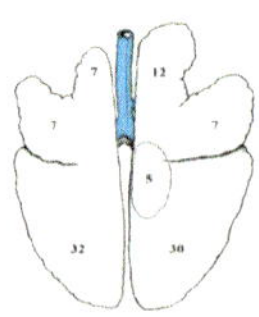	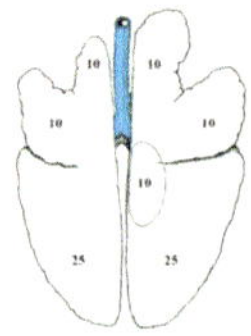	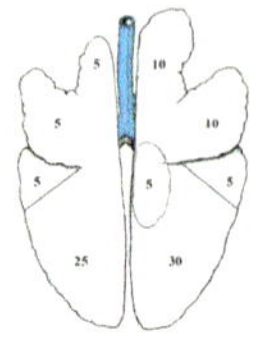
Max. Score	100%	100%	100%	100%
Method	Goodwin et al. (1969)	Madec and Kobisch (1982)	Hannan et al. (1982)	Sibila et al. (2014)
Scoring system	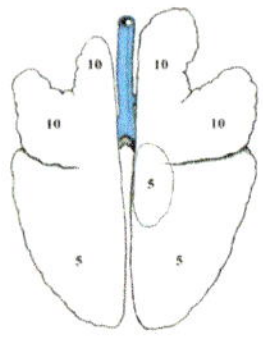	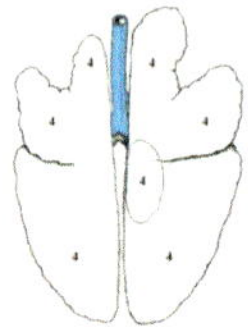		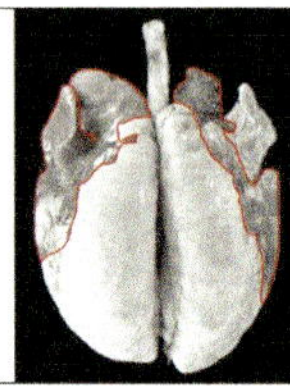
Max. Score	55 points	28 points	35 points	100%

Ph. Eur: European pharmacopeia

Figure 5.8. A review of cranioventral pulmonary consolation methods. (Adapted from Garcia-Morante et al., 2016)

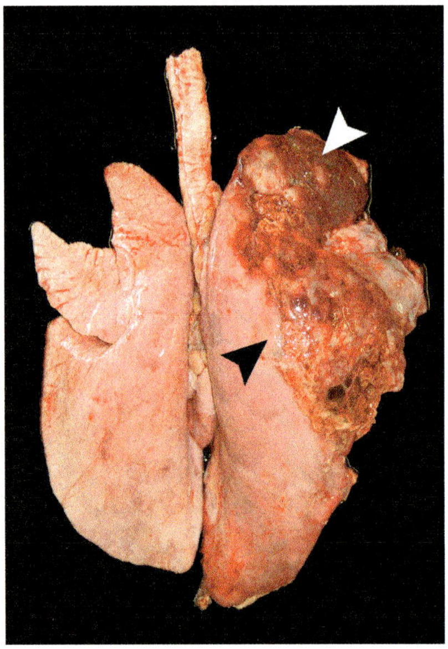

Figure 5.9. Lung showing cranioventral pulmonary consolidation (CVPC) (white arrow) together with fibrinous pleurisy lesions (black arrow) in apical and middle right lung lobes. In this case, CVPC on the right lung is not possible to be assessed (Source: CReSA-IRTA)

From a pathological point of view, slaughter pigs (Figure 5.7) or pigs subjected to necropsy can also be visually inspected by the veterinarian and the presence of *M. hyopneumoniae*-like gross lesions assessed. However, examination of lung pathology is not suited to establish an etiologic diagnosis. Anyway, different lung lesion score systems have been developed as a support for the clinical diagnosis of MP and EP. Several lung lesion scoring systems to assess the prevalence and extension of such gross lesions are described in the literature. Almost all these methods estimate the area of lung tissue affected by CVPC based on visualization and palpation. These data can be normalized by the size or the relative weight of each lobe. Nevertheless, the outcome of the main lung lesion scoring systems used in the peer-reviewed literature show good correlation between them (García-Morante et al., 2016a) (Figure 5.8). Noteworthy, presence of concomitant lesions such as fibrinous pleurisy, may obscure CVPC observation and scoring (Figure 5.9).

Abattoir examinations are challenging due to the difficulties in record taking. Therefore, a fast scoring system, in accordance with the speed of the slaughter line, is preferred. Using a voice activated recording device greatly simplifies records and allows the hands to manipulate the pluck. Photographs may be ideal, but they need to be reviewed later, which is time consuming and slaughterhouses may refuse to allow a camera into the premise. On the slaughter line, the lungs are removed from the carcasses and it is vital that the observer positively identifies the origin of the lungs under examination. It is also important that the clinician examines a statistically significant number of representative lungs from the farm. In EP positive farms, evaluation of the lungs at the abattoir may allow correlation of the cough at the herd level with the degree of consolidation in the slaughterhouse. However, as most infected farms are vaccinating against *M. hyopneumoniae*, the coughing as well as the slaughterhouse examination is influenced by the vaccination.

The value of the clinical-pathological monitoring is high since it provides insight into the severity degree of the *M. hyopneumoniae* infection. However, monitoring assessment must be coupled with laboratory detection of the bacterium or evidence of seroconversion (see Chapter 8) to confirm the specific etiology.

CHAPTER 6

IMMUNE RESPONSES AGAINST PORCINE *MYCOPLASMA* INFECTIONS

Artur Summerfield[1]

1 Institute of Virology and Immunology, Faculty of Veterinary Medicine, University of Bern, Switzerland

6.1 Introduction

The immune responses against *Mycoplasma* infections are relatively inefficient at clearing the infection and there is a tendency for mycoplasmas to establish chronic infections. Nevertheless, *Mycoplasma* infections are associated with strong inflammatory responses, which in cases of clinical diseases tend to cause chronic inflammation with immune responses that are inefficient and can even participate in disease pathogenesis (Dobbs et al., 2009; Jones and Simecka, 2003). At the level of the epithelia, hyperplasia of epithelial cells - perivascular and peribronchiolar accumulation of mononuclear cells, is observed. In addition, it appears that mycoplasmas cause ciliostasis and cilia damage thereby severely affecting first line immunological defense mechanisms (DeBey and Ross, 1994). Lung infections by *Mycoplasma* are associated with a heavy infiltration of lymphoid cells, in particular T helper cells, which appear to have a pathogenic role (Jones and Simecka, 2003; Dobbs et al., 2009). In addition, natural infection appears to confer resistance against reinfection, suggesting the existence of protective adaptive immune responses (Kobisch and Friis, 1996). However, in an experimental study, no protection was induced by a primary infection with low virulent *M. hyopneumoniae*, and even signs of enhanced disease were observed (Villarreal et al., 2009). The reason for this discrepancy is unclear but one possibility is the poor immunogenicity of low virulent strains or the timing of the challenge in this experiment. Early studies in mice which have a defective adaptive immune response indicate that innate immunity has a major role to play in controlling *M. pulmonis* (Cartner et al., 1998). A central theme of *Mycoplasma* infections is that protective immune responses should be viewed in the light of a balance between beneficial responses and those contributing to disease pathogenesis. Importantly, these can be the same immunological mechanisms (Hickman-Davis, 2002; Hodge and Simecka, 2002; Dobbs et al., 2009).

This chapter will give an overview of the current knowledge on immune responses against *M. hyopneumoniae* by integrating the knowledge acquired from other *Mycoplasma* infections, including mouse models.

6.2 Innate immune responses

6.2.1 Overview of innate immune responses

Innate immunity represents the first line of defense that is absolutely essential to prevent or control early pathogen replication before adaptive immune responses are induced. For a pathogen like *M. hyopneumoniae* which colonizes the epithelium of the respiratory tract, relevant immune responses start at the surfaces of the respiratory tract, including physical protection of the epithelium by mucus, local production of opsonins like collectins, local production of antimicrobial peptides, ciliary activity to physically remove any attached bacteria (Abbas et al., 2016), and possibly also the secretion of polyreactive immunoglobulin (Ig)A (Mouquet and Nussenzweig, 2012). A second line of defense is initiated after contact of the mycoplasmas with epithelial cells and cells of the immune system present locally in the airways, the epithelium itself, the *lamina propria* or deeper submucosal areas. These can be macrophages, neutrophils and dendritic cells (DC) each playing different roles in the innate immune response. Macrophages play an important role in tissue homeostasis and regulation of immune responses but are highly phagocytic and after activation acquire potent antimicrobial activity. In contrast, neutrophils are typically attracted in large numbers in the course of inflammation. These cells represent an absolutely essential component of the innate immune response against bacteria. They are also highly phagocytic but possess much more and dangerous weapons against bacteria, including a large array of aggressive enzymes and reactive oxygen species. Finally, DC represent the professional antigen presenting cell required for the induction of naïve T-cell responses. They also participate in innate immune defenses through their strong ability to sense invading pathogens and to respond with an array of different cytokines and chemokines. They play a major role in directing the type of immune response. The contact of mycoplasmas with these cells is expected to result in their activation through particular pathogen recognition receptors which recognize microbial-associated molecular patterns (MAMPS, Chapter 6.2.2), and results in the release of pro-inflammatory mediators. In this respect, it is important to understand that the most important part of an innate immune response to a bacterial infection represents inflammation (Abbas et al., 2016). This inflammation is absolutely essential for the control of the bacterial infection and also for induction of adaptive immune responses. It will guide the type of adaptive immune response induced. Direct effector functions of the innate immune response are: (i) phagocytosis with destruction of the pathogen in the phagolysosome by reactive nitrite, reactive oxygen metabolites as well as enzymes; (ii) complement-mediated killing of the bacteria; (iii) killing by released antimicrobial peptides and other products such as proteases and reactive oxygen species (Abbas et al., 2016).

6.2.2 Inflammatory responses after infection of pigs with *Mycoplasma hyopneumoniae*

After experimental infection of pigs, *M. hyopneumoniae* can be first detected at the luminal surface of bronchial and bronchiolar epithelial cells at seven days post inoculation (DPI). Later during the infection, mycoplasmas are also in association with alveolar and interstitial macrophages as well as type I pneumocytes in the alveoli (Kwon et al., 2002). This colonization of the lower respiratory tract is accompanied with a strong inflammatory response with the induction of many pro-inflammatory cytokines and chemokines such as interleukin (IL)-1, IL-6, IL-8, tumor necrosis factor (TNF), IL-10 and prostaglandin E2, which are all detectable in bronchoalveolar lavage fluid (Asai et al., 1993; Asai et al., 1994; Harding et al., 1997) and in lung tissue (Choi et al., 2006; Ahn et al., 2009; Woolley et al., 2013). Another study also demonstrated a strong increase in cyclooxygenase-2 expression relating to the strong prostaglandin-mediated inflammatory response (Andrada et al., 2014). Parts of this innate immune response are illustrated in Figure 6.1. The inflammation and cytokine responses are associated with lung lesions and peribronchiolar lymphoid cuffing (Harding et al., 1997; Kwon et al., 2002; Choi et al., 2006; Lorenzo et al., 2006). In addition to the pro-inflammatory mediators, also the T-cell cytokines such as IL-2 and IL-4 can be detected in bronchoalveolar exudates. Immunohistochemistry showed that cytokines are associated with mononuclear cells of the alveolar septa, as well as macrophages and lymphocytes of the bronchus-associated lymphoid tissue (BALT), from one-week post infection onwards (Lorenzo et al., 2006). These responses are typically associated with intense lymphoid hyperplasia (Rodriguez et al., 2004), and are characterized by an accumulation of IgG- and IgA-expressing plasma cells, $CD4^+$ T cells, macrophages and DC in the BALT in inflamed lung tissue (Sarradell et al., 2003).

Although the above responses appear worrisome considering the pathogenesis of *Mycoplasma* infection, studies in mice have demonstrated a striking protective role of innate immune responses. These studies employed highly susceptible C57BL/6 and resistant C3H/He mice, both with a severe combined immunodeficiency (SCID) which is associated with a highly defective adaptive immune response. Despite this lack of adaptive immunity, the C3H/He SCID mice were able to control *M. pulmonis* infection (Cartner et al., 1998).

6.2.3 Sensing of *Mycoplasma* by the innate immune system

The above innate and inflammatory responses can only be induced if the innate immune system can sense invading *Mycoplasma* through specific receptor-ligand interactions. The receptors of the innate immune system termed pathogen recognition receptors (PRR), can be divided into five families which are the Toll-like receptors (TLR), nucleotide-binding oligomerization domain (NOD)-like receptors (NLR), C-type lectin like receptor (CLR) as well as the nucleic acid sensing cytoplasmic retinoic acid-inducible gene-I-(RIG-I) like receptors (RLR) and the cytosolic DNA sensors (CDS) (Abbas et al., 2016). TLR2 was demonstrated to play an important role in *Mycoplasma* sensing (Lien et al., 1999). This receptor recognizes bacterial lipoproteins if associated with TLR6 as a heterodimer (Takeda et al., 2002; Takeuchi and Akira, 2007). These *in vivo* mouse studies were confirmed *in vitro* for *M. hyopneumoniae* and porcine TLR2 and TLR6 (Muneta et al., 2003). Also, the lipopeptide FSL-1 was described as a ligand for TLR2/TLR6 (Okusawa et al., 2004). On the TLR side also TLR1 as a heterodimer with TLR6 was shown to be involved in the recognition of dipalmitoylated lipoproteins from *M. pneumoniae* (Shimizu et al., 2005). *M. pneumoniae* also promotes inflammatory responses through TLR4, autophagy and through a toxin-dependent activation of the inflammasome mediated by several NLRs, such as NLRP3 (Shimizu, 2016). This inflammasome activation is responsible for IL-1β and IL-18 secretion detected following infection (Khare et al., 2012; Segovia et al., 2018), and is also observed in pigs following *M. hyopneumoniae* infection (Muneta et al., 2006).

While these inflammatory responses certainly contribute to disease and lung lesions, they clearly are also important in protection. In a mouse model using *M. pulmonis*, it was shown that TLR2 played a major role in innate resistance (Love et al., 2010). This was also confirmed by the lack of control in MyD88 deficient mice, in which TLR2 signaling is abrogated (Lai et al., 2010). TLR2-mediated sensing is probably important in activating innate immune defenses such as infiltration of myeloid cells (Chapter 6.2.4) or release of antimicrobial peptides (Chapter 6.2.6). Furthermore, mucin formation following TLR2 activation by *M. pneumoniae* was found to prevent *Mycoplasma* from adhering to the epithelium (Chu et al., 2005).

6.2.4 Contribution of myeloid cells

Innate protection against *M. pulmonis* in the mouse model is dependent on macrophages (Lai et al., 2010). In fact, mycoplasmas activate alveolar macrophages to produce several of the above mentioned pro-inflammatory cytokines. In the mouse model, nitric oxide was identified to be important for killing of *Mycoplasma* by macrophages (Hickman-Davis, 2002), although porcine macrophages seemed unable to produce nitric oxide (Sautter et al., 2018). Interestingly, *M. hyopneumoniae* appears to evade phagocytic uptake by alveolar macrophages, at least *in vitro* (Deeney et al., 2019).

There are only a few and conflicting reports with respect to the role of neutrophils in protection. Lai et al. (2010) reported that *M. pulmonis* in the mouse model was dependent on macrophages but not on neutrophils. On the other hand, it was demonstrated that T_h17 responses (see Chapter 6.4.2) are protective in a neutrophil-dependent fashion (Mize et al., 2018; Sieve et al., 2009). The latter would be expected considering the central importance of neutrophils in controlling bacterial infections, in particular through their high phagocytic activity combined with a most active arsenal of antimicrobial enzymes stored in their granules. In addition, neutrophils possess very potent anti-microbial activities through release of reactive oxygen species. The latter was demonstrated for several *Mycoplasma* species (Peterhans et al., 1984; Alabdullah et al., 2015; Di Teodoro et al., 2018). This oxidative burst can be induced by *Mycoplasma* lipoproteins resulting in the formation of neutrophil extracellular traps (NETs), which represent DNA complexed with antimicrobial peptides and proteases to trap and kill bacteria. This was detected in the mammary gland of sheep infected with *M. agalactiae* (Cacciotto et al., 2016). Interestingly, *M. bovis* possesses a nuclease that can degrade these neutrophil extracellular traps (Gondaira et al., 2017; Mitiku et al., 2018). Considering the lack of nitric oxide as a potent antimicrobial activity in porcine macrophages, future studies should address the role of neutrophils in the innate immune responses against *M. hyopneumoniae*, which is likely to be important. The contribution of myeloid cells is schematically represented in Figure 6.1.

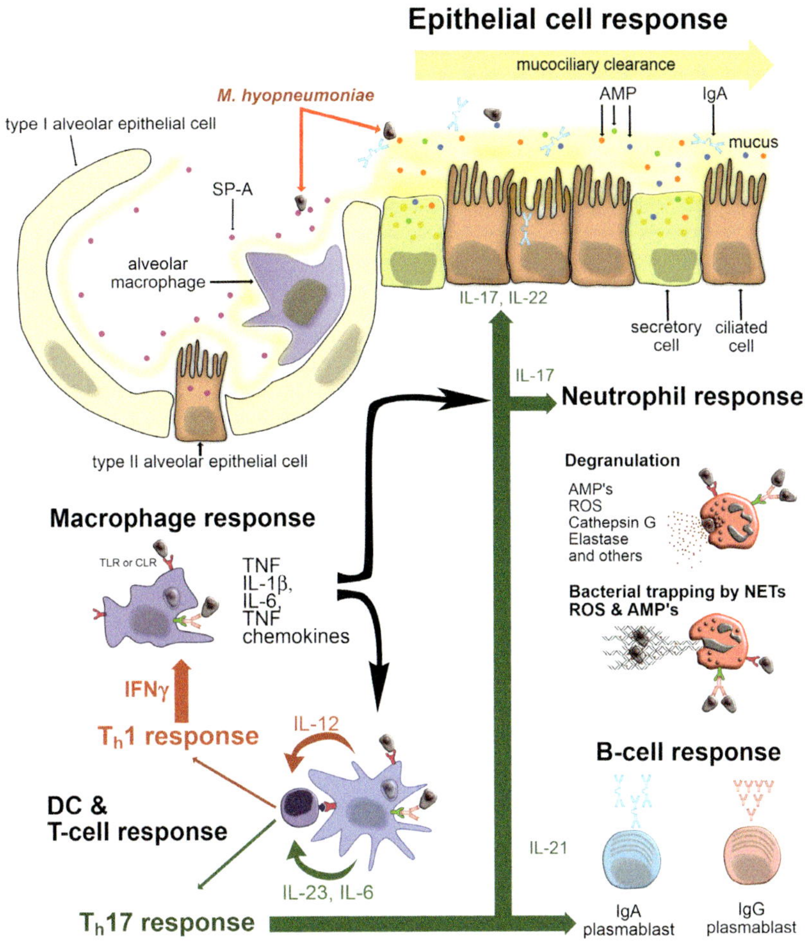

Figure 6.1. Scheme of protective immune responses against *Mycoplasma hyopneumoniae*. At the level of the epithelium important protective elements are the mucociliary clearance, the production of mucus and anti-microbial proteins (AMP) by secretory cells in the epithelium, of surfactant protein A (SP-A) by type II alveolar epithelial cells, and secretory IgA transcytosed through the epithelium. Macrophages in the alveoli, the *lamina propria* and submucosa will be activated by *M. hyopneumoniae* to produce pro-inflammatory cytokines and chemokines. This will result in activation of the endothelium and recruitment of neutrophils to enhance the inflammatory response and clear the infection. DC will also be recruited, activated and take up the antigen and migrate to the lymph node for antigen presentation to T cells. The antibacterial neutrophil response is composed of the secretion of AMP, release of reactive oxygen species (ROS) and enzymes following degranulation. In addition, neutrophil extracellular traps (NETs) may participate in the anti-mycoplasma response. After initiation of the T-cell response in the draining lymph node, activated T_h1 and T_h17 cells will migrate to the sites of inflammation. A major component of T_h1 responses is the IFNγ-mediated macrophage activation which strongly enhances their anti-microbial activity, while T_h17 response promote the above mentioned functions of the epithelium, the neutrophil as well as the IgA response. IgG in the mucosal may participate in the anti-mycoplasma response through opsonization for enhanced phagocytosis or through complement activation although this has not been demonstrated. (Source: A. Summerfield)

6.2.5 Complement and other opsonins

Older studies indicate that mycoplasmas appear to be relatively inefficient in activating the alternative complement pathway (Howard, 1980) and killing of mycoplasmas by the membrane attack complex may not be functional (Webster et al., 1988). This could represent an important immune evasion strategy, also because mycoplasmas appear to resist phagocytosis *in vitro* in absence of any complement or other opsonins (Hickman-Davis, 2002).

Collectins are multimeric proteins recognizing microbial carbohydrates. They are present in the plasma, produced during inflammation, and certain are secreted at mucosal surfaces. Through their ability to bind to many microbial structures such as mannose and N-acetylglucosamin, they play an important role as opsonins. In particular the surfactant protein SP-A has been studied as an important factor for control of *M. pulmonis* in the mouse model (Hickman-Davis, 2002), and was also identified to inhibit the growth of *M. pneumoniae* (Piboonpocanun et al., 2005).

Unfortunately, there are no studies addressing the role of complement or other opsonins in *Mycoplasma* infections of pigs. Specific IgG certainly represents a powerful opsonin resulting in *Mycoplasma* phagocytosis by macrophages, but will only be effective during the adaptive phase of the immune response. Nevertheless, mycoplasmas were reported to be bound by polyreactive antibodies (Ben-Aissa-Fennira et al., 1998), representing part of the natural antibody repertoire expressed early in life, which can also mediate bacterial opsonization (Gunti and Notkins, 2015).

6.2.6 Antimicrobial peptides

Cationic antimicrobial peptides such as HG1 and LL-37 were detected to be active against *M. pulmonis* (Park et al., 2013). Similarly, the murine cathelicidin cathelin-related antimicrobial peptide (CRAMP), related to the human cathelicidin LL-37, had potent antimicrobial activity against *M. pneumoniae*. It was detected in the BALF of infected mice and released from neutrophils in contact with *Mycoplasma* (Tani et al., 2011). In addition to cathelicidins, human β-defensin-2 and -3 were identified to have a strong antimicrobial activity against *M. pneumoniae* (Kuwano et al., 2006). Also, for antimicrobial peptides information is lacking for *M. hyopneumoniae*.

6.3 Antibody response against *Mycoplasma hyopneumoniae*

6.3.1 General considerations of antibody responses against *Mycoplasma*

During infections or vaccinations, antibodies can be induced against any non-self-antigen and therefore antibodies *per se* are not necessary involved in protective immune responses. For the latter, antibody functions need to be taken into consideration. For *Mycoplasma* infections, antibodies could already act at the first line of immune defenses, which is on the mucosal surfaces in form of secretory IgA. Such IgA would ideally prevent bacterial adhesion to epithelial cells and inhibit colonization of the respiratory tract. Importantly, secretory IgA is typically not induced by parenteral vaccination but requires delivery of a vaccine to mucosal surfaces. However, antigens on mucosal surfaces are usually not immunogenic unless a strong adjuvant or a replicating agent with a certain degree of pathogenicity is employed.

In addition to secretory IgA, antibodies of IgM, IgG, and IgA isotypes will be present in the interstitial lung tissue. These can be locally produced in germinal centers of the BALT or can diffuse from blood into tissues. While such antibodies cannot prevent adhesion of *Mycoplasma* to the apical epithelial surface, they have important functions in terms of opsonization of bacteria for phagocytosis by macrophages and neutrophils. These cells express Fc receptors binding the Fc portion of the antibody, and this results in phagocytosis, destruction of the pathogen, and activation of myeloid cells. A second important effector function of antibodies is the activation of the classical complement cascade resulting in the release of very potent chemoattractant factors for myeloid cells, as well as the killing of bacteria by the final product of the complement cascade representing the membrane attack complex (Abbas et al., 2016).

6.3.2 Kinetics of antibody responses after *Mycoplasma hyopneumoniae* infection

Infection of hysterectomized piglets kept in an SPF environment made antibodies 3 - 4 weeks post inoculation with *M. hyopneumoniae*, peaking at 11 - 12 weeks. A booster reaction was detected after secondary infection, and these animals were protected against pneumonia (Kobisch et al.,

1993) demonstrating that adaptive immune responses can be protective. Nevertheless, another study found no protection after previous infection of piglets with a low virulent *M. hyopneumoniae* strain and secondary challenge with a virulent strain (Villarreal et al., 2009). A possible reason for the discrepancy between the two studies is the time between first and second infection which was four months versus one month.

The kinetics of antibody responses will also depend on the route of administration. For example, after intravenous inoculation, seroconversion was found around 10 days in all animals, after intraperitoneal inoculation seroconversion occurred after 13 days but only in 24% of the animals, after subcutaneous inoculation 29% of the animals seroconverted at 21 days and after intranasal inoculation the first responses were detected at 14 days, with all animals seroconverting (Lloyd et al., 1987). Another study compared endotracheal, intranasal and aerosol inoculation of pigs and found endotracheal administration of the bacteria to be most efficient in terms of inducing colonization of lung with *M. hyopneumoniae,* seroconversion and mucosal IgA (Garcia-Morante et al., 2016b). Pigs in contact with animals with *M. hyopneumoniae* caused pneumonia first and had antibody in their sera at a mean of 27 days (Lloyd et al., 1987). In another study, seroconversion was observed even beyond 60 days post infection (Gomes-Neto et al., 2014). It is important to note that the kinetics and rate of seroconversion will strongly depend on the *M. hyopneumoniae* strain, the age at inoculation, co-infection with other swine pathogens and animal source. Also, not all animals within a group might seroconvert (Garcia-Morante et al., 2017a).

The development of Ig^+ cells and plasma cells in retropharyngeal and bronchial lymph nodes, in lung tissue, and in nasal mucous membrane was followed during an experimental infection of pigs with *M. hyopneumoniae*. A significant increase of antigen-specific plasma cells in lymph nodes and lung tissue as well as antibodies in tracheobronchial secretions was demonstrated two weeks after infection followed by a marked increase of Ig-expressing cells in lymph nodes and lung tissue (Suter et al., 1985). The authors discussed the data in terms of autoimmune reactions in the pathogenesis of enzootic pneumonia (EP), a concept that had been previously proposed (Baumgartner and Nicolet, 1984; Biberfeld, 1979; Roberts and Little, 1970), but later this hypothesis disappeared from the scientific literature on *M. hyopneumoniae*. It is possible that these reactions are caused by the induction of polyreactive antibodies following infection. At least *in vitro*, we found that *M. hyopneumoniae* activates innate-like B cells, recently described in pigs (Braun et al., 2017) to produce polyreactive antibodies (Trueeb, Braun, Auray, Kuhnert and Summerfield, unpublished data). It is clear that *M. hyopneumoniae* has a potent effect on B cells *in vivo*, as infection of pigs results in a 25-fold increase of the B cell population in the lung parenchyma at 28 days post infection (Walker et al., 1996). *In vitro* stimulation of antibody production against *M. hyopneumoniae* was analyzed using peripheral blood mononuclear cells. *M. hyopneumoniae*-specific B cells, were detected seven days after the primary immunization and reached their highest levels one week later (Wallgren et al., 1992).

The kinetic and levels of antibody responses after experimental infection through the upper respiratory tract will certainly depend on the ability of the mycoplasmas to colonize and cause lesions in the respiratory tract. For instance, pigs inoculated with *M. flocculare* did not develop any lesions and also showed a weak antibody response detected at 6 - 8 weeks post-infection. This contrasted with *M. hyopneumoniae* infected animals, which seroconverted 4 - 5 weeks post infection and had greater antibody responses (Strasser et al., 1992). For pigs infected by *M. hyopneumoniae* both under experimental and field conditions, a positive association between the presence of antibodies (both systemic and local) and lung lesions was found (Garcia-Morante et al., 2016b; Garcia-Morante et al., 2017b).

6.3.3 Role of antibodies in protection against *Mycoplasma*

In the mouse model of *M. pulmonis* infection, early studies indicated an important role for antibodies, as transfer of serum from immune mice was protective, while transfer of naïve spleen cells even had a negative effect on lung lesions (Cartner et al., 1998). For *M. hyopneumoniae* infection of pigs, no data on antibody transfer have been published, but data published already in 1969, would indicate that antibodies are not protective. In fact, following vaccination, no correlation of antibody levels with reduction of lung lesions and no protection of piglets from vaccinated sows was observed (Goodwin et al., 1969). This was confirmed later by others (Djordjevic et al., 1997; Thacker et al., 1998). In addition, a longitudinal study including 825 pigs in eight chronically infected herds identified no overall relationship between average daily weight gain and the serological response to *M. hyopneumoniae* (Andreasen et al., 2001).

However, IgG antibodies have been demonstrated to opsonize mycoplasmas for phagocytosis and killing of bacteria, as shown for *M. bovis* with bovine macrophages (Howard, 1984) and at least in cattle local lung IgG together with macrophages was proposed to be important in recovery (Howard et al., 1987). This would be in line with the described difficulty in phagocytosis of non-opsonized mycoplasmas (Deeney et al., 2019) and the weak activation of the alternative complement pathway (Chapter 6.2.5). Furthermore, as described in the next section, there is evidence of a role of maternally derived immunity to protect piglets, which could be mediated by antibodies.

6.3.4 Maternally-derived antibodies

If piglets receive colostrum containing *M. hyopneumoniae*-specific antibodies, these maternally derived antibodies (MDA) will be detectable in the piglet's blood until 9 weeks of age (Wallgren et al., 1998). There are conflicting reports with respect to two very important questions concerning MDA, one being their protective values, the other if they interfere with piglet vaccination. While one study found little or no impact on vaccine-induced antibody responses when piglets were vaccinated at 7 days of age (Martelli et al., 2006), others demonstrated such interference (Hodgins et al., 2004; Grosse Beilage and Schreiber, 2005; Bandrick et al., 2014a). This is not too surprising because interference by MDA will depend on the type of vaccine, the adjuvant, the antigen dose, and the levels of maternal antibodies present at the time of vaccination. Independent of a possible interference with antibody responses, vaccination of piglets in presences of maternal antibodies induced protective immunity (Wilson et al., 2013).

The other important question whether maternally derived antibodies are protective has also not been definitively answered. While the beneficial effect of sow vaccination has been reported (Sibila et al., 2008; Arsenakis et al., 2019), it is unclear if this is mediated by antibodies or by cell mediated immunity since functional maternal lymphocytes specific for *M. hyopneumoniae* are transferred by colostrum (Bandrick et al., 2014b).

6.3.5 Mucosal antibody responses

In mouse models, it was demonstrated that local immune responses are more effective at controlling mycoplasmas compared to systemic responses (Hodge and Simecka, 2002). This seems logical considering the pathogenesis of *Mycoplasma* infections and is probably also correct for *M. hyopneumoniae*.

In fact, *M. hyopneumoniae* adheres to ciliated epithelial cells and this adherence can be inhibited by low dilutions of sera from convalescent pigs (Zielinski and Ross, 1993). Nevertheless, the same authors found no effect with lavage fluids that contained *M. hyopneumoniae*-specific IgA, possibly because of too low concentrations (Zielinski and Ross, 1993). After infection of pigs with *M. hyopneumoniae*, specific mucosal antibodies of the IgG and IgA isotypes were detected mainly in pigs in which an intensive *M. hyopneumoniae* colonization and lung lesions were found, indicating that a relatively high antigenic load and possibly disruption of the respiratory barrier function is required to induce mucosal antibodies (Garcia-Morante et al., 2017b). Considering that *M. hyopneumoniae* are chronic infections, this would indicate that once the infection is established, the locally produced antibodies are relatively inefficient at clearing the infection. Nevertheless, an older study indicates that resolution of pneumonia appears to correlate with antibody responses (Messier et al., 1990). Parenteral vaccination with bacterin vaccines does not induce such responses but appear to prime the immune system for higher IgG and IgA responses in the respiratory tract following challenge (Djordjevic et al., 1997). This could possibly be mediated by induction of *Mycoplasma*-specific memory T_h cells but information on mucosal T-cell responses to *M. hyopneumoniae* are missing. Clearly future studies are required to address this important area.

6

6.4 T-cell mediated immune responses against *Mycoplasma*

6.4.1 General considerations of T-cell responses against *Mycoplasma*

From the two major subsets of T cells, the $CD4^+$ T-helper (T_h) and the $CD8^+$ cytotoxic T cells (T_c), one would assume that only the T_h cells are important for immunity against mycoplasmas. This is because T_c are involved in recognition and killing of host cells infected by intracellular pathogens through antigenic peptide presentation in the context of major histocompatibility complex (MHC) class I. In contrast, T_h cells recognize peptides presented after endo- and phagocytosis of extracellular antigens by antigen presenting cells in the context of MHC class II. Dendritic cells are professional antigen presenting cells required for activation of immunologically naïve T cells, while macrophages and B cells typically present antigen during effector phases of the immune response or for induction of recall responses. The interaction between macrophages and T_h cells results also in a strong activation of macrophages, which is central to many antibacterial responses. Also, antigen presentation by B cells to T_h cells provides a very strong signal to B cells resulting in antibody production, Ig isotype switching, and affinity maturation. These are the two most crucial reasons for the importance of T_h cells in antibacterial immunity.

It is important to consider that T cells, when compared to antibodies, have a high potential to cause immunopathological effects. For T_h cells, this is mainly induced indirectly through the release of cytokines and chemokines, which strongly enhance inflammatory responses by innate immune cells. Therefore, physiological T-cell responses must be tightly regulated by cells and factors to prevent over-activation. This process is termed immunoregulation and involves anti-inflammatory cytokines like IL-10 and transforming growth factor-β, as well as so-called regulatory T cells (T_{reg}) representing a T-cell subset specialized in dampening immune responses (Abbas et al., 2016).

Several subsets of T_h cells have been described in the last 30 years of immunological research. Originally, these were T_h1 and T_h2 cells only, but it was then realized that this dichotomy is not sufficient to describe and understand immune responses. In most immunology textbooks, a description of four T_h subsets will be found: the T_h1, T_h2, T_h17 and the above mentioned T_{reg}. Considering that T_{reg} have a "special" function in immunoregulation, immune responses to pathogens can be seen as tripartite, which are now often called "type 1", "type 2" and "type 3" immunity. Type 1 immunity is typically induced by intracellular pathogens such as viruses and intracellular bacteria. These responses are characterized by a dominant T_h1 response with IFN-γ as a key cytokine. Type 2 immunity is typically induced by parasitic infections, corresponding to T_h2 responses with IL-4, and IL-13 as the key T-cell cytokines. Finally, type 3 immunity will be mainly induced by extracellular microbes such as many bacteria and fungi. These responses are known to be associated with a strong T_h17 response with cytokines of the IL-17 family and IL-22 being the key cytokines. Type 3 immunity is particularly important to protect mucosal surfaces from extracellular pathogens as it strongly promotes epithelial cell regeneration, mucous and antimicrobial protein production, and release and neutrophil recruitment.

6.4.2 Role of T cells in protective immunity against *Mycoplasma* infections

In the literature, there are several reports indicating that T cells would not be important in the control of *Mycoplasma* infection and even cause disease exacerbation. Examples are that *M. pneumoniae* also does not cause more severe pneumonia in human patients with T-cell deficiencies (Hickman-Davis, 2002). Furthermore, in a hamster model, T-cell depletion resulted in less severe infections following *M. pneumoniae* infection (Taylor et al., 1974). Even studies using thymectomized pigs indicated that T cells participate in the development of pneumonic lesions after *M. hyopneumoniae* infection (Tajima et al., 1984). In fact, lung infections with mycoplasmas are typically associated with a heavy infiltration of lymphoid cells, in particular T_h cells, which appear to have a pathogenic role (Jones and Simecka, 2003; Dobbs et al., 2009). Such lymphoid hyperplasia was also demonstrated in pigs infected with *M. hyopneumoniae* (Sarradell et al., 2003; Rodriguez et al., 2004).

6

However, mouse studies in which various T-cell subsets were depleted using monoclonal antibody prior to *M. pulmonis* infection, indicate that T_h cells do play an important role also for resistance against the pathogen. This was confirmed by adoptive transfer of T_h cells from immune to naïve mice (Jones et al., 2002; Dobbs et al., 2009). For pigs infected with *M. hyopneumoniae*, resolution of pneumonia correlated with development of increased sensitization of lymphocytes, but also with IgG in the lung (Messier et al., 1990). Interestingly, lymphocytes isolated from pigs immune to *M. hyopneumoniae* were found to cross-react with *M. hyorhinis* and *M. flocculare* but restimulation with *M. hyopneumoniae* was superior (Kishima et al., 1985).

6.4.3 Role of different types of T-cell responses

Using IL-4 and IFNγ knockout mice, it was demonstrated that IFN but not IL-4 has a protective role against *M. pulmonis*, at least concerning pathogenic effects in the lower respiratory tract (Woolard et al., 2004). Altogether, studies in the mouse model indicate that T_h1 responses are required for protection, although they can also be involved in pathogenic events (Dobbs et al., 2009). The idea that T_h1 responses, which are known to be associated with a significant pro-inflammatory potential, can play an important role in such pathogenic responses was also confirmed by a study in which T_{reg} were depleted. This resulted in a dramatic increase in disease severity in the

M. pulmonis mouse model. Interestingly, it was associated with enhanced humoral immune responses, enhanced T_h1 and T_h17 responses, but with no impact on pathogen loads, demonstrating the delicate balance that is required for the immune response in controlling *Mycoplasma* infections without causing detrimental tissue damage in the respiratory tract (Dobbs et al., 2009; Odeh and Simecka, 2016). Interestingly, also $CD8^+$ T cells were reported to dampen exaggerated inflammatory responses in the *M. pulmonis* mouse model (Dobbs et al., 2009). A most crucial mechanism of T_h1 responses that is likely to be important against *Mycoplasma* infections is the very potent effect of IFN to activate macrophages for high antimicrobial activity.

Not surprisingly, T_h17 responses also play an important role during *Mycoplasma* infections. Using a *M. pneumoniae* mouse model, it was shown that local T_h17 response in the lung which possibly contribute to lung inflammation and pathology are induced (Kurata et al., 2014). This was confirmed in children suffering from *M. pneumoniae* infection. This study indicated that T_h17 responses could be associated with clearance of infection but could also contribute to immunopathological damage (Wang et al., 2016b). However, using mouse models infected with *M. pulmonis*, treatment with an anti-IL17A antibody or IL17 receptor knockout mouse, clearly demonstrated the protective role of IL-17 and that this function is neutrophil-dependent (Mize et al., 2018; Sieve et al., 2009). Priming of such T_h17 responses could be mediated by membrane-derived lipoproteins which are able to induce the IL-17-promoting IL-23 in DC (Goret et al., 2017).

A number of interesting studies were published which made an immunological comparison of two pig lines, one showing increased immunological resistance for producing lung lesions after *M. hyopneumoniae* infection. These pigs had higher serum levels of IL-17, IFN-γ and IL-10 but lower antibody responses post-vaccination. They also had relatively more circulating granulocytes compared to lymphocytes (Borjigin et al., 2016a, 2016b; Sato et al., 2016; Shimazu et al., 2013; Shimazu et al., 2014). These studies support the importance of both T_h1 and T_h17 responses in the protective adaptive immune response of pigs against *M. hyopneumoniae*, in analogy to the results obtained in the mouse model. We have recently developed a microparticle-based vaccine that induced no antibodies, no T_h1 but T_h17 responses (Matthijs et al., 2019b). This vaccine provided a good level of protection, again supporting that targeting T_h17 responses for *M. hyopneumoniae* represents a promising approach (Matthijs et al., 2019b).

6.5 Conclusions

While basic information on the immune responses against *M. hyopneumoniae* is available there are still many gaps that have prevented the understanding of the pathogenesis of EP, and following of a more rational design of vaccines. It seems clear that the innate immune system efficiently senses *M. hyopneumoniae,* resulting in a strong pro-inflammatory response, but it remains unclear if this response is more protective or pathogenic. In addition, information on an innate mechanism with direct antimicrobial effects such as on phagocytosis and destructive pathways in macrophages and neutrophils, complement and other soluble opsonins and antimicrobial peptides are rare or completely missing. The literature on porcine *Mycoplasma* infections coupled with information from mouse models indicates that adaptive immune responses against *M. hyopneumoniae* are clearly complex and represent a fragile balance between pathogenic and protective T_h-cell responses. Despite so many years of research, we are only starting to understand the contribution of antibodies and the various subsets of T_h cells for protective immunity against *M. hyopneumoniae*. Antibody responses to infection are rather slow to develop and appear to require colonization of the respiratory tract and lung with mycoplasmas, which are associated with inflammation and lesions. Altogether, it seems clear that systemic antibodies do not correlate with protection. The role of mucosal antibodies in protection remains to be demonstrated. Maternally-derived immunity has been reported but it is unclear if this is mediated by antibodies or transfer of maternal lymphocytes. The literature supports the view that T_h1 and maybe even more T_h17 are important in controlling *M. hyopneumoniae*.

Taken together, future research is required to fill important knowledge gaps, but despite this, the cumulated knowledge on the immunology of *M. hyopneumoniae* infections should be exploited to make better vaccines. This may include the development of better adjuvants and vaccine delivery systems, and also the development of a live attenuated vaccine that can be applied via the mucosal route to directly promote IgA-mediated protection at mucosal surfaces.

CHAPTER 7

INTERACTIONS OF *MYCOPLASMA HYOPNEUMONIAE* WITH OTHER PATHOGENS AND ECONOMIC IMPACT

Corinne Marois-Créhan[1]
Joaquim Segalés[2]
Derald Holtkamp[3]
Chanhee Chae[4]
Céline Deblanc[5]
Tanja Opriessnig[3, 6]
Christelle Fablet[7]

1 Ploufragan-Plouzané-Niort Laboratory, French Agency for Food, Environmental and Occupational Health & Safety (Anses), Ploufragan, France
2 Facultat de Veterinària (Universitat Autònoma de Barcelona), Centre de Recerca en Sanitat Animal (CReSA)-Institut de Recerca i Tecnologia Agroalimentàries (IRTA), Campus de la Universitat Autonoma de Barcelona, Bellaterra, Spain
3 College of Veterinary Medicine, Iowa State University, Ames, USA
4 College of Veterinary Medicine, Seoul National University, Seoul, Republic of Korea
5 Swine Virology and Immunology Unit, French Agency for Food, Environmental and Occupational Health and Safety (Anses), Ploufragan, France
6 The Roslin Institute and The Royal (Dick) School of Veterinary Studies, University of Edinburgh, UK
7 Ploufragan-Plouzané-Niort Laboratory, French Agency for Food, Environmental and Occupational Health and Safety (Anses), Ploufragan, France

7.1 Introduction

Under field conditions, it is common that multiple pathogens circulate in a herd at a given time leading to co-infections or combinations of sequences of infections that may in turn create opportunities for pathogen interactions. In the respiratory system, *Mycoplasma hyopneumoniae* (*M. hyopneumoniae*) may be in association with various infectious pathogens also damaging the respiratory tract including bacteria, viruses, or parasites. The term 'porcine respiratory disease complex (PRDC)' is currently used to describe mixed respiratory infections of multiple aetiologies where *M. hyopneumoniae* is considered to be one of the key agents.

From a diagnostic point of view, it is important to differentiate a respiratory disease caused by a specific pathogen from PRDC. This latter clinical condition is polymicrobial in nature (Sibila et al., 2009; Opriessnig et al., 2011a), and a wide variety of bacterial and viral pathogens have been detected in cases of PRDC. Early reports indicated *M. hyopneumoniae* and porcine reproductive and respiratory syndrome virus (PRRSV) as the most important contributors to the syndrome (Thacker et al., 1999), but other respiratory pathogens such as swine influenza A viruses (IAV-S), porcine circovirus 2 (PCV-2), *Actinobacillus pleuropneumoniae* (*A. pleuropneumoniae*) and *Pasteurella multocida* (*P. multocida*) have been involved in PRDC cases (Opriessnig et al., 2011a). In fact, each herd tends to have a unique combination of pathogens and non-infectious risk factors.

The individual pathogenesis of specific respiratory diseases is relatively well known. While viruses mainly act by disrupting the muco-ciliary apparatus and/or affecting functionality or numbers of pulmonary alveolar and intravascular macrophages, bacteria recruit inflammatory cells through enhanced cytokine responses (Opriessnig et al., 2011a).

In the context of PRDC, combination of several pathogens may cause additive or synergistic effects at the tissue level. However, literature is still scarce of sound data on such interactions and mechanisms by which clinical signs and lesions are potentiated. Such knowledge is paramount to design proper PRDC control and prevention strategies, especially in the context of *M. hyopneumoniae* interaction with viruses, bacteria and parasites.

The purpose of this chapter is to explore the technical and economic impacts of *M. hyopneumoniae* interactions with other pathogens, as well as to analyze the interactions of this bacterium with respiratory bacterial and viral pathogens, parasitic infections and mycotoxins. The effects of those combinations on the clinical outcome and the underlying mechanisms of the interactions are detailed.

7.2 Impact of *Mycoplasma hyopneumoniae* interactions with other pathogens on production and economic performance

Economic losses attributed to *M. hyopneumoniae* occur primarily in the wean-to-finish phase of production. Economic losses occur because of poor productivity, high animal health costs, and increased variation in pig growth. Productivity losses include reductions in growth rates, increase in feed conversion, and increases in mortality and culling rates. Various approaches have been used to estimate productivity losses and other economic impacts due to *M. hyopneumoniae* infections under field conditions. Dykhuis-Haden et al. (2012) used historical production data from a single large production company in the Midwest of the United States. Groups were retrospectively classified as affected if any routine diagnostic information suggested exposure to various respiratory pathogens. Groups affected by *M. hyopneumoniae* alone had an absolute increase in mortality of 2.2% and a reduction in average daily gain (ADG) of 18 g/day. Groups affected by *M. hyopneumoniae* in combination with PRRSV and IAV-S experienced more severe losses with a 5.4% and 3.5% absolute increase in mortality and a reduction in ADG of 64 and 82 g/day, respectively. The economic value of those losses was estimated to be US$0.63 per head placed in groups affected by *M. hyopneumoniae* alone, increasing to US$9.69 and US$10.12 per head placed in groups affected by *M. hyopneumoniae* in combination with PRRSV and IAV-S respectively. Silva et al. (2019) compared groups of finishing pigs sourced from sow farms that were positive for *M. hyopneumoniae* to groups of pigs from sow farms where *M. hyopneumoniae* had been eliminated and thus were negative for *M. hyopneumoniae.* Groups from positive sow farms grew 36 g/day slower (4.2% reduction), had a 0.018 increase in feed conversion (0.6% increase) and had a 1.26% higher mortality rate (24.3% increase) compared to groups of pigs from negative sow farms. In addition, the cost of antimicrobial treatments was higher by US$1.21 per pig marketed and additional US$0.25 per pig was spent on *M. hyopneumoniae* vaccines compared to groups from negative sow farms. The economic impact of the poorer productivity and higher antimicrobial treatment and vaccine costs were estimated to increase the cost of production by an average of US$7.00 per pig marketed in sow farms that were positive for *M. hyopneumoniae*. Using a different approach, Rautiainen et al. (2000) compared the performance of individual pigs within groups based on their serological status and presence of lung lesions at slaughter. Pigs affected by *M. hyopneumoniae*, based on the detection of specific antibodies and lung lesions at slaughter, grew from 75 to 176 g/day slower (10.2% to 17.5% reduction) than pigs unaffected. Similarly, Regula et al. (2000) reported a 38g/day reduction in average daily gain in pigs that were seropositive for *M. hyopneumoniae* compared to those that were negative.

7.3 Interactions of *Mycoplasma hyopneumoniae* with bacteria involved in lung diseases

Many bacterial species can be detected in pig lung lesions (pneumonia, pleurisy, abscesses) and may interact with *M. hyopneumoniae* due to this shared niche. They act as primary and/or secondary pathogens depending on whether they can damage the lung tissue by themselves or need other co-infecting pathogens or cofactors to induce substantial lesions in the respiratory tract. *M. hyopneumoniae* and *A. pleuropneumoniae* are considered two main primary bacterial pathogens (Fraile et al., 2010). *Bordetella bronchiseptica* (*B. bronchiseptica*) is also recognized as a primary pathogen of PRDC. The other bacterial species considered to be secondary or opportunistic respiratory pathogens include mainly *P. multocida*, *M. hyorhinis*, *Glaesserella* (*Haemophilus*) *parasuis*, α hemolytic *Streptococcus* sp. (with *Streptococcus suis*), *Actinobacillus suis*, *Trueperella pyogenes* and *Salmonella enterica* serovars Choleraesuis and Typhimurium. Recently, Oliveira Filho et al. (2018) suggested to classify some *P. multocida* strains as primary pathogens since they caused necrotic bronchopneumonia and diffuse fibrinous pleurisy and pericarditis in pigs without any other recognized co-factor. However, these last results should be confirmed by future studies.

7.3.1 Interaction with *Actinobacillus pleuropneumoniae*

A. pleuropneumoniae is the etiologic agent of swine pleuropneumonia (Gottschalk, 2012a). In its most virulent form, *A. pleuropneumoniae* induces severe rapidly fatal, fibrino haemorrhagic and necrotizing pleuropneumonia in naïve swine of all ages, especially in growing pigs. The *A. pleuropneumoniae* infection may become chronic or subclinical without previous signs of disease, and outbreaks may suddenly appear or subclinical infections may remain silent. The bacterium is able to persist in pig tissues, particularly in tonsillar crypts and in sequestered necrotic lungs. In this bacterial species, 15 serotypes are described and their prevalence and virulence vary among countries: serotypes 2 and 9 are the most prevalent in European countries and particularly in France, and serotypes 1 and 5 are the most frequently detected in North America (Gottschalk, 2012a). Virulence factors involved in the colonization (fimbriae type IV, lipopolysaccharides) and the lysis of alveolar epithelial cells (four exotoxins: Apx I, II, III and IV) were described (Frey, 1995; Gottschalk, 2012a). The toxins Apx I, II and III, in addition to lyse respiratory epithelial cells, have lytic properties also on macrophages and neutrophils, involved in the innate immune response (Dom et al., 1992; Cullen and Rycroft, 1994). The lytic properties of the toxin disappear when the toxin concentrations decrease; nevertheless, these low concentrations and/or the *A. pleuropneumoniae* capsular polysaccharides can still alter the chemotactic activity of macrophages and the phagocytosis (Inzana et al., 1988; Tarigan et al., 1994). Porcine immunoglobulins IgA and IgG can also be degraded by a protease produced by *A. pleuropneumoniae* (Negrete-Abascal et al., 1994; Negrete-Abascal et al., 1998). These modulations of immune responses could explain why the infected host remains contagious for an extended period of time and characterize *A. pleuropneumoniae* as a long-cycle pathogen (as *M. hyopneumoniae*).

M. hyopneumoniae and *A. pleuropneumoniae* co-infection is a common cause of PRDC (Thacker, 2006). For example, in France, *M. hyopneumoniae* and *A. pleuropneumoniae* were detected in 69.3% and 20.7% of 3731 lungs from 125 herds, respectively and, at the pig level, a significant association was found between PCR-positive results for *M. hyopneumoniae* and *A. pleuropneumoniae* (Fablet et al., 2012a). Experimental studies demonstrated that *M. hyopneumoniae* may potentiate the severity of *A. pleuropneumoniae*-induced lesions in mixed-infected pigs (Yagihashi et al., 1984; Marois et al., 1989a,b; Marois et al., 2009; Lee et al., 2014). Pigs experimentally infected with *M. hyopneumoniae* and 7, 16 or 28 days later with *A. pleuropneumoniae* were particularly affected. They showed severe clinical signs (hyperthermia, coughing and death), lung lesions (corresponding to those induced by the two bacterial strains combined) and reduced growth (Yagihashi et al., 1984; Marois et al., 1989a,b; Marois et al., 2009). This *M. hyopneumoniae*-*A. pleuropneumoniae* co-infection is responsible for immune response modulations: *M. hyopneumoniae* may influence phagocytic responses, especially in pigs secondarily infected with *A. pleuropneumoniae*, *e.g.* by decreasing the phagocytosis capacity of macrophages thus enabling more tissue damage by *A. pleuropneumoniae* (Caruso and Ross, 1990; Ciprián et al., 1994).

7.3.2 Interaction with *Bordetella bronchiseptica*

B. bronchiseptica is widespread and plays multiple roles in respiratory diseases. In addition to being the primary etiological agent of non-progressive atrophic rhinitis and promoting nasal colonization by toxigenic strains of *P. multocida* (which leads to severe progressive atrophic rhinitis), it is a primary cause of bronchopneumonia in young pigs and contributes to the PRDC in older pigs (Palzer et al., 2008; Brockmeier et al., 2012). However, the frequency of combinations of *B. bronchiseptica* and *M. hyopneumoniae* on farms and the impacts on the respiratory tract are not well-known. Very few field prevalence studies involving both pathogens are reported in the literature and to our knowledge, no experimental co-infection study has been performed. In the USA, *B. bronchiseptica* has a relatively low prevalence in diagnostic submissions, whereas *M. hyopneumoniae* is the second most common microorganism found in field respiratory cases following *P. multocida* (Opriessnig et al., 2011a), thus indicating few cases of potential interactions. In more recent Austrian studies, *B. bronchiseptica* was also detected in a relatively low frequency in lungs affected by interstitial pneumonia or *Pneumocystis spp.* positive lung samples (Kureljušić et al., 2016; Weissenbacher-Lang 2016). In these studies, *M. hyopneumoniae* was also the second most frequent agent. No *in vitro* cellular study has been published to understand the mechanisms involved during a *M. hyopneumoniae*/*B. bronchiseptica* co-infection. Nevertheless, an *in vitro* cytotoxic effect of *B. bronchiseptica* on macrophages has been described, which probably has an important role in the colonization and persistence of this pathogen and probably other respiratory pathogens, as *M. hyopneumoniae* (Brockmeier and Register, 2007; Brockmeier et al., 2012). Yim et al. (2017) found that *B. bronchiseptica* antigens may increase the production of *M. hyopneumoniae* antigen-specific antibodies. This opens the door for the use of *B. bronchiseptica* antigens as a vaccine adjuvant or as a vaccine antigen.

7.3.3 Interaction with *Pasteurella multocida*

P. multocida is responsible for pneumonic pasteurellosis that is frequently the final stage of enzootic pneumonia or PRDC. Septicemic pasteurellosis, pleurisy and abscessation can be observed in severe cases (Register et al., 2012; Oliveira Filho et al., 2018). This bacterium has a significant impact on swine health worldwide, because of induced pneumonia but also progressive atrophic rhinitis, caused by toxigenic *P. multocida* either alone or in combination with *B. bronchiseptica* (Register et al., 2012; Park et al., 2016b). The role of the dermonecrotic toxin in pneumonic pasteurellosis has been under debate (Pijoan et al., 1984; Register et al., 2012) even if it has been emphasized by Ahn et al. (2008). Infections with *P. multocida* induce a cranioventral pulmonary consolidation due to suppurative bronchopneumonia, which is microscopically characterized by neutrophil infiltration of bronchial and alveolar spaces and interstitial thickening (Register et al., 2012). In France, *P. multocida* was detected in 36.9% of lungs and associated with pneumonia on both pig and herd scales. Moreover, at this level, a significant association was found between a PCR-positive result for *P. multocida* and DNA detection of *M. hyopneumoniae* (Fablet et al., 2012a). Similar outcomes were reported in Germany (Palzer et al., 2008).

Five capsular serotypes, A, B, D, E, and F were described in *P. multocida* with A and D comprising most swine isolates (Register al., 2012). Serotype A is most commonly cultured from pneumonic lungs while most progressive atrophic rhinitis isolates are serotype D, but either serotype may be isolated from either condition (Register al., 2012).

Experimental studies were performed in order to examine the role of *P. multocida* as a secondary agent (Ciprián et al., 1988; Amass et al., 1994; Andreasen et al., 2000; Stipkovits et al., 2001; Eamens et al., 2012; Tocqueville et al., 2017). All these studies highlighted more severe clinical signs (coughing, hyperthermia) and lesions (pneumonia) when pigs were infected with *M. hyopneumoniae* and then *P. multocida* when compared to animals single infected. The damaged ciliated epithelium and suppressed immunity by the *M. hyopneumoniae* infection are likely the main factors underlying the secondary *P. multocida* infection. Moreover, *P. multocida* has a type A polysaccharide capsule that aids in resisting phagocytosis. The pathogenesis mechanisms to the co- or subsequent infections with *M. hyopneumoniae* and *P. multocida* are not clearly elucidated. Nonetheless, a recent experimental study showed that *M. hyopneumoniae* alters the *in situ* composition of fucosyl glycoconjugates to enhance adherence of *P. multocida* type A to the bronchial and bronchiolar epithelial cells (Park et al., 2016b). No study of immune response mechanisms induced by *M. hyopneumoniae* and *P. multocida* co-infection has been reported in the literature.

7.3.4 Interaction with other *Mycoplasma* species

Two other *Mycoplasma* species are detected in lungs with *M. hyopneumoniae*: *M. hyorhinis* and *M. flocculare* (Fourour et al., 2018). *M. hyorhinis* was the first mycoplasma described in pig farming in 1955, since its growth in culture medium is relatively fast and less difficult compared to other *Mycoplasma* species of pig, including *M. hyopneumoniae* (Switzer, 1955; Gois et al., 1968; Friis, 1971a; Gois et al., 1971). Initially, *M. hyorhinis* was described as a commensal agent of the respiratory system because its presence was not systematically associated with a clinical picture (Thacker and Minion, 2012). Although studies have confirmed that some infections may be subclinical, others have demonstrated its involvement in different types of pathologies such as polyserositis (pericarditis, pleurisy, peritonitis), arthritis, ear infections, conjunctivitis and sepsis (Thacker and Minion, 2012). The involvement of *M. hyorhinis* in pneumonia remains controversial in the literature. However, its role in PRDC seems likely since although it is more often detected in pneumonic than in healthy pigs (Palzer et al., 2008; Hansen et al., 2010; Luehrs et al., 2017).

M. flocculare has been the subject of little research and limited information is available. Most authors describe this *Mycoplasma* as common in pig farms, but no epidemiological studies have been conducted (Kobisch and Friis, 1996; Thacker and Minion, 2012). Usually described as commensal of the respiratory system, *M. flocculare* may be found in healthy lungs or with pneumonia, but no pathogenicity is directly associated with it.

The mixed infection with *M. hyopneumoniae* and *M. flocculare* does not seem to influence the clinical or the pathological picture of the disease (Strasser et al., 1992). However, proliferations of lympho-histiocytic-like cells in the lamina propria and lesions of the epithelial layer of the nasal mucosa or lung tissue have been observed during experimental infections on specific-pathogen-free (SPF) pigs (Kobisch and Friis, 1996). More recently, an investigation of the impact of secondary *M. flocculare* or *M. hyorhinis* infection in SPF pigs pre-infected with *M. hyopneumoniae* was performed (Fourour et al., 2019b). Pigs infected with *M. hyopneumoniae* and *M. hyorhinis* or *M. flocculare* showed lower growth rate during the third week of the experiment and higher haptoglobin concentrations in contrast to pigs infected with *M. hyopneumoniae* alone. In pigs infected with *M. hyopneumoniae* and *M. hyorhinis*, (i) *M. hyorhinis* was detected in bronchi (lack in mono-infected pigs with *M. hyorhinis*), (ii) *M. hyopneumoniae* was detected in polyserositis lesions and (iii) there was a slight delay in the production of total anti-

M. hyopneumoniae IgG. Without significant differences in terms of the extent of pneumonia, this experimental trial suggested that the *M. hyopneumoniae* and *M. flocculare* or *M. hyorhinis* association could contribute to the overall deterioration in pig health, by inducing an increase of the inflammatory status of pigs and probably involving the impairment of the immune system (Fourour et al., 2019b). Co-stimulation of porcine bone-marrow-derived dendritic cells (BM-DCs) with *M. hyopneumoniae* and *M. hyorhinis* or *M. flocculare* showed a decreased BM-DCs production of TNF-α (Fourour et al., 2019a). Moreover, *M. flocculare* and *M. hyopneumoniae* association increased BM-DCs ability to secrete IL-10. Therefore, the combination of *M. hyorhinis* or *M. flocculare* with *M. hyopneumoniae* seems to be able to participate to the modulation of the immune response and to affect the severity of the disease (Fourour et al., 2019a).

7.3.5 Interaction with other bacterial species

G. (Haemophilus) parasuis, *S. suis*, *A. suis*, *T. pyogenes* and *S. enterica* serovars Choleraesuis and Typhimurium may induce septicemia and therefore reach the lungs and have a possible role in bronchopneumonia (Palzer et al., 2008; Opriessnig et al., 2011a; Aragon et al., 2012; Carlson et al., 2012; Gottschalk, 2012b; Taylor, 2012; Jarosz et al., 2014; Maes et al., 2018). These pathogens have not been experimentally studied in association with *M. hyopneumoniae*. *T. pyogenes* and *S. enterica* are seldom. In Germany, α-haemolytic streptococci were the most frequently detected bacterial pathogens from 239 bronchoalveolar lavage fluids collected in pigs (from 93 farms) with clinical signs of respiratory disease. They were associated with the presence of *M. hyorhinis*, *P. multocida* and *B. bronchiseptica*, and also occurred more often in pigs that were already infected with other pathogens (Palzer et al., 2008). Two hypotheses were issued: 1) the proliferation of α-hemolytic streptococci could affect the lung surface, making it more receptive to colonization by other secondary or opportunistic pathogens; and 2) the colonization of lungs by streptococci is enhanced after an infection with these pathogens (Palzer et al., 2008). One pathogenic α-hemolytic streptococcus in the pig is *S. suis*, causing meningitis, arthritis, pericarditis, polyserositis and septicemia and has a possible role in bronchopneumonia (Gottschalk, 2012b). The mechanism by which *S. suis* induces pneumonia is unknown. In France, DNA of *S. suis* and *H. parasuis* was not as frequently detected as *M. hyopneumoniae* DNA in pig lungs, with 6.4%, 0.99% and 69.3% of PCR positive lungs collected at the slaughterhouse, respectively (Fablet et al., 2012a). However, a

7

significant positive association was found between a PCR-positive result for *M. hyopneumoniae* and DNA detection of *H. parasuis* at the lung level (Fablet et al., 2012a). The exact pathogenesis and host response are still unknown for the co-infections with *M. hyopneumoniae* and the opportunistic bacteria described in this paragraph, which also brings into question the role of these bacterial pathogens, like *H. parasuis*, in PRDC.

A summary of the knowledge currently available on the effect(s) of *M. hyopneumoniae* interaction with bacteria is presented in Table 7.1.

Table 7.1. Overview of the main results on the effect of *M. hyopneumoniae* interactions with bacteria.

	Clinical signs		Lung lesions	
Interaction with	**Synergistic effects**	**Additive effects**	**Synergistic effects**	**Additive effects**
Actinobacillus pleuropneumoniae		Yagihashi et al. 1984 Marois et al. 1989 Marois et al. 2009		Yagihashi et al. 1984 Marois et al. 1989 Marois et al. 2009
Bordetella bronchiseptica	Undocumented	Undocumented	Undocumented	Undocumented
Pasteurella multocida		Ciprièn et al. 1988 Amass et al. 1994 Andreasen et al. 2000 Stipkovits et al. 2001 Eamens et al. 2012 Tocqueville et al. 2017		Ciprièn et al. 1988 Amass et al. 1994 Andreasen et al. 2000 Stipkovits et al. 2001 Eamens et al. 2012 Tocqueville et al. 2017
Mycoplasma hyorhinis		Fourour et al. 2019		
Mycoplasma flocculare		Fourour et al. 2019		
Other bacterial species	Undocumented	Undocumented	Undocumented	Undocumented

7.4 Interactions of *Mycoplasma hyopneumoniae* with viruses involved in lung diseases

Understanding how *M. hyopneumoniae* and respiratory viruses interact with each other can be critical in the effort to control PRDC. The primary respiratory viruses include PRRSV, IAV-S, PCV-2 and pseudorabies virus (PRV). Their interactions with *M. hyopneumoniae* are described thereafter.

7.4.1 Interaction with PRRSV

PRRSV is a swine-specific enveloped RNA virus and a member of the family *Arteriviridae*. Infection with PRRSV is characterized by reproductive failure in pregnant sows, high pre-weaning mortality in piglets infected *in utero* and respiratory signs in growers and finishers (Done et al., 1996; Kranker et al., 1998; Rossow, 1998). Interacting with other infectious pathogens, the virus might be involved in cases of PRDC, which often exhibit concurrence with *M. hyopneumoniae* (Chae, 2016).

7

The interaction of *M. hyopneumoniae* with PRRSV within the porcine respiratory tract is complex. Under experimental conditions, when both pathogens are present, the sequence of infection can have very different outcomes. *M. hyopneumoniae* infection followed by PRRSV three weeks later, had a potentiating effect on the PRRSV-induced pneumonia (Figure 7.1.A; Thacker et al., 1999). On the other hand, inoculation with PRRSV followed by *M. hyopneumoniae* 10 days later had no effect on the severity of pneumonia (Figure 7.1.B; Thacker et al., 1999). Studies with concurrent infections of *M. hyopneumoniae* and PRRSV at either 3 or 6 weeks of age also did not result in a synergistic effect (Figures 7.1.C and 7.1.D; Van Alstine et al., 1996; Thacker et al., 1999; Bourry et al., 2015). Indeed, pigs dually infected with *M. hyopneumoniae* and PRRSV showed a combination of clinical signs characteristic of each pathogen but no significant exacerbation of pathogenicity. The mechanisms by which both pathogens may interact are not yet clearly and deeply disentangled. Experimental studies indicate that the potentiation and prolongation of PRRSV-induced pneumonia by *M. hyopneumoniae* rely on the inflammatory processes elicited in the course of both mycoplasmal and PRRSV pneumonia (Thanawongnuwech et al., 2001; Thanawongnuwech et al., 2004). Recruitment of macrophages and lymphocytes in the lung parenchyma following *M. hyopneumoniae* infection may provide higher numbers of cells susceptible to PRRSV infection, thus potentiating and prolonging the viral pneumonia (Opriessnig et al., 2011a). Furthermore, the inflammatory responses induced by both pathogens may further weaken the lung immune response so that PRRSV clearance decreases (Thanawongnuwech et al., 2001; Thanawongnuwech et al., 2004). Therefore, *M. hyopneumoniae* might play an initial role in inflammatory modulation by inducing pro-inflammatory cytokines and recruitment of macrophages, thus enabling the animal to be more susceptible to PRRSV infection.

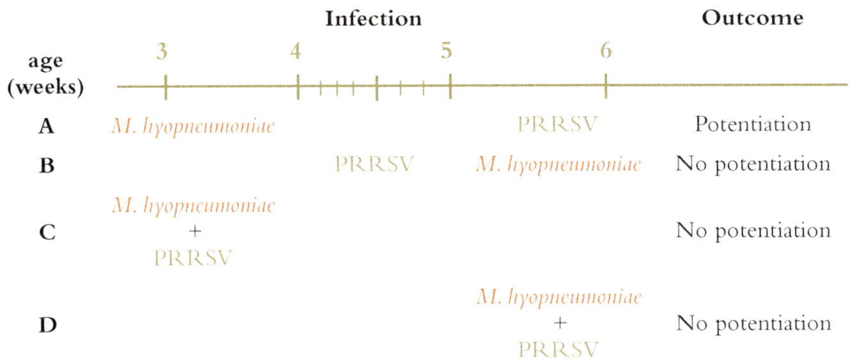

Figure 7.1. Interaction of *Mycoplasma hyopneumoniae* with porcine reproductive and respiratory syndrome virus (PRRSV). *M. hyopneumoniae* infection prior to PRRSV infection potentiates PRRSV-induced pneumonia (A). No potentiating effect is observed on PRRSV-induced pneumonia when *M. hyopneumoniae* infection follows PRRSV infection (B) or in the case of co-infection at either 3 (C) or 6 weeks old (D) (Source: Van Alstine et al., 1996; Thacker et al., 1999).

Understanding this interaction is also important for vaccination. Since prior infection with *M. hyopneumoniae* can result in increased severity of PRRSV-induced pneumonic lesions (Thacker et al., 1999), vaccination against *M. hyopneumoniae* alone could significantly reduce these signs. Two studies have already shown this effect, as *M. hyopneumoniae* vaccines were able to neutralize the effect that *M. hyopneumoniae* has on PRRSV infection potentiation (Thacker et al., 2000b; Park et al., 2014). These results are also consistent with a field study in which vaccination against *M. hyopneumoniae* only improved the growth performance under co-infection of pigs with *M. hyopneumoniae* and PRRSV (Moreau et al., 2004). Taken together, these data suggest that vaccination of pigs against *M. hyopneumoniae* is particularly important if pigs are co-infected with *M. hyopneumoniae* and PRRSV (Chae, 2016).

In summary, there are three important things to consider when designing a vaccine regimen involving these agents. First, a previous infection with *M. hyopneumoniae* increases the severity of PRRSV-induced pneumonia. Second, vaccination with *M. hyopneumoniae* alone appears to alleviate the severity of PRRSV-induced pneumonia and, in consequence, *M. hyopneumoniae* vaccination should be the first priority.

7.4.2 Interaction with PCV-2

PCV-2, a small, non-enveloped, single-stranded DNA virus is one of the most economically important viruses in modern pig production when uncontrolled. PCV-2 is commonly associated with lymphoid depletion and immune suppression in infected pigs (Segalés et al., 2004) and therefore frequently associated with other pathogens. Early studies indicated that, in porcine circovirus disease (PCVD) clinical cases, *M. hyopneumoniae* can be found at a rate of 35.5% in the USA (Pallarés et al., 2002) or 31.4% in Korea (Kim et al., 2003). Furthermore, pigs that were infected with PCV-2 were 3.77 times as likely to be also infected with *M. hyopneumoniae* compared to PCV-2 negative pigs (Dorr et al., 2007), suggesting an important role of *M. hyopneumoniae* on the development of (PCVD).

Like PCV-2, infection with *M. hyopneumoniae* is commonly associated with immunomodulation (stimulation or suppression; Kishima and Ross, 1985; Kishima et al., 1985; Maes et al., 1996), but *M. hyopneumoniae* also has non-specific mitogenic effects on swine lymphocytes (Messier and Ross, 1991) or cytokines secreted by activated macrophages (Messier and Ross, 1991; Asai et al., 1993; Thanawongnuwech et al., 2001), and therefore, could potentiate the effects of PCV-2 infection *via* these routes. Under *in vitro* conditions using the PK-15 cell line, PCV-2 replication was enhanced by subsequent co-inoculation with *M. hyopneumoniae* in a time and dose dependent method, but not by prior or simultaneous co-inoculation (Wang et al., 2016a), possibly highlighting important *in vitro* and *in vivo* infection dynamic differences. To further investigate the relationship of PCV-2 and *M. hyopneumoniae*, conventional pigs were experimentally infected with *M. hyopneumoniae* at 4 weeks of age followed by experimental infection with PCV-2 two weeks later (Opriessnig et al., 2004). Dually infected pigs had decreased growth rates and moderate dyspnea. *M. hyopneumoniae* potentiated the severity of PCV-2 associated lung and lymphoid lesions, increased the amount and prolonged the presence of PCV-2 antigen, and increased the incidence of PCVD in pigs (Opriessnig et al., 2004). Similar results were obtained using the same infection interval obtained by another group (Seo et al., 2014). In contrast, simultaneous PCV-2 and *M. hyopneumoniae* coinfection did not potentiate disease in conventional pigs (Sibila et al., 2012). *Mycoplasma hyopneumoniae* needs time to establish infection and to induce lymphoid hyperplasia which once present could provide an important first site of PCV-2 replication. This could explain the differences in results in simultaneously infected pigs versus pigs infected with PCV-2 two weeks after *M. hyopneumoniae* infection.

Besides the importance of concurrent *M. hyopneumoniae* and PCV-2 co-infection in young pigs, this particular association is also important in high health breeding herds and boar studs or in incoming young gilts or boars with insufficient immunity against these pathogens. Gilts or young boars that enter breeding herds or boar studs are often exposed to infectious pathogens such as *M. hyopneumoniae* or PCV-2 for the first time, are applied multiple adjuvanted vaccines, and are exposed to other stressors including transportation and regrouping. Such gilts and boars are more prone to develop clinical disease associated with *M. hyopneumoniae* and PCV-2 coinfection as demonstrated by case reports (Opriessnig et al., 2006). To investigate this under controlled conditions, young boars were experimentally co-infected with both pathogens and the results previously obtained in growing pigs were repeatable in this young boar model (Opriessnig et al., 2011b). When the cytokine and chemokine mRNA expression profiles in trachea-bronchial lymph nodes from pigs concurrently infected with PCV-2 and *M. hyopneumoniae* were investigated, *M. hyopneumoniae* potentiated PCV-2 infection by increasing interferon-γ and IL-10 mRNA expression levels (Zhang et al., 2011). The increase of IFN-γ and chemokines and the decrease of IFN-α in single PCV-2 and dually infected pigs were associated with increased severity of lymphoid lesions and the presence of PCV-2 antigen (Zhang et al., 2011).

7.4.3 Interaction with swine influenza A viruses

Swine influenza A viruses are enveloped, single-stranded RNA viruses belonging to the family *Orthomyxoviridae* and are etiological agents of swine influenza, an acute respiratory disease characterized by hyperthermia, loss of appetite, lethargy, and respiratory problems. They are enzootic in areas densely populated with pigs all around the world and are recognized as major pathogens of the PRDC when associated with other virus(es) and/or bacteria (Opriessnig et al., 2011a).

Detection of an IAV-S in combination with another pathogen is very common. For example, in the USA, analyses of 2,872 lungs collected from pigs with respiratory problems showed that IAV-S's were associated with another respiratory pathogen in 19% of cases, whereas only 3.1% of samples were infected by IAV-S only (Choi et al., 2003). The *M. hyopneumoniae*-IAV-S association represented 22.3% of these co-infection cases (*i.e.* 4.3% of total cases) and was the second most frequent association involving IAV-S (the first one being *P. multocida*-IAV-S).

Several experimental studies with pigs previously infected or not with *M. hyopneumoniae* have been conducted to investigate the impact of *Mycoplasma* on the outcome of a subsequent IAV-S infection and to understand the mechanisms of this interaction. The timing between both infections appeared to be an important factor influencing the exacerbation, or not, of influenza. Indeed, clinical signs and pneumonic lesions were exacerbated in *M. hyopneumoniae*-IAV-S infected animals when the pathogen inoculations occurred 21 days apart but not 7 days (Thacker et al., 2001; Yazawa et al., 2004). In addition, whether CDCD pigs or conventional pigs are used, this underlines that *Mycoplasma* needs to be already established in the respiratory tract at the time of IAV-S infection to potentiate it. The virus subtype seems to be also an important factor in PRDC development since *M. hyopneumoniae* infection increased the severity of a subsequent infection with IAV-S H1N1, but not H1N2 (Deblanc et al., 2012). Animals co-infected with *M. hyopneumoniae* and H1N1 presented a weight loss during 4 days following the IAV-S infection, a prolonged hyperthermia, and extended pneumonia lesions as compared to H1N1 single infected piglets.

Mechanisms underlying the influenza exacerbation are not fully understood yet, but it appeared that several modifications induced by *M. hyopneumoniae* would be at the origin of the increased pathogenesis and recovery delay. Firstly, it was shown that *Mycoplasma* induced an oxidative stress that influenced the severity of the subsequent influenza infection and that co-infection induced major changes in nutrient metabolism (Deblanc et al., 2013; Le Floc'h et al., 2014). Secondly, *M. hyopneumoniae* infection affected neither the influenza replication nor the IFN-induced antiviral response in the lung at early time post-infection (≤ 2 days; Deblanc et al., 2016), but it would promote the persistence of the virus in the lung. Indeed, the viral genome was still present in the diaphragmatic lobes of co-infected animals 7 days after H1N1 infection, but not in animals infected only with H1N1 (Deblanc et al., 2012). Given the fact that *M. hyopneumoniae* infection induces a functional decrease of phagocytic cells (Thacker and Minion, 2012; Deeney et al., 2019), it may be partly responsible for the delay in the elimination of the virus in the lung. Finally, *Mycoplasma* infection predisposed the animals to a higher inflammatory response following H1N1 infection (Deblanc et al., 2016). Co-infected pigs had an increased production of pro-inflammatory cytokines (IL-6, IL-1β and TNF-α) and a greater infiltration of macrophages and neutrophils into the lung compared to pigs infected only with H1N1. This marked inflammatory state would play an important role in exacerbating the tissue injuries and clinical signs subsequent to H1N1 infection and prolonging them over time. *M. hyopneumoniae* infection induces a chronic pulmonary inflammation peaking 21 days after infection (Choi et al., 2006; Redondo et al., 2009). Therefore, it is supposed that the response to IAV-S infection have an additive, but not synergistic, effect on the inflammation induced by *M. hyopneumoniae* already in place.

7.4.4 Interaction with other viruses

Concomitant presence of antibodies against *M. hyopneumoniae* and viruses such as Pseudorabies (Aujeszky's disease) virus (PRV), Porcine respiratory coronavirus (PRCV – designated as transmissible gastro-enteritis virus in early studies), Hemagglutinating encephalomyelitis virus, Enteroviruses and Porcine cytomegalovirus (PCMV) has been widely described (Marois et al., 1989a,b; Maes et al., 1999b). Moreover, *Mycoplasma* has also been detected concurrently with viruses such as PCMV (Hansen et al., 2010) and *Torque teno sus* viruses (Rammohan et al., 2012; Weissenbacher-Lang et al., 2016) under natural conditions. However, the concomitant presence of *M. hyopneumoniae* and other viruses or their antibodies does not imply the existence of interaction with them.

Besides the well-known effects of the experimental infection of *M. hyopneumoniae* with the viruses mentioned in previous sections, the interaction between *Mycoplasma* and PRV, PRCV and swine adenovirus were explored. The inoculation of *M. hyopneumoniae* one week before PRV showed more extensive lung lesions (cranio-ventral pulmonary consolidation) in the double challenged group as well as higher load of *Mycoplasma* in broncho-alveolar lavage fluids compared with the group that received *M. hyopneumoniae* alone. These observations were consistent at both 2 and 4 weeks post-*M. hyopneumoniae* inoculation (Shibata et al., 1998). Similarly, the simultaneous co-inoculation of *Mycoplasma* with PRCV resulted in an increased severity of lung lesions when compared with the solely *M. hyopneumoniae* inoculated group 2 weeks after challenge (Marois et al., 1989a,b). The co-infection of *M. hyopneumoniae* with swine adenovirus in gnotobiotic pigs also resulted in more severe lesions than inoculation of both pathogens separately (Kasza et al., 1969). In consequence, the limited and dated information available with these viruses points to a potentiation synergy with *M. hyopneumoniae* regarding lung lesion extension, although involved mechanisms of interaction are virtually unknown. No recent information is available regarding the co-infection of *M. hyopneumoniae* with other viruses.

The current knowledge on the effect(s) of *M. hyopneumoniae* interaction with viruses is presented in Table 7.2.

Table 7.2. Overview of the main results on the effect of *M. hyopneumoniae* interactions with viruses.

	Clinical signs		Lung lesions	
Interaction with	Synergistic effects	Additive effects	Synergistic effects	Additive effects
PRRSV			Thacker et al., 1999	
PCV-2*	Opriessnig et al., 2004		Opriessnig et al., 2004	Seo et al., 2014
Swine influenza A viruses		Thacker et al., 2001 Yazawa et al., 2004 Deblanc et al., 2012		Thacker et al., 2001 Yazawa et al., 2004 Deblanc et al., 2012 Deblanc et al., 2016
PRV				Shibata et al., 1998
PRCV				Marois et al., 1989
Adenovirus				Kasza et al., 1969

*In Sibila et al. (2012) the simultaneous PCV-2 and *M. hyopneumoniae* coinfection of conventional pigs did not produce any synergistic or additive effect.

7.5 Interactions of *Mycoplasma hyopneumoniae* with parasitic infections and mycotoxins

Parasitic infections or feed contaminated with mycotoxins can negatively affect the host ability to respond to respiratory infectious pathogens (Tjørnehøj et al., 1992; Oswald, 2011). Early in the 75s, Flesja and Ulvesaeter (1980) observed that the extent of pneumonia was significantly associated with the presence of liver white spot lesions due to migrating *Ascaris suum* (*A. suum*) larvae. This study provided the first support for putative relationships between parasites and infectious pathogens involved in pneumonia. The effect of *A. suum* infection on *M. hyopneumoniae* vaccination efficacy was then explored in an experimental trial that is described in the chapter on vaccination of the current book (Steenhard et al., 2009).

Toxic agents such as mycotoxins may also compromise pig health and the immune responses following an infectious challenge. Mycotoxins are produced by fungi before, during, or after harvest and may be present in various raw materials commonly used for pig feeds and with various levels of contamination. Studies assessing the relationship between *M. hyopneumoniae* and mycotoxins are scarce but the results indicate that the effects of toxic agents on *M. hyopneumoniae* infection may differ according to the toxin. While experimental exposure of pigs to feed artificially contaminated by Fumonisin B was found to exacerbate the course of *M. hyopneumoniae* infection and associated lung lesions (Posa et al., 2013; Posa et al., 2016), exposure to feed contaminated by the mycotoxin deoxynivalenol (DON) did not enhance the severity of *M. hyopneumoniae* infection when compared to pigs fed with non-contaminated stuffs (Michiels et al., 2018).

7.6 Conclusions

In conclusion, *M. hyopneumoniae* generally acts as a primary pathogen and interaction with one or more bacterial and/or viral pathogens generally lead to more severe disease. The effects of the interaction on the clinical outcome and lung lesions may depend on the strain of the pathogens involved, the infectious dose, infection time, infection route and the sequence of infections. The understanding of the interactions between several infectious agents, their mechanism and effects is often limited to the experimental approaches used, which do not always represent the full diversity and complexity of infections and interactions involved in natural infections.

However, in order to develop appropriate and therefore more effective control strategies against *M. hyopneumoniae* infection, it seems essential to better identify all the pathogen associations, including perhaps the marginal ones, involved in pulmonary lesions in each particular case and to understand the pathophysiological mechanisms, as well as host immunity responses. New laboratory approaches such as 'omics' techniques will certainly be a source of great scientific progress in the coming years to better disentangle the interactions between *M. hyopneumoniae* and the associated infectious agents to reduce the animal health and welfare issues of respiratory diseases, as well as the consequences on the production performances and in turn on the economic impacts for the farmers.

CHAPTER 8

DIAGNOSIS OF *MYCOPLASMA HYOPNEUMONIAE* INFECTION AND ASSOCIATED DISEASES

Chanhee Chae[1]

João Carlos Gomes-Neto[2]

Joaquim Segalés[3]

Marina Sibila[4]

1 College of Veterinary Medicine, Seoul National University, Seoul, Republic of Korea
2 Nebraska Innovation Campus, University of Nebraska-Lincoln, Lincoln, USA
3 Facultat de Veterinària (Universitat Autònoma de Barcelona), Centre de Recerca en Sanitat Animal (CReSA)-Institut de Recerca i Tecnologia Agroalimentàries (IRTA), Campus de la Universitat Autonoma de Barcelona, Bellaterra, Spain
4 Centre de Recerca en Sanitat Animal (CReSA), Institut de Recerca i Tecnologia Agroalimentàries (IRTA), Campus de la Universitat Autònoma de Barcelona, Bellaterra, Spain

8.1 Introduction

Antemortem diagnosis of *Mycoplasma hyopneumoniae (M. hyopneumoniae)* infection can be difficult mainly due to the inaccessibility of the microorganism site of action, its capabilities to evade the host immune system (leading to a delayed seroconversion), and by the specific isolation requirements. All these points make the confirmation of *M. hyopneumoniae* "free" status of a swine herd challenging. This scenario is worsened by the fact that the outcome of the infection probably depends on many factors, such as infectious pressure, time to exposure, co-infections, and presumably the number and virulence of *M. hyopneumoniae* strains involved. As a consequence of these many influencing factors, *M. hyopneumoniae* infection course can vary from a subclinical infection to a mild respiratory disease in an endemic scenario or to an epidemic severe respiratory outbreak in naïve populations (Maes et al., 2018). Therefore, the diagnostic approach should be decided depending on the objective pursued: *M. hyopneumoniae* involvement in respiratory disease or lesions, detection of subclinically infected pigs in an eradication program, and/or determining the negativity of replacement stock. In this chapter, the different diagnostic techniques available alongside their advantages and limitations are reviewed.

8.2 Clinical-pathological diagnosis

The first indication of respiratory disease attributable to this pathogen is the observation of a certain number of growing and/or finishing animals showing an intermittent non-productive cough of variable severity (see Chapter 5). The diagnostic value of this clinical finding is limited as it can be triggered by many environmental factors as well as caused by other infectious pathogens (Maes et al., 2018). Additionally, subclinically infected animals are likely to go unrecognized if diagnosis is determined only on clinical respiratory signs. Hence, the association of these clinical signs with *M. hyopneumoniae* needs laboratory confirmation by pathogen and/or antibody detection/isolation. A quantitative assessment of coughing bouts (coughing index: total number of coughing bouts / number of examined pigs x total time of observation) in finishing pigs has been described (Nathues et al., 2012). However, its low capacity to determine the etiology of this clinical sign encourages the use of this method after training sessions to differentiate dry versus moist cough in combination with antibody and pathogen detection by ELISA or PCR, respectively (Nathues et al., 2012).

Macroscopically, *M. hyopneumoniae* produces a red-to-purple area of consolidation (cranioventral pulmonary consolidation, CVPC). These areas are normally distributed bilaterally in apical, intermediate, accessory and cranial portions of diaphragmatic lobes of distribution (Garcia-Morante et al., 2016a) (see Chapter 5). Lung lesion assessment during necropsy or slaughter checks is frequently employed to evaluate the extension of the lesions presumably associated with *M. hyopneumoniae* infection, since it is usually correlated with the severity of clinical signs at herd level. Lung inspections can also identify infected animals without clinical signs but suffering from incipient (subclinically infected animals) or healing (chronically infected animals) lung lesions. More information on *M. hyopneumoniae* gross lung lesions' characteristics and assessment methods is given in Chapter 5.

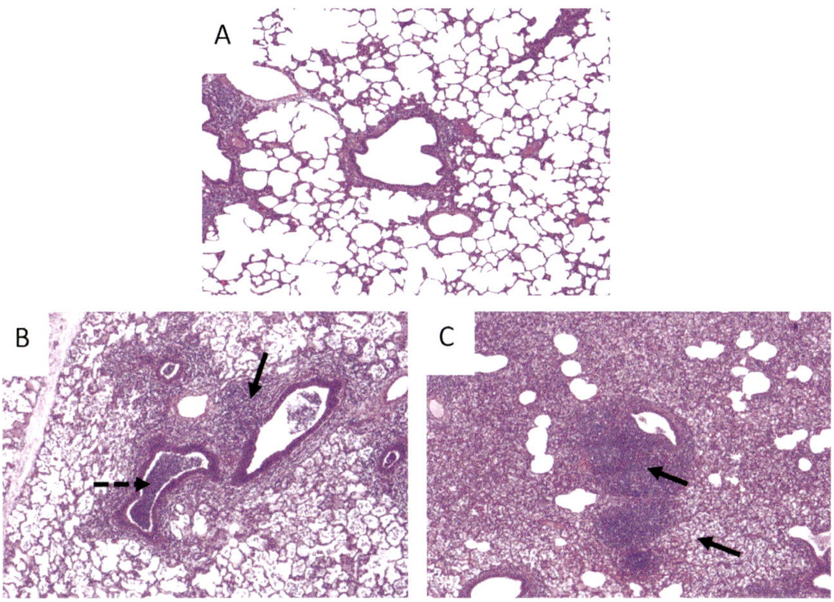

Figure 8.1. **(A)** Normal lung. **(B)** Bronchointerstitial pneumonia with hyperplasia of bronchus-associated lymphoid tissue (BALT, black arrow) induced by *M. hyopneumoniae* infection; the lung has also suppurative bronchopneumonia (discontinuous black arrow) produced by a concomitant bacterial infection. **(C)** Bronchointerstitial pneumonia with severe BALT hyperplasia with formation of lymphoid follicles (black arrow). (Source: CReSA-IRTA)

If microscopic analysis is desired or required, lung tissue portions taken from CVPC affected areas of different lobes should be immersed in 10% formalin. Microscopically, these lesions are described as broncho-interstitial pneumonia (BIP) with hyperplasia of the bronchus-associated lymphoid tissue (BALT) (Figure 8.1.B). Evaluation of microscopic lesions helps in differentiating gross lesions due to pulmonary collapse (atelectasis) or to score the BIP lesion severity (Sibila et al., 2012). Additionally, microscopic analysis allows the detection of incipient lesions not yet observable at macroscopic level. Presence of lymphocytes and plasma cells within the interstitium surrounding the airways forming peribronchiolar and perivascular lymphoid nodules are indicative of more severe lesions (Calsamiglia et al., 2000) (Figure 8.1.C). While some neutrophils may be present, concurrent bacterial infections are often found and lesions of suppurative bronchopneumonia are predominant (Figure 8.1.B) in such multi-etiological scenarios.

8.2.1 Differential diagnosis

Importantly, clinical, gross and microscopic lesions caused by *M. hyopneumoniae* infection are not exclusive of this bacterium. All potential causes (infectious and non-infectious) of coughing and dyspnea should be included in the clinical differential diagnoses (Ramirez, 2019). Moreover, CPVC due to *M. hyopneumoniae* could resemble the lesions caused by other bacterial (*Pasteurella multocida, Bordotella bronchiseptica, Glaesserella (Hemophilus) parasuis, S. suis)* or viral (swine influenza A virus [IAV-S]) pathogens (Figure 8.2). Similarly, animals infected by IAV-S infection or parasitized by *Metastrongylus* spp. can also display peribronchiolar and perivascular lymphoid hyperplasia. Consequently, the etiological differential diagnosis of these pathological findings should ideally include other microorganisms, with confirmation via ancillary diagnostics.

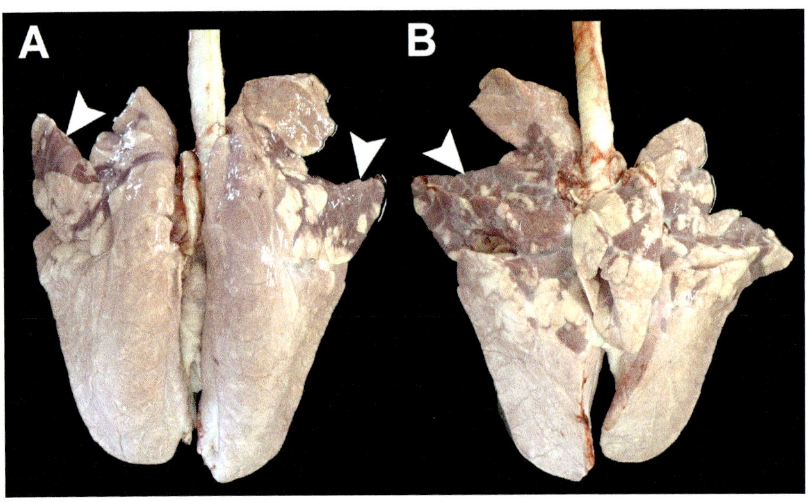

Figure 8.2. Dorsal and ventral vision (left and right, respectively) of a lung showing cranioventral pulmonary consolidation (CVPC) lesions (white arrows) caused by an experimental challenge of swine influenza A virus. (Source: CReSA-IRTA)

8.3 Detection of the pathogen

8.3.1 Isolation and culturing

Bacterial isolation was traditionally considered the "gold standard" technique to demonstrate *M. hyopneumoniae* presence in lung tissue. However, culturing this microorganism is difficult and time-consuming due to the specific nutritional requirements and the optimal incubation conditions needed. This implies that a negative culture result is not a conclusive test. Additionally, contamination with *M. hyorhinis*, a less fastidious, faster growing and common inhabitant of porcine respiratory tract, is quite frequent. Consequently, *M. hyoneumoniae* isolation is seldom performed as a diagnosis technique. This technique is normally performed only when the isolate is required (Pieters and Maes, 2019).

Mycoplasmas' genome contains a limited number of genes, implying the need to obtain some essential nutrients from an external source. Under laboratory conditions, these limitations are overcome by cultivating these microorganisms in a rich medium supplemented with serum (Beier et al., 2018). The most frequently utilized medium to grow mycoplasmas was developed by Niels Friis (1975) and is still the basis for some of the variants used currently. There are other media available for commercial purchase that support *M. hyopneumoniae* growth, but their detailed formula is sometimes not in the public domain (Cook et al., 2016). Friis medium contains, among other components, fresh yeast extract, brain heart infusion broth, thallium acetate, phenol red as a pH indicator, as well as swine and horse serum (Amstrong et al., 1994). Friis medium, as well as most variants for *M. hyopneumoniae* culture from field samples, is a liquid broth. In general terms, the growth of this microorganism in solid agar is very difficult and may need several weekly passages in liquid medium until a colony is formed in solid agar. This limited growth in solid medium has been linked to the agar purity and its capability to inhibit the growth or to retain some nutrients (Cook et al., 2016).

To prevent *M. hyorhinis* contamination, Friis recommended the use of 5% hyperimmune anti-*M. hyorhinis* serum together with 500 μg/ml cycloserine (Kobisch and Friis, 1996). More recently, Cook et al. (2016) described a selective medium for *M. hyopneumoniae* culture by adding kanamycin to suppress *M. hyorhinis* growth.

The samples of choice for *M. hyopneumoniae* isolation would be lung tissue, or even bronchial swab or broncho-alveolar lavage fluid, from an animal in the acute phase of the disease. The *M. hyopneumoniae* bacterial load present in chronic lesions is assumed to be low. Isolation should be attempted, preferably, from tissue obtained from the transition area between healthy and affected tissue (Amstrong et al., 1994). In order to avoid contamination from other bacteria, filtration of this suspension through 0.45 μm is recommended (A. Pérez de Rozas, personal communication). This filtered suspension should be diluted in broth medium and incubated from days to weeks at 37°C in an aerobic atmosphere (Pieters and Maes, 2019). The incubation period for the medium is normally determined by the change in color of the phenol red indicator. Turbidity is not observed in liquid media growth. In solid agar medium, *M. hyopneumoniae* colonies observation using a phase contrast stereomicroscope may take from days to weeks (Cook et al., 2016; García-Morante et al., 2018). Nevertheless, identification of colonies by a *M. hyopneumoniae*-specific PCR is recommended (Pieters and Maes, 2019). Therefore, colony forming units (CFU) is, for this microorganism, usually an unsuitable titration method. Instead, color changing units (CCU) is the most frequent titration method used. However, this ten-fold dilution approach is an indirect, subjective, and low accuracy method (Calus et al., 2010). ATP luminometry, flow cytometry and/or qPCR detection techniques can lend support to CCU titration method (Calus et al., 2010; Assunção et al., 2005a; García-Morante et al., 2018).

8.3.2 Detection and localization of *Mycoplasma hyopneumoniae* in tissues

Common detection methods of *M. hyopneumoniae* in tissues are immunofluorescence assay (IFA), immunohistochemistry (IHC), and *in situ* hybridization (ISH) (Doster and Lin, 1988; Kwon et al., 2002; Opriessnig et al., 2004; Sarradell et al., 2003). Both IFA and IHC are methods of detecting specific target proteins while ISH detects target DNA. The tissue samples should be frozen for IFA or fixed for IHC and ISH. The standard fixation method uses formalin and a subsequent embedding of tissues in paraffin. Formalin fixed samples are often the only sample available in pathology laboratory archives because fresh and frozen tissue samples from clinical cases are rarely saved for extended periods. Samples fixed in formalin are tested by IHC and ISH (Figure 8.3) to determine the extent of colonization and *M. hyopneumoniae* tissue localization. Frozen tissues are usually employed in IFA to determine *M. hyopneumoniae* tissue localization. Noteworthy, these detection methods require sampling of actively infected lung tissue, as colonization of airways is not homogenous across all sections of lung. Thus, samples could still be negative in an infected animal, even with a sensitive detection method.

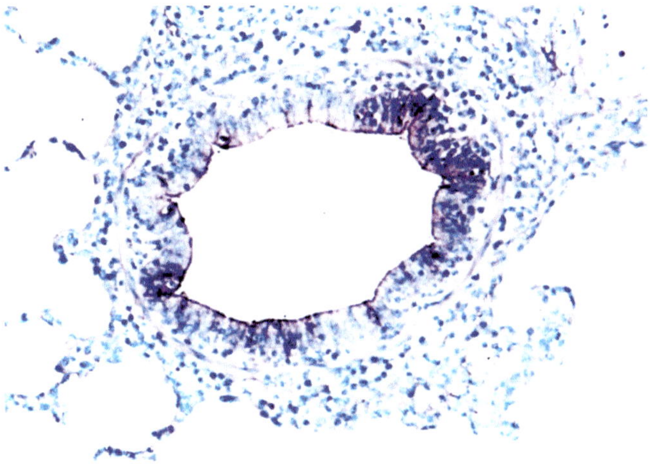

Figure 8.3. *M. hyopneumoniae* DNA present on the luminal surface (purple reaction) of bronchiolar epithelial cells. *In situ* hybridization. (Source: C. Chae)

The specificity and sensitivity of the IFA and IHC are highly dependent on the quality of the antibody employed. Monoclonal and polyclonal antibodies have been utilized for the detection of *M. hyopneumoniae* in formalin-fixed, paraffin-embedded tissues (Doster and Lin, 1988; Opriessnig et al., 2004; Sarradell et al., 2003). In general, a monoclonal antibody has a higher specificity compared to a polyclonal antibody due to the fact that it targets a single epitope (Haines and Chelack, 1991). Nevertheless, depending on the size and position of the epitope, the binding can be limited resulting in reduced signal. Polyclonal antibodies tend to be less specific due to non-specific binding from other antibodies in the serum produced by previous immunizations; however, they have a higher sensitivity because they target more epitopes, increasing the binding and the overall signal. This is the reason why polyclonal antibody-based IHC is a more practical routine diagnostic method for the detection of *M. hyopneumoniae* in tissues due to the increased sensitivity.

In situ hybridization is a method capable of detecting small quantities of *M. hyopneumoniae* DNA (Kwon, et al., 2002). ISH protocol development can be more practical than IHC because preparation of *M. hyopneumoniae* specific DNA probes by PCR is easier compared to the generation of monoclonal or polyclonal antibodies. The increased availability of probes and sequence databases also makes the process less time-consuming, rendering the technique more applicable in diagnosis and research areas.

It is important to note, however, that there are no perfect methods that allow detection of *M. hyopneumoniae* in tissues. All methods described above have advantages and disadvantages. IFA and IHC can be limited because *M. hyopneumoniae* shares antigenic determinants with two other porcine mycoplasmas such as *M. flocculare* and *M. hyorhinis*, which can cause false positives (Armstrong et al., 1983; Bolske et al., 1987). Such cross-reactivity can be avoided using ISH with a *M. hyopneumoniae*-specific DNA probe (Kwon et al., 2002). However, routine use of ISH may be restricted for some diagnostic laboratories because the procedure is technically more complex, cumbersome, and expensive when compared to IHC. The use of non-radioactive probes labeled with digoxigenin facilitates the application of this method in diagnostic laboratories that already perform IHC.

8.3.3 Detection of the pathogen by PCR

8.3.3.1 Sample selection

M. hyopneumoniae attaches to ciliated epithelial cells of the trachea, bronchi and larger bronchioles. In consequence, detection of its DNA can be performed in different specimens collected along the respiratory airways. Samples more frequently used to assess *M. hyopneumoniae* infection are airway tract swabs (nasal, tonsillar, laryngeal, tracheal and/or bronchial), oropharyngeal brushing, tracheobronchial mucus (TBM), lavage fluids (tracheobronchial or bronchoalveolar), lung (homogenate) tissue and oral fluids (rope collection) (Kurth et al., 2002; Sibila et al., 2004a; Fablet et al., 2010; Vangroenweghe et al., 2015a; Hernandez-García et al., 2017; Pieters et al., 2017).

Several studies report that *M. hyopneumoniae* DNA detection rates vary in different types of samples collected from the same animal (Kurth et al., 2002; Sibila et al., 2004c; Fablet et al., 2010; Vangroenweghe et al., 2015a; Pieters et al., 2017). These intra-animal comparative studies pointed out that the most sensitive samples to detect *M. hyopneumoniae* infection by PCR methods in naturally infected live *(antemortem)* pigs are laryngeal swabs (Pieters et al., 2017), tracheobronchial swabbing/washing (Kurth et al., 2002; Fablet et al., 2010), deep tracheal catheters (Sponheim et al., 2019) and/or TBM (Vangroenweghe et al., 2015a). Although nasal swab was initially proposed as a suitable sample to monitor *M. hyopneumoniae* infection (Calsamiglia et al., 1999), its limited sensitivity has been described in different studies (Kurth et al., 2002; Sibila et al., 2004c; Marois et al., 2007). On the contrary, *postmortem* samples should be taken from CVPC affected areas (when present), being the tracheobronchial or bronchial swabs and/or lung homogenate (Burrough et al., 2018; Rawal et al., 2018) the most frequently employed. Remarkable, lung tissue may in some occasions result in false negative samples or in higher cycle threshold (Ct) values compared to bronchial swabs (Burrough et al., 2018; M. Pieters personal communication). The material used for sampling can also influence the PCR results obtained. According to Takeuti et al. (2017a), nylon-flocked swabs showed higher absorption capacity and higher detection rates than rayon-bud ones. Differences in Ct values, however, were only 0.5 to 1.7.

Alternative approaches to individual specimens include pools of individual samples or aggregated samples (such as oral fluids). Pooling strategies in *M. hyopneumoniae* diagnoses has been poorly investigated. Published available data indicate that pooling TBM samples from five 12-25 days-old piglets per litter resulted in higher sensitivity, when tested by real-time PCR, than pooling 10 samples (Vangroenweghe et al., 2015a). However, information on the effect of pooling on the estimation of *M. hyopneumoniae* load is still lacking. Oral fluid sampling offers the advantage of a non-invasive collection approach but its sensitivity and consistency for *M. hyopneumoniae* DNA detection seem to be low and limited, respectively (Hernandez-García et al., 2017; Pieters et al., 2017; Arsenakis et al., 2019), and need further optimization procedures.

Besides respiratory tract samples, *M. hyopneumoniae* DNA has been occasionally detected, under experimental conditions, in different types of non-respiratory samples obtained from internal organs such as spleen, kidney and/or liver (Buttenschøn et al., 1997; Marois et al., 2007; Woolley et al., 2012). The biological relevance and occurrence of such findings in naturally infected pigs is largely unknown. In addition, *M. hyopneumoniae* DNA has also been detected in processing fluids (Vilalta et al., 2019). Usefulness of these types of samples for *M. hyopneumoniae* infection diagnosis or monitoring is still under study and not yet validated.

8.3.3.2 DNA extraction, amplification and quantification

Almost no information is available on the effect of the DNA extraction methods on *M. hyopneumoniae* PCR detection outcome. Indeed, there is only one study reporting differences in Ct values obtained from DNA extracted from TBM samples depending on the extraction method practiced (Vangroenweghe et al., 2015a). These results encourage further investigation of the best extraction method, especially in those incipient infections with low *M. hyopneumoniae* loads.

The first *M. hyopneumoniae* PCR assays were set up in early 90s, all of them conventional reactions. A few years later, development of nested PCR (nPCR) based on two amplification runs significantly increased the assay sensitivity and specificity (reviewed by Sibila et al., 2009). Nonetheless, these one or two-step end-point PCR assays have several limitations such as cross-contamination between samples (during the amplicon processing and/or analysis) and the qualitative (presence/absence) and visually estimated outcome. In order to improve these two aspects, several real-time/quantitative PCR (rt-PCR/qPCR) methods have been developed (Dubosson et al., 2004; Strait et al., 2008; Marois et al., 2010; Sibila et al., 2012). These fluorescence-based assays allow the real-time measuring of the amplification reaction kinetics without post-processing steps. Moreover, these systems enable the use of reference genes as internal control for extraction and amplification steps as well as for data normalization. The outcome of this real-time amplification can be obtained in Ct values or in copy numbers extrapolated from a standard curve (constructed with a log-ten dilution of a positive control of known concentration). This latest option is utilized to calculate the analytic sensitivity of such quantitative tests. It has been described that 1 *femto*gram (*fg*) of DNA corresponds to 978 kb. Therefore, in those cases in which the target is a non-repetitive element and considering an 897,405 nt genome size, 1 *fg* of *M. hyopneumoniae* DNA would correspond to 1 genome equivalent (geq; Kurth et al, 2002; Strait et al., 2008; Fourour et al., 2018).

Although most of the PCR-based assays are designed to detect one target (monoplex), others are set up to differentially detect multiple *M. hyopneumoniae* targets (Caron, et al., 2000; Marois et al., 2010) or multiple agents (Stakenborg et al, 2006a; Lung et al., 2017; Fourour et al., 2018). This former option is used for assessing the presence of different porcine mycoplasmas during isolation methods (Stakenborg et al., 2006a) or in enzootic pneumonia compatible lung lesions (Fourour et al., 2018). Similarly, Lung et al. (2017) described the usefulness of a multiplex PCR to detect simultaneously and rapidly common bacterial pathogens of porcine respiratory disease complex cases.

Sensitivity of all these PCR methods and their variants is highly variable and may depend on the extraction method, type of the assay, primer target(s), and origin of the sample to be processed. Additionally, sensitivity is frequently reported in different units (CFU, CCU, cells, *fg*, or *geq*) per reaction/volume, complicating the comparison between assays. Specificity is considered very high, 100% in most reported methods. However, when a panel of different *M. hyopneumoniae* strains were tested, some false negative (failure in detection of some *M. hyopneumoniae* strains) results were obtained (Dubosson et al., 2004; Strait et al., 2008; Marois et al., 2010; Sibila et al., 2012). This situation is probably explained by the great variety of chosen targets. While some of the assays are targeting very conserved regions, such as 16S rRNA (Sibila et al., 2009), others are focused on highly variable genes, such as the adhesin p97 (*mhp183*; Strait et al., 2008; Marois et al., 2010). Indeed, in this latter case, the specificity of the assay varies depending on the region targeted within the gene. For instance, it was reported that one of these regions (R2) was lacking in different field strains as well as in the J-strain (Marois et al., 2010). Contrarily, detection of false-positive PCR results (detection of mycoplasmas other than *M. hyopneumoniae*) has not been reported in the literature. Notwithstanding, this possibility cannot be completely ruled out for those assays targeting very conserved regions, unless a specificity assay includes a wide and complete list of different mycoplasmas and related acholeplasmas of unknown prevalence (Strait et al., 2008).

Apart from these multiple and varied PCR based methods to detect *M. hyopneumoniae* DNA available in the peer-reviewed literature, several *M. hyopneumoniae* specific commercial kits/reagents are available on the market. Whereas some of these products are based on the PCR/qPCR mentioned before, others utilize proprietary primers and probe sequences. In addition, some of them are intended for research purposes only.

8.4 Detection of antibodies against *Mycoplasma hyopneumoniae* infection

M. hyopneumoniae is typically considered a ubiquitous bacterium in modern swine production systems worldwide, and in consequence, any direct and directional interpretation simply combining diagnostics and disease occurrence data can be confounded by a plethora of factors (Gomes-Neto et al., 2014; Maes et al., 2018). As for antibody responses, diagnostic limitations are partially a result of the inability to repeatedly and reliably determine timing of causative events leading to measurable immunological alterations in post-natural exposure with or without concurrent herd vaccination scenarios. Nonetheless, antibody-based assays have been and will continue to be useful, adding value when utilized in combination with other molecular tools commonly employed for the investigation of the spread and prevalence of this pathogen.

Under natural conditions, seroconversion against *M. hyopneumoniae* is very variable as the individuals within a population may get infected at different moments. Additionally, the impact of physical proximity to infectious animals may be synergistically affected by the exposure time, passive immunity (Martelli et al., 2006; Bandrick et al., 2011), and perhaps transmissibility or even strain virulence (Vicca et al., 2003), which may increase the variability in the proportion of seropositive pigs, and in the time to seroconversion (*i.e.*, for both direct and indirect contact animals). In contrast, under experimental conditions, time to seroconversion is much more predictable as all the animals are infected at the same time. In this scenario, *M. hyopneumoniae* has been detected in pigs for more than 200 days post-exposure (Pieters et al., 2009), and yet, seroconversion may be delayed by direct or indirect contact with neighboring shedders (Fano et al., 2005; Pieters et al., 2010).

Typically, the ELISA kits commercially available detect the most common isotype of antibody which is the anti-*M. hyopneumoniae* immunoglobulin (Ig)-G molecule. Although the specificity of IgG may vary, one of the main targets of this immunoglobulin are surface proteins produced by the pathogen. IgG-based responses are commonly measured in the serum of animals, first detectable as early as 21 dpi (Pieters et al., 2017). Other antibody isotypes to be employed for detection of *M. hyopneumoniae* exposure include IgM (under experimental conditions detected at 9 dpi but showing a rapid decline) and IgA (*i.e.*, proxy for mucosal immune activation) (Gomes-Neto et al., 2014; Pieters et al., 2017). Detection of IgA antibody response has been assessed in different samples such as nasal swab, pen-based oral fluid, tracheobronchiolar or bronchoalveolar lavage fluid samples (Garcia-Morante et al., 2017a; Pieters et al., 2017; Bai et al., 2018; Bjustrom-Kraft et al., 2018). Each of these sample types offer different advantages and limitations, but considering the ambiguous results obtained across the different studies, further work remains to be done to develop robust protocols and assays. Colostrum is also a known sample type that can be utilized for measuring antibodies against *M. hyopneumoniae* and may be used to assess the exposure status of sows across different parities, in addition to the presence of passive immunity available to piglets (Jenvey et al., 2015). Importantly, quantifying humoral immune responses at different levels does not equate to predicting protective immunity (Zinkernagel, 2012).

Up to now, antibody response derived from *M. hyopneumoniae* natural exposure is indistinguishable from the antibodies elicited by the available commercial vaccines. Interestingly, IgA detection has recently been shown to discriminate between infected versus vaccinated pigs, without an apparent interference of passive immunity (Bai et al., 2018).

Indirect or blocking ELISAs are available as commercial kits. Both detect primary antibodies present in the sera or other sample specimens of pigs, with the blocking assay attempting to improve specificity by preventing cross-reactivity with undesirable targets. Gomes-Neto et al. (2014) also tested the efficacy of different commercially available ELISA kits and showed varying degree of performances over time as previously demonstrated (Erlandson et al., 2005; Fano et al., 2012; Gomes-Neto et al., 2014), and the presence of cross-reactivity against other mycoplasmas at the serum level for some of the assays tested. These specificity problems, however, were not observed when using pen-based oral fluids. It is important to disclaim that this particular study had a small sample size (Gomes-Neto et al., 2014). Although cross-reactivity with other mycoplasmas continues to be an issue, the potential for more specific assays targeting surface proteins exists (Petersen et al., 2016, 2019). At the moment, any attempt to estimate the specificity or sensitivity of any given antibody-based assay comes with limitations because of the difficulties with sampling issues (*e.g.*, sample size) and the determination of true positive status of the animals. Importantly, the degree of cross-reactivity among mycoplasmas may be affected by the variation in immunogenicity/antigenicity across different strains of non-*M. hyopneumoniae* species, which remains to be studied.

Similarly, there is a lack of understanding on whether the genetic diversity of *M. hyopneumoniae* population impacts the overall accuracy and predictive value of ELISA tests. Perhaps some antibody isotypes are less prone to such a confounding effect, or certain antigens/immunogens are ubiquitously present across all strains preventing this from being an issue for diagnosis. Nonetheless, there is demonstrable variation in the *M. hyopneumoniae* population structure across herds (Michiels et al., 2017; Garza-Moreno et al., 2019b). Other epidemiological factors may also contribute to the variation in antibody responses such as vaccine-driven cross-reactivity (Bai et al., 2018), the impact of host genetics (Ruiz et al., 2002), or maybe even the composition of the swine gut microbiome (Schachtschneider et al., 2013). All these factors should be considered while trying to understand the kinetics of humoral immune responses to this pathogen.

8.5 Selecting an adequate sample size

Number and type of samples depend on the diagnosis objective. If the goal is establishing the causality of a clinical disease, respiratory tract samples of acutely and non-treated (when possible) affected animals (coughing) should be taken. *In vivo* diagnosis mostly utilizes PCR methods on respiratory tract airways samples whereas *postmortem* specimens should be collected from lower respiratory airways (area/s of lung where CVPC is observed) and PCR or IHC/ISH can be used. If the goal is monitoring or surveillance programs, sample size calculation should be based on the selected laboratory technique (sensitivity and specificity), the expected prevalence of infection in a population and the desired confidence level (Burrough et al., 2019). This prevalence depends on the different parameters such as the age of the animals, the stage of infection (subclinical, acute or chronic) and/or the epidemiological scenario. As for instance, eradication or replacement selection programs require more extensive surveillance and proper detection of a smaller amount of pathogen when compared to field monitoring approaches. Therefore, in order to calculate the proper sample size with a desired confidence, statistical tables should be consulted (Burrough et al., 2019).

8.6 Conclusions

M. hyopneumoniae infection is widespread in the worldwide domestic pig population. In consequence, confirmation of *M. hyopneumoniae* infection/seroconversion does not imply the observation/presence of clinical signs and/or lung lesions. Likewise, presence of animals showing non-productive coughing and/or CVPC lesions in lungs is not equivalent to *M. hyopneumoniae* infection and involvement in the respiratory problems observed. Antibody detection methods (via ELISA tests) are the most common and affordable method for herd monitoring but do not allow the differentiation between natural exposure and vaccination antibody responses. PCR testing methods can detect DNA (from live or dead bacteria) in infected animals even prior to seroconversion or appearance of clinical signs. Therefore, implication of *M. hyopneumoniae* in a respiratory outbreak should be established using a multiple test approach within a differential diagnosis frame. Similarly, confirmation of *M. hyopneumoniae* infection herd status, as per eradication or replacement selection programs, requires testing lower respiratory tract samples complemented, when possible, with serum samplings taken in a longitudinal fashion and assessing lung lesions' inspections at abattoir. Since diagnosis of *M. hyopneumoniae* at herd level can be confusing, a farm classification according to *M. hyopneumoniae* clinical-pathological parameters, infection (PCR) and seroconversion status has been suggested (Garza-Moreno et al., 2018). Hence, farms can be classified as negative, provisionally negative, and positive. Among these latter ones, farms can be subdivided into subclinically infected I, subclinically infected II and clinically affected (Table 8.1).

Table 8.1. Farm classification according to *M. hyopneumoniae* health status proposed by Garza-Moreno et al., 2018.

Classification		Clinical Signs	Lung Lesions	ELISA Results[a]	PCR Results
Negative		Not observed	Not observed	Negative	Negative
Provisional Negative		Not observed	Not observed	Positive	Negative
Positive	Subclinical infected I	Not observed	Not observed	Positive/Negative	Positive
	Subclinical infected II	Not observed	Observed	Positive/Negative	Positive
	Clinical affected	Observed	Observed	Positive/Negative	Positive

[a] ELISA results (negative/positive) could depend on infection pattern in the farm and sampling time point.

In conclusion, diagnosis of *M. hyopneumoniae* infection at herd/animal level can be complex due to its intrinsic properties, co-factors influencing the outcome of the disease, and/or diagnostic technique limitations. Bearing in mind this assumption, a combination of different diagnostic parameters would allow covering all these aspects.

CHAPTER 9

GENERAL CONTROL MEASURES AGAINST *MYCOPLASMA HYOPNEUMONIAE* INFECTIONS

Enric Marco[1]

Paul Yeske[2]

Maria Pieters[3]

1 Marco VetGrup SLP, Barcelona Spain
2 Swine Vet Center, P.A., St. Peter, USA
3 College of Veterinary Medicine, University of Minnesota, St. Paul, USA

9.1 Introduction

The control of *Mycoplasma hyopneumoniae (M. hyopneumoniae)* infections in the field has been, for many decades, based on the use of antibiotic compounds and also on the application of vaccine products, which started in the early 90s. Disease control using antibiotics can be accomplished, but implies their use for long periods of time, with the most commonly used antibiotics being tetracyclines, macrolides, pleuromutilin or quinolones, among others (del Pozo, 2014). However, the risk of development of antimicrobial resistance is a growing public health threat of broad concern to countries and multiple sectors. In fact, the WHO recommendation is to replace the use of antimicrobials with other strategies, especially when referring to antimicrobials that have been classified as critically important (WHO, 2015). In this new scenario, other disease control measures in addition to vaccines have become more critical, especially for indoor housing and intensive production. Different conditions in density, herd size, climate, and health status of the source of pigs can dramatically influence the manifestation of *M. hyopneumoniae* infection. Thus, production system, replacement acclimatization, pig flow, herd management, climate and housing conditions are probably those other main factors that could be modified with the aim of reducing the severity of *M. hyopneumoniae* associated disease. Moreover, in this section, the focus of *M. hyopneumoniae* disease control is placed on factors different from antibiotic treatment and vaccine application, as those are covered in Chapters 10 and 11.

An overview of the most prominent factors affecting *M. hyopneumoniae* disease development and clinical presentation are listed in Table 9.1 and are discussed in detail in different sections of this chapter.

Table 9.1. Factors affecting *M. hyopneumoniae* associated disease.

Factor		Reference(s)
Production system		
Herd size	Smaller farms vs. larger	Hurnik et al., 1994
	Number of farrowing crates/ room 16 or less	Nathues et al., 2013b
Number of pig sources	One health source	Hurnik et al., 1994
		Stärk, 2000
Pig flow and batching	Multiple site vs. 1 or 2 sites	Sibila et al., 2004a
	Five-week vs. 4-week batching system	Vangroenweghe et al., 2012
	Batching systems other than 2 and 4 weeks	Nathues et al., 2013b
	Parity segregation	Moore, 2005
Replacement animals		
Gilts	Low replacement rate	Nathues et al., 2013b
	Similar *M. hyopneumoniae* health status	Fano et al., 2006
Management		
Prior to weaning	Colostrum intake	Bandrick et al., 2008
	Cross-fostering	van der Peet-Schwering et al., 2008
	Weaning age	Maes et al., 2008
		Pieters et al., 2014
All-in/all-out	Al-in/all-out	Scheidt et al., 1995
Stocking density	Density in natural ventilated buildings	Pointon et al., 1995
Stocking rates	Space allowance	Hyun et al., 1998 Jang et al., 2017
Group size	Pigs per air space	Tielen et al., 1978
	Pigs per nursery pen	Stärk et al., 1998
Other disease management	Controlling other pathogens including parasites	Maes et al., 2008
	Mycotoxins in feed	Taranu et al., 2005 Pierron et al., 2016
Climate and housing conditions		
Seasonality	Season of the year	Stärk et al., 1992 Elbers et al., 1992 Maes et al., 2001 Segalés et al., 2012 Scollo et al., 2017 Vangroenweghe et al., 2015b
Thermal conditions	Temperature and humidity	Gordon, 1963
	Temperature	Geers et al., 1989
	Air speed	Scheepens, 1996
Air contaminants	Air contaminants level	Donham, 1991
		Cargill et al., 1998
		Jolie et al., 1999

9

investigations on the influence of temperature and humidity on the incidence of pneumonia showed lower incidence and less severe lesions in the hot-humid environment (Gordon 1963). Work by Geers et al. (1989) indicated a similar relationship, with less coughing in higher temperatures. Based on these findings the "sweat house" concept was developed in UK farming during the 70s and 80s. Pigs raised in this type of facilities showed less incidence of lung lesions than those in conventional housing. It is assumed that the major factor in reducing lung disorders in the "sweat house" conditions is the reduced number of airborne particles within the 1-3 microns range (Thomas, 2013). A strategic increase in temperature and humidity in the indoor stable climate is still used successfully as a corrective measure to prevent respiratory disorders in some parts of the world such as Japan (H. Ishikawa, personal communication). Housing systems should guarantee that air speed does not exceed 0.15 m/s at pig level and environmental temperature should guarantee thermo-neutrality (Scheepens, 1996).

9.5.3 Air contaminants

Intensive pig farming concentrates large numbers of animals in a small volume of air. This fact may result in a worsening of the air quality breathed by the pigs, especially when the air exchange is reduced. Measurements in 44 finishing farms in Denmark resulted in high levels of some important air contaminants (Table 9.4).

Table 9.4. Measurements of air quality and contaminants in finishing farms (Baekbo, 1990).

Parameter	Min	Mean	Max
Dust (mg/m^3)	0.04	2.11	9.46
Respirable (< 5 microns)	0.01	0.49	1.87
Ammonia (ppm)		8.7	21.0
Endotoxin ($x10^3$ EU/m^3)	1.0	38	240
Total microbes (c.f.u./m^3)	0.15	$20.1x10^5$	171
Carbon dioxide (ppm)	200	1,381	2,800

The relationship between air quality and respiratory disease in pigs has been investigated. Several air contaminants (*e.g.* dust, ammonia carbon dioxide, and microbes) have been observed to be correlated with pneumonia and pleurisy detected at the slaughterhouse (Donham, 1991). In Australia, the severity of pneumonia and prevalence of pleurisy in pigs examined at slaughter was reduced in pigs reared in barns with acceptable air quality, compared with pigs reared in barns with poor air quality (Cargill et al., 1998). Pigs from the same farm (with high prevalence of pneumonia) raised in an isolated facility, where the airborne contamination was lower than in the farm environment, did not develop macroscopic pneumonia lesions indicative of *M. hyopneumoniae* infection (Jolie et al., 1999).

Among air contaminants, ammonia is the air contaminant detected in highest concentration in closed swine houses. Consequently, and because it is an irritant, ammonia has been studied more than other air pollutants. Pigs exposed to atmospheric ammonia (50 ppm or higher) at young ages develop inflammatory changes in the wall of the respiratory tract, as well as reduced bacterial clearance from lungs (Drummond et al., 1978, 1980). Exposure to a combination of non-commensal bacteria (alpha-hemolytic cocci) and ammonia stimulate the immune system in a progressive manner, and there were direct effects of immune stimulation on food intake and growth resulting in poorer feed-conversion efficiency, even though the effects remained subclinical (Murphy et al., 2012). Investigations on the simultaneous influence of particulate matter and ammonia on lung lesions and the presence of *M. hyopneumoniae* in finishing pigs identified a positive association with pleurisy prevalence at slaughterhouse, but only a trend, causing a higher number of *M. hyopneumoniae* nested-PCR positive pigs in the finishing phase. Particulate matter appeared to have a greater impact than ammonia for the prevalence of pneumonia and pleurisy (Michiels et al., 2015). Particulates in the air of animal houses originate from feed, debris, manure, and animals themselves, and may contain a great variety of bacteria and microorganisms (Zhang, 2004).

Although no direct correlation has been observed between *M. hyopneumoniae* infection and carbon dioxide concentration in the air, carbon dioxide may be a good indicator for other air contaminants (Donham, 1991). Pigs are responsible for producing most of the carbon dioxide present in the air of barns, and levels will vary according to stocking density and ventilation rates, as well as pig activity (Pedersen et al., 2002). Generally, the levels of carbon dioxide are used as indicators of the level of ventilation (Van't Klooster and Heitlager, 1994). The required level of ventilation can be determined by the kg of pig/m^3. High levels of carbon dioxide (> 2000 ppm) indicate insufficient ventilation and therefore might be associated with an increase of other air pollutants and microbial load. Maximal safe concentrations of air contaminants were estimated on the basis of dose-response correlation to swine health or human health problems (Table 9.5).

Table 9.5. Recommended maximal concentrations of air contaminants in pig barns (Donham, 1991).

Dust (mg/m^3)	2.4
Ammonia (ppm)	7
Endotoxin (mg/m^3)	0.08
Total microbes (c.f.u./m^3)	10^5
Carbon dioxide (ppm)	1,540

Housing conditions influence the quality of air breathed by pigs, especially in naturally ventilated buildings. In these cases, working with the proper stocking density (< 33.3 kg of pig per cubic meter) can be essential to improve the control of pneumonia caused by *Mycoplasma*. However, when mechanical ventilation is in place, stocking densities are less critical. Mechanical ventilation systems allow to achieve the desired climate (temperature, humidity, and air speed), and to obtain air renewed in such a way to maintain the levels of contaminants within the recommended limits (Table 9.5).

9.5.4 Improving air quality

Ventilation helps in reducing airborne ammonia concentration and dust, but this only applies when a high standard of surface hygiene is maintained, and stocking rates are optimal. In sheds with dirty floors, ventilation rates cannot compensate for sub-standard hygiene (Banhazi et al., 2004). Provided good dunging patterns are maintained, a high standard of air hygiene is also maintained, but dunging patterns can be easily altered by increasing stocking rates (kg pig/m^2), presence of leaking pipes, draughts, and poor floor maintenance (Cargill et al., 2002). Waiting for pens to be perfectly dry before re-stocking, when all-in/all-out management is performed, can have an influence over the dunging pattern of pigs up to 8 weeks after re-stocking (Murphy, 2011). A good management of manure or a slurry handling system will reduce airborne ammonia concentration. Barns with deep pits and partially slatted floors perform best in terms of air quality as long as the solid parts of the floor stays dry (Murphy, 2011). Dust concentration reduction can be achieved by increasing humidity or by adding fat to the feed (Pedersen et al., 2000). Spraying a mixture of oil and water directly onto pen floors reduces significantly the concentration of both inhalable and respirable airborne particles (Banhazi, 2007).

CHAPTER 10

ANTIMICROBIAL TREATMENT OF *MYCOPLASMA HYOPNEUMONIAE* INFECTIONS

Anne V. Gautier-Bouchardon[1]

1 Ploufragan-Plouzané-Niort Laboratory. French Agency for Food, Environmental and Occupational Health and Safety (Anses), Ploufragan, France

10.1 Introduction

Control of *Mycoplasma hyopneumoniae* (*M. hyopneumoniae*) infections is primarily based on (i) optimization of management practices, biosecurity measures and housing conditions to reduce the occurrence and transmission of this pathogen and (ii) vaccination, which is widely used worldwide (Maes et al., 2008; Simionatto et al., 2013). However, vaccination provides only partial protection and does not prevent infection (Maes et al., 2008; Pieters et al., 2010). The use of antimicrobial agents can be necessary in cases of clinical outbreaks, to improve health, welfare of pigs and to reduce performance losses due to bacterial infections. Tetracyclines, β-lactams, sulfonamides with trimethoprim and macrolides are the most frequently used antibiotic classes for respiratory infections in pigs (Chauvin et al., 2002; Timmerman et al., 2006). Tetracyclines and macrolides are more specifically used to control and treat *M. hyopneumoniae* infections since mycoplasmas are intrinsically resistant to β-lactams and potentiated sulfonamides (del Pozo Sacristán, 2014; Maes, 2008; Gautier-Bouchardon, 2018). Other antimicrobial families such as lincosamides, pleuromutilins, amphenicols, aminoglycosides, aminocyclitols and fluoroquinolones are also potentially active against *M. hyopneumoniae* (Gautier-Bouchardon, 2018).

There is increasing concern worldwide about antimicrobial use in livestock and selection of antibiotic resistant bacteria (Collignon & McEwen, 2019; Scott et al., 2019). Antibiotic resistance can lead to treatment failures and tend to increase disease severity as well as incidence and costs of infection in herds. There is therefore a need for surveillance of susceptibility levels of the target pathogen towards antimicrobials. Selection of *M. hyopneumoniae* strains with reduced antimicrobial susceptibility has been described *in vitro* for tylosin and oxytetracycline (Hannan et al., 1997b) and *in vivo* during experimental infection for marbofloxacin (Le Carrou et al., 2006). Strains with reduced susceptibility levels have also been isolated from clinical outbreaks (Felde et al., 2018a; Inamoto et al., 1994; Klein et al., 2017; Qiu et al., 2018; Tavio et al., 2014; Thongkamkoon et al., 2013; Vicca et al., 2004).

This chapter reviews the antimicrobials and antimicrobial treatments used to control *M. hyopneumoniae* infections. It also gives an overview of *in vitro* and *in vivo* activities of these antimicrobials and of resistance mechanisms described for *M. hyopneumoniae* towards these antimicrobials.

10.2 Antimicrobial treatments

10.2.1 Antimicrobials

Without a cell wall, *M. hyopneumoniae* is unaffected by antibiotics that target cell-wall synthesis such as β-lactams, glycopeptides, and fosfomycin. All *Mycoplasma* species are also naturally resistant to polymyxins, sulfonamides, trimethoprim, rifampicin and first generation quinolones such as nalidixic acid (Schultz et al., 2012; Taylor-Robinson and Bebear, 1997). *M. hyopneumoniae* is also insensitive to 14-membered lactone ring macrolides such as erythromycin (Gautier-Bouchardon, 2018). Mechanisms responsible for these intrinsic resistances are described in more details elsewhere in this chapter.

The most widely used antimicrobial agents in pigs against *M. hyopneumoniae* infections are tetracyclines and 15- and 16-membered lactone ring macrolides. Pleuromutilins, amphenicols, aminoglycosides and aminocyclitols, lincosamides and fluoroquinolones are less frequently used but are active against *M. hyopneumoniae* (del Pozo Sacristán, 2014; Gautier-Bouchardon, 2018; Maes et al., 2008; Maes et al., 2018).

10.2.1.1 *Tetracyclines* (tetracycline, oxytetracycline, doxycycline, chlortetracycline)

Tetracyclines represent a large group of bacteriostatic antibiotics that inhibit protein synthesis by binding reversibly to the 30S subunit of ribosomes and thus blocking their protein synthesis by competing with transfer RNA for the same binding site (Baietto et al., 2014; Brodersen et al., 2000). This prevents the formation of new proteins, and therefore bacterial growth. Tetracyclines are widely used in the treatment of respiratory infections in pigs due to their broad-spectrum activity on Gram-negative and Gram-positive bacteria both aerobic and anaerobic, but also on *Mycoplasma* species and even several protozoa (del Pozo Sacristán, 2014;Vicca, 2005).

Tetracycline, chlortetracycline and oxytetracycline are naturally-occurring first generation tetracyclines whereas doxycycline is a second generation semi-synthetic analog with improved pharmacokinetic and chemical properties (Chopra and Roberts, 2001). Oral bioavailability of tetracyclines in pigs is low (3 to 18%) and can be lowered in fed compared to fasted pigs (Nielsen and Gyrd-Hansen, 1996), their intestinal absorption being reduced up to 50% by formation of complexes with cations (calcium, magnesium,

10

iron and aluminum). However, in contrast to mammalian cells, cells of most bacterial species actively concentrate tetracyclines (Chopra and Roberts, 2001). Their bacteriostatic activity is optimal in an acid environment and they diffuse easily through biological barriers and cell membranes.

Common side effects of tetracyclines are diarrhea and increased phototoxicity but the absence of major side effects has led to their extensive use in swine (Chopra and Roberts, 2001; Karriker et al., 2012).

10.2.1.2 *Macrolides* (tylosin, tilmicosin, tylvalosin, tulathromycin, tildipirosin and gamithromycin) and lincosamides (lincomycin)

Macrolides and lincosamides have very close antimicrobial activities. They are bacteriostatic antibiotics that inhibit protein synthesis by binding reversibly to the peptidyl transferase component of the 50S subunit of ribosomes and thus blocking the growth of the nascent peptide chain (Leclercq, 2002; Schwarz et al., 2016). However, bactericidal activity was also reported for some macrolides (Tavio et al., 2014). The spectrum of activity of macrolides covers Gram-positive bacteria, some Gram-negative bacteria, intracellular bacteria and mycoplasmas (depending on the size of the macrolide lactone ring for several *Mycoplasma* species) (Gautier-Bouchardon, 2018; Leclercq, 2002). *M. hyopneumoniae* is insensitive to 14-membered lactone ring macrolides such as erythromycin, due to a mutation in the domain V of its 23S rRNA (Gautier-Bouchardon, 2018). Some 16-membered lactone ring macrolides (such as tylosin, tilmicosin and tylvalosin) or semi-synthetic macrolides (tulathromycin, tildipirosin and gamithromycin) are dedicated to veterinary use. Lincosamides have a more limited spectrum of activity and are mainly active on Gram-positive bacteria, mycoplasmas and some Gram-negative anaerobic bacteria such as *Lawsonia* and *Brachyspira* (Gautier-Bouchardon, 2018; Schwarz et al., 2016). Macrolides and lincosamides are recommended to treat *M. hyopneumoniae* infections. Macrolides are also used in pigs to treat respiratory infections caused by other bacteria, such as *Actinobacillus pleuropneumoniae*, *Bordetella bronchiseptica*, *Glaesserella (Haemophilus) parasuis*, and *Pasteurella multocida* (del Pozo Sacristán, 2014). In addition to their antibacterial properties, anti-inflammatory effects have been described in macrolides (Crosbie and Woodhead, 2009; Moges et al., 2018). These immunomodulatory properties can lead to superior clinical benefits during bacterial infections.

Macrolides and lincosamides have a high lipid solubility and tend to be characterized by higher oral bioavailability than tetracyclines, with variations due in part to their stability at low pH, depending on chemical structures and formulations (Baietto et al., 2014; Fohner et al., 2017). After oral administration, they are well absorbed from the gastrointestinal tract and have an extensive tissue penetration: in the spleen, liver, kidneys, and particularly the lungs. They accumulate within many cells, including macrophages and neutrophils (Leclercq, 2002), in which they may be found at a concentration more than 20 times the plasma concentration (Baietto et al., 2014): because of their weakly basic character, ion-trapping of macrolides and lincosamides occurs inside cells since intracellular pH is lower than extracellular fluid pH (Maddison et al., 2008).

Toxicity and adverse effects are uncommon for most macrolides. Diarrhea has sometimes been described after oral administration and tissue irritation, causing pain, and inflammation may occur when some macrolides are administered parenterally (Karriker et al., 2012; Giguère, 2013). Tilmicosin may cause cardiac toxicity after parenteral administration to different animal species including pigs (Papich, 2017).

10.2.1.3 *Pleuromutilins* (tiamulin, valnemulin)

Pleuromutilins are semisynthetic bacteriostatic antibiotics that were exclusively used in veterinary medicine until recently. Like macrolides, they inhibit protein synthesis by binding to the peptidyl transferase component of the 50S subunit of ribosomes (Paukner and Riedl, 2017). Their activity is close to the antimicrobial spectrum of macrolides but they are particularly active against anaerobic bacteria, mycoplasmas and Gram-positive bacteria such as streptococci. In pigs, tiamulin and valnemulin are commonly used for the treatment of respiratory infections caused by *A. pleuropneumoniae*, *M. hyopneumoniae*, *Brachyspira* sp. and *Lawsonia intracellularis* (Schwarz et al., 2016). In some countries, it is recommended to reserve pleuromutilins mainly to treat *Brachyspira* infections, since high prevalence of resistance to other antimicrobial classes exists (van Duijkeren et al., 2014; Card et al., 2018).

Pleuromutilins are characterized by a good absorption after oral administration, a high bioavailability in tissues, with high drug concentrations achieved in macrophages (del Pozo Sacristán, 2014; Paukner and Riedl, 2017). On rare occasions, adverse reactions such as erythema or mild edema of the skin may occur in pigs following the use of tiamulin, but they are usually mild and transient (Karriker et al., 2012). Lethal side effects may be observed in combination with ionophores (Hampson, 2012).

10.2.1.4 *Phenicols* (florfenicol)

Florfenicol is a fluorinated derivative of thiamphenicol exclusively approved for use in animals (Schwarz et al., 2016). This bacteriostatic antimicrobial binds to the 50S ribosomal subunit, inhibiting the peptidation reaction and the translation of bacterial mRNA. Florfenicol is a broad-spectrum antimicrobial agent, active against Gram-positive and Gram-negative bacteria, as well as mycoplasmas (Shin et al., 2005). The main target bacteria in pigs are *P. multocida*, *A. pleuropneumoniae*, *B. bronchiseptica*, *Streptococcus suis* and *M. hyopneumoniae* (Ciprián et al., 2012; del Pozo Sacristán, et al., 2012b; Priebe and Schwarz, 2003).

Florfenicol is characterized by rapid absorption and distribution after oral or parenteral administration, with a wide tissue distribution, and a slow elimination (Liu et al., 2003). Rectal eversion and perianal edema may occur during treatments (Karriker et al., 2012).

10.2.1.5 *Fluoroquinolones* (enrofloxacin, marbofloxacin)

Fluoroquinolones are bactericidal antimicrobials developed synthetically from the quinolone class of antimicrobials (Baietto et al., 2014). Fluoroquinolones have affinity for DNA gyrase and/or topoisomerase IV (depending on Gram-positive or Gram-negative species) and prevent DNA replication of bacteria by binding to the complex of each of these enzymes and blocking progress of the DNA replication enzyme complex: the collision of the replication fork with these trapped complexes results ultimately in cell death (Hooper, 2001). Fluoroquinolones are indicated for treatment of respiratory infections in pig. Prudent use of these antimicrobials, listed as critically important antibiotics, is recommended to prevent selection of resistant bacteria from food-producing animals (Collignon and McEwen, 2019; Scott et al., 2019).

Fluoroquinolones can be administered by oral or parenteral routes, with a fast and high absorption after oral administration, high tissue distribution with a good penetration through biological barriers (Inui et al., 1998;

Nielsen and Gyrd-Hansen, 1997). Cases of lameness have been reported after injection of fluoroquinolones, but these antimicrobials are generally considered as safe in swine (Karriker et al., 2012).

10.2.1.6 *Aminoglycosides and aminocyclitols* (spectinomycin)

Aminoglycosides and aminocyclitols are bactericidal antimicrobials that act mostly as protein synthesis inhibitors by interfering with the proofreading process at the 30S ribosomal subunit level, causing increased rate of error in synthesis with premature termination and inhibition of ribosomal translocation. They can also disrupt the integrity of bacterial cell membrane (Baietto et al., 2014; Brown and Riviere, 1991). They are active against many Gram-positive and Gram-negative aerobic bacteria and are indicated in the treatment of *M. hyopneumoniae* infections.

Tissue diffusion of aminoglycosides after parenteral administration is poor since these antimicrobials have a low lipophilic degree and are too polar to cross mammalian cell membranes, but there is a good diffusion into extracellular fluids such as peritoneal, pleural and pericardial fluid and synovial effusions (Brown and Riviere, 1991). They also easily penetrate in lung parenchyma and bronchial secretions (Goldstein et al., 2002). Oral administration is not recommended to treat respiratory infections because of their poor absorption from the gastrointestinal tract (Baietto et al., 2014; Brown and Riviere, 1991). Cases of nephrotoxicity have been described with prolonged parenteral use of aminoglycosides (Karriker et al., 2012).

10

10.2.2 Administration routes of antimicrobials for treatment of respiratory diseases in pigs

For reasons of practicality and efficiency, antimicrobials are often administered to pigs through feed or water to entire groups (*e.g.*, pens of pigs) in intensive farming, either for prophylaxis (to healthy animals during high-risk periods for disease to prevent a possible clinical outbreak) or metaphylaxis (mass medication to healthy or subclinically-infected animals in the same group as diseased animals) (Burch, 2012; del Pozo Sacristán, 2014). The critical period for the transmission of *M. hyopneumoniae* was shown to be around the beginning of the finishing period, after the transfer of the animals to the finishing houses at 10 weeks of age (Leon et al., 2001). Antimicrobial curative treatment of enzootic pneumonia (EP) is often applied by oral administration to the whole affected group of pigs, since a fast diffusion and action of the antimicrobial is not as important as for acute infections. Antimicrobials are used therapeutically through parenteral administration to treat individual clinically sick animals suffering from acute respiratory tract disease (Vicca, 2005). A French study performed between 2010 and 2013 showed that routes of administration of antimicrobials in pigs were mostly oral, through feed and water, and that the parenteral route represented only 8% of the prescriptions in 2013 (Hémonic et al., 2018): treatments for respiratory diseases represented 23.7% of the prescriptions. Another study, performed in Belgium in 2010, showed that prophylactic group treatment was applied in 98% of the visited fattening pig herds and evidenced a drastic decrease of the proportion of metaphylactic group treatments (Callens et al., 2012). However, these results cannot be extrapolated globally since antimicrobial use can be different depending on regions of the world (Collignon and McEwen, 2019).

In-feed medication is the most common method of administering antimicrobials at group level but presents several disadvantages: its relatively high cost (since diseased but also healthy pigs receive medication), low dose accuracy (depending on feed-intake, especially for diseased pigs) and slow implementation compared to parenteral administration (del Pozo Sacristán, 2014). Achieving sufficiently high antibiotic levels in the lung tissue can be difficult (due to poor absorption, lowered bioavailability or poor diffusion in tissues for some antibiotics): a close follow-up of the feed intake is required to achieve appropriate therapeutic levels and to avoid toxic effects due to overdosing or treatment failure and increased resistance due to underdosing (Burch, 2012; Vicca, 2005).

In-water medication, which also allows administration of antimicrobials at group level, leads to a faster implementation than the in-feed medication, with a lower cost, and is indicated for chronic and acute diseases (del Pozo Sacristán, 2014). Major disadvantages are the high water spillage (by pigs playing with the drinking nipples or by some drinking systems), instability or poor solubility of some antimicrobials in water, reaction with the water pipe material (*e.g.* tetracyclines with iron) and seasonal variations of the water intake by pigs (Vicca, 2005).

Oral treatments are often underdosed (Callens et al., 2012). However, even if the intake and absorption of the drug is intermittent and occurs irregularly over a 24 hour period, the flow and absorption along the gastrointestinal tract is prolonged: depending on the compound, plasma and lung concentrations may therefore be relatively constant and achieve correct therapeutic doses, in comparison with a parenteral administration which may result in higher concentrations during a more limited period (Burch, 2012). The treatment's outcome will also depend on the mode of action of the antimicrobial used (time- or concentration-dependent, with or without a post-antibiotic affect). The main antimicrobial classes used in pigs to treat *M. hyopneumoniae* infections (tetracyclines, macrolides, fluoroquinolones and aminoglycosides) are mainly concentration-dependent with moderate to prolonged post-antibiotic effect. Their efficacy is determined more by concentrations achieved in tissues than by time of exposure (Levison and Levison, 2009).

Pulse-medication is a variant of in-feed or in-water medication, with a short period of treatment alternated with a longer period without medication to allow the development of an active immune response (Kavanagh, 1994; Le Grand and Kobisch, 1996; Vicca, 2005). Prophylactic and long-term use of antimicrobials should be discouraged because it increases the risk for development of antimicrobial resistance.

Parenteral administration can be used for group or individual treatments (del Pozo Sacristán, 2014). Callens et al. (2012) stated that, in Belgium, injectable group treatments were mainly used in suckling pigs as a single administration of new long-acting and more potent antimicrobials. Individual treatments by parenteral administration in growing and fattening periods are mainly used for pigs with acute disease, during onset and first days of clinical signs and can be followed by an oral group treatment (Vicca, 2005). Parenteral treatments are often overdosed. This can be due, at least in part, to misevaluation of the bodyweight at the moment of administration, or to a lack of precision of dosing (Callens et al., 2012).

10.2.3 Efficacy of several antimicrobials against *Mycoplasma hyopneumoniae* infections under experimental and field conditions

A review of several studies conducted to assess the efficacy of various antimicrobials used for the prevention and/or the treatment of *M. hyopneumoniae* infections under experimental and field conditions is presented in Table 10.1. However, due to considerable variations in study design (time and duration of treatment, dose of antimicrobial and route of administration, observation periods after treatment and parameters followed to measure the treatment efficacy), comparison of results is difficult. For example, the three studies on in-feed chlortetracycline treatments after experimental infection were performed with different doses (from 50 to 500 mg/kg feed), with treatments starting three or seven days before infection or 21 days after infection and implemented during 14, 18 or 33 to 41 days (Huhn, 1971a, 1971b; Thacker, Thacker, and Wolff, 2006) (Table 10.1).

However, it can be concluded that, for the majority of antimicrobials tested, an improvement of economic (average daily weight gain, feed conversion ratio) and clinical parameters (clinical signs, lung lesions) were observed (Table 10.1). Some studies performed with multi-bacterial experimental infection or in herds with EP or respiratory disease evidenced an effect on concomitant bacteria (Binder et al., 1993; Hannan et al., 1982) or mortality rate (Jouglar et al., 1993; Martineau, 1980; Nutsch et al., 2005; Stipkovits et al., 2003). These results underline the fact that it may be important to choose a treatment with an antimicrobial active against *M. hyopneumoniae*, but also on secondary bacteria (such as *P. multocida*, *S. suis*, *H. parasuis* and/or *A. pleuropneumoniae*) that can be, at least in part, responsible for the clinical signs observed and for the decrease of the herd's performance parameters. However, antimicrobial treatments also have a negative effect on beneficial microbiota (*e.g.* gut and nasal microbial communities), which participate to the correct development of the immune system and resistance against pathogens colonization (Correa-Fiz et al., 2019).

Several studies evidenced a persistence of *M. hyopneumoniae* after administration of tetracyclines (Huhn, 1971b; Thacker et al., 2006), macrolides (Vicca et al., 2005), fluoroquinolones (Hannan and Goodwin, 1990; Le Carrou et al., 2006), amphenicols (Ciprián et al., 2012), pleuromutilins (Goodwin, 1979; Ross and Cox, 1988; Schuller, Neumeister, and Vogl, 1977) and antimicrobial combinations (Le Grand and Kobisch, 1996) (Table 10.1). Moreover, the lesion reduction was not maintained with time after treatment in several cases, suggesting new lesion development once medication had been withdrawn, presumably because the *Mycoplasma* started to multiply again and active immunity had not been fully established during the treatment and observation period (Burch, 2012).

Table 10.1. Effect of different antibiotic regimens for treatment of *M. hyopneumoniae* infections in experimentally infected pigs or in field conditions.

Antimicrobials[a]	EI or FC[b]	Regimen[c, d]	Effects[e]	References
Tetracyclines				
Chlortetracycline	EI	50, 100 or 200 ppm for 33 to 41 days (starting 7 days before EI and until necropsy)	LL ↓	Huhn, 1971a
	EI	400 ppm in-feed for 18 days (starting 21 days after EI)	LL ↓; reisolation of MHp	Huhn, 1971b
	EI	500 ppm in-feed for 14 days (3 days before EI)	CS ↓, LL ↓and MHp load ↓; reisolation of MHp	Thacker et al., 2006
	FC	300 ppm in feed for 7 days	Better ADWG and FCR	Burch et al., 1986
	FC	500 ppm in-feed for 2 consecutive weeks (at onset of CS)	No beneficial effect	del Pozo Sacristán et al., 2012a
	FC	500 ppm in-feed for 2 non-consecutive weeks (at onset of CS)	CS ↓, LL↓ and performance losses ↓	del Pozo Sacristán et al., 2012a,b
Doxycycline	FC	800 ppm in-feed for 3 consecutive weeks	CS ↓	Ganter et al., 1995
	FC	11 ppm in-feed for 8 days (fattening period)	CS ↓, better ADWG and cure rate	Bousquet et al., 1998

10

Antimicrobials[a]	EI or FC[b]	Regimen[c, d]	Effects[e]	References
Macrolides				
Tylosin	EI	782 mg/L in-water for 33 to 41 days (starting 7 days before EI and until necropsy)	No beneficial effect	Huhn, 1971a
	EI	100 ppm in-feed for 21 days (12 days after EI)	CS ↓and LL ↓; no beneficial effect on ADWG; reisolation of MHp	Vicca et al., 2005
	FC	4 mg/kg by IM for 3 days (after birth and/or at weaning)	Improvement of ADWG; no beneficial effect on FCR and CS	Kunesh, 1981
	FC	7 mg/kg by IM twice a day for 5 days	CS ↓; faster elimination of MHp DNA	Qiu et al., 2018
	FC	100 ppm in-feed for 3 consecutive weeks (at onset of CS)	No beneficial effect	del Pozo Sacristán, et al., 2012a
Tilmicosin	EI	363 ppm in-feed for 21 days (starting 7 days before EI)	CS ↓; reduction of MHp seroconversion; no beneficial effect on ADWG or LL	Clark et al., 1998
	FC	100, 200 or 300 ppm in-feed for 9 or 14 days	CS ↓and better ADWG only with 300 ppm; less concomitant bacteria	Binder et al., 1993
		200 ppm in-feed for 3 weeks (after weaning: 34-55 days of age) and for 2 weeks (at 77-98 days of age)	Better ADWG, FCR; CS ↓, LL ↓	Mateusen et al., 2001
Tulathromycin	EI	2.5 mg.kg by single IM (5 days after EI)	CS ↓ and LL ↓; better WG	McKelvie et al., 2005
	FC	2.5 mg.kg by single IM	CS ↓ and better ADWG	Nanjiani et al., 2005
	FC	2.5 mg.kg by single IM (pigs with SRD)	Better cure and mortality rate	Nutsch et al., 2005
Tylvalosin	FC	2.5 mg.kg in-water for 7 days (at weaning and in fattening period: 63-70 days of life	Better ADWG and homogeneity, LL prevalence and severity ↓	Pallares et al., 2015

Lincosamides				
Lincomycin	FC	5 mg/kg by IM for 3 days (after birth and/or weaning)	No beneficial effect on ADWG, FCR and CS	Kunesh, 1981
	FC	5 mg/kg by IM for 3 days (at 1 and 39 days)	Improvement of ADWG, FCR and CS ↓	Lukert & Mulkey, 1982
	FC	200 ppm in-feed for 3 weeks (fattening period)	Improvement of ADWG, FCR; CS ↓ and LL ↓	Mateusen et al., 2002
Fluoroquinolones				
Enrofloxacin	EI	5 mg/kg by IM for 3 days (5 days after EI)	CS ↓, LL ↓ and better WG	McKelvie et al., 2005
	FC	5 mg/kg *per os* for 3 weeks (after birth)	CS ↓	Frank, 1989
	FC	5 mg/kg by IM every 3 days for 3 weeks (after weaning)	CS ↓	Frank, 1989
	FC	5 mg/kg by IM every 2 days for 1 week (after birth)	CS ↓	Frank, 1989
Marbofloxacin	EI	1 or 2 mg/kg by IM for 3 days (starting 27 days after 1st EI and 4 days after 2nd EI)	No beneficial effect; reisolation of MHp	Le Carrou et al., 2006
	FC	2 mg/kg by IM for 3 or 5 days (fattening period)	Improvement of ADWG and cure rate	Thomas et al., 2000
Norfloxacin	EI	100 ppm in-feed for 3 consecutive weeks (1 month after EI)	Better ADWG and FCR; reisolation of MHp	Hannan & Goodwin, 1990
6-chloro analog of Norfloxacin	EI	200 ppm in-feed for 3 weeks (1 month after EI)	Better ADWG and FCR; reisolation of MHp	Hannan & Goodwin, 1990
	EI	400 ppm in-feed for 3 weeks (1 month after EI)	Better ADWG, FCR and LL ↓; reisolation of MHp	Hannan & Goodwin, 1990

10

M. hyopneumoniae are often referring to the IRPCM recommendations (Felde et al., 2018a; Klein et al., 2017; Tavio et al., 2014; Thongkamkoon et al., 2013; Vicca et al., 2004) and CLSI guidelines (Klein et al., 2017) as a basis for MIC determinations.

An inoculum of 10^3 to 10^5 color changing units (CCU)/mL in broth medium or 10^3 to 10^5 colony-forming units (CFU)/mL on agar medium is recommended by IRPCM guidelines (Hannan, 2000) whereas CLSI recommends 10^4 to 10^5 CFU/mL for broth and agar assays (CLSI, 2011). MIC is generally defined as the lowest antibiotic concentration that inhibits growth when growth is observed in the control without antibiotic for broth dilution MIC testing (CLSI, 2011; Hannan, 2000; Waites et al., 2012). However, results are also sometimes reported by authors as final MIC values after longer incubation periods (Felde et al., 2018a; Ter Laak et al., 1991). Agar dilution MIC testing is not recommended for *M. hyopneumoniae* since colonies are very small and difficult to observe under a stereomicroscope but this method has already been used (Friis and Szancer, 1994; Hannan et al., 1989). For agar dilution, MIC is defined as the lowest concentration of antibiotic that inhibit colony formation when colonies are observed on the control plate without antibiotic (CLSI, 2011; Waites et al., 2012). However, MIC has sometimes been wrongly defined as the concentration resulting in reduction of 50% or more in colony size or number (Hannan, 2000; Hannan et al., 1989; Ter Laak et al., 1991).

Quality control (QC) reference strains are available for MIC assays with human mycoplasmas (Waites et al., 2012) but not for *M. hyopneumoniae*. To validate the results obtained, it is therefore important to perform several independent MIC determination assays on separate occasions and to include strains with known MIC values for the antibiotics studied, when available.

Two types of criteria can be used for interpretation of results of antimicrobial susceptibility testing to categorize bacteria: clinical breakpoints, used to guide a therapeutic approach, and epidemiological cut-off (ECOFF) values, used to point out non wild-type bacteria harboring at least one phenotypically-detectable acquired resistance mechanism (Schwartz et al., 2010; Toutain et al., 2017). Interpretative criteria for several antimicrobials and bacteria are available, *e.g.* from CLSI (CLSI, 2013, 2015), the European Committee on Antimicrobial Susceptibility Testing (EUCAST, 2019) and its veterinary subcommittee (VetCAST) and from the Comité de l'Antibiogramme de la Société Française de Microbiologie (CA-SFM) and its veterinary subcommittee (CA-SFM, 2019). The absence of specific clinical breakpoints for mycoplasmas makes it difficult to evaluate the likely *in vivo* therapeutic efficacy from MIC data established *in vitro*. MIC results are often compared to breakpoints given for other bacteria (CLSI, 2013, 2015; EUCAST 2019; CA-SFM, 2019) or to breakpoints suggested by IRPCM (Hannan, 2000) or Ter Laak and collaborators (Ter Laak et al., 1991). Neither ECOFF values have been defined, due to the lack of MIC data for most *Mycoplasma* species, since a minimum of 100 MIC values in the putative wild-type population is needed (Toutain et al., 2017). The presence of a resistance mechanism can however be supposed when a non-monomodal Gaussian distribution of MICs is observed when testing a sufficient number of strains (Vicca, 2005; Toutain et al., 2017). This approach may allow determining which *M. hyopneumoniae* isolates do not belong to the wild-type population (Schwartz et al., 2010).

10.3.2 Minimal bactericidal concentration (MBC) determination

Based on their antimicrobial action, antibiotics can be classified into bactericidal (which kill bacteria) and bacteriostatic (which prevent the growth of bacteria) agents. The bactericidal activity of an antimicrobial agent is generally related to its mechanism of action: antibiotics acting on cell wall or interfering with essential bacterial enzymes are likely to be bactericidal, whereas antibiotics inhibiting ribosomal function and protein synthesis are likely to be bacteriostatic (French, 2006).

The MBC is defined as the lowest concentration needed to reduce the viability of the initial inoculum by ≥99.9% after incubation for a fixed length of time (CLSI, 1999). Antibacterial agents are usually classified as bactericidal if the MBC is no more than four times the MIC. However, these categories are not absolute, since the killing effect of the antimicrobial agent can vary with the species and the method used (CLSI, 1999). The practical relevance of this MBC value is limited since no pharmacokinetic/pharmacodynamic (PK/PD) integration is performed and no clinical breakpoints nor ECOFF values based on MBC are available.

a G2057A (*E. coli* numbering) substitution in the domain V of the 23S rRNA (Stakenborg et al., 2005a). Additional, acquired A2058G, A2059G and A2064G point mutations were found in the domain V of 23S rRNA of field strains with acquired resistance to 16-membered macrolides such as tylosin and to lincosamides (Felde et al., 2018a; Qiu et al., 2018; Stakenborg et al., 2005a) (Table 10.2). No mutations in domain II of 23S rRNA or in proteins L4 and L22 have been described in *M. hyopneumoniae* so far.

10.5.2 Resistance to fluoroquinolones

Fluoroquinolones kill dividing bacteria by inhibiting the topoisomerases II (DNA gyrase) and IV enzymes required for DNA replication (Redgrave et al., 2014). Chromosomally mediated resistance to fluoroquinolones in several *Mycoplasma* species is due to alterations in the quinolone resistance-determining regions (QRDR) of the *gyr*A and *gyr*B genes encoding DNA-gyrase and the *par*C and *par*E genes encoding topoisomerase IV. The targeting of either DNA-gyrase or topoisomerase IV by fluoroquinolones varies with bacterial species and specific fluoroquinolone (Redgrave et al., 2014). No active efflux mechanism or other plasmid-mediated resistance have been described so far in *M. hyopneumoniae*.

M. hyopneumoniae strains with reduced susceptibility to marbofloxacin were isolated from infected pigs after *in vivo* marbofloxacin treatments (Le Carrou et al., 2006). This reduced susceptibility was associated with a Ser80Phe or an Asp84Asn in the ParC QRDR (Table 10.2). A Ser80-Tyr substitution was also evidenced in the ParC QRDR of five field strains isolated from pig herds in Belgium and harboring reduced susceptibility to flumequine and enrofloxacin (Vicca et al., 2007). An extra mutation Ala83-Val, leading to further increase of the enrofloxacin MIC, was also evidenced in GyrA for one of these strains (Table 10.2). These four point mutations were also evidenced in a more recent study, and two additional mutations were found in the GyrA QRDR: Gly81Ala and Glu87Gly (Felde et al., 2018a). These studies suggested that ParC could be the primary target of fluoroquinolones in *M. hyopneumoniae* (first mutations observed in strains with moderate reduced susceptibility). Double substitutions in the *parC* and *gyrA* genes are associated with higher MIC values (Felde et al., 2018a; Vicca et al., 2007). These results are in accordance with previous results obtained in *M. gallisepticum*, suggesting that alteration of the primary target site can be followed by secondary mutations in lower-affinity binding sites and that highly resistant organisms will typically carry a combination of mutations within DNA-gyrase and topoisomerase IV (Reinhardt et al., 2002a; Reinhardt et al., 2002b).

10.5.3 Resistance to other antimicrobials

Strains resistant to tetracyclines have been reported (Inamoto et al., 1994). However, so far, no mechanism of resistance to tetracyclines, aminoglycosides/aminocyclitols, pleuromutilins or amphenicols has been described in *M. hyopneumoniae*. Mutations or resistance determinants responsible for decreased susceptibility to these antimicrobial families have been described in other *Mycoplasma* species (Gautier-Bouchardon, 2018).

10.6 Conclusions

Several studies have evidenced the efficacy of various antimicrobials for the treatment and control of *M. hyopneumoniae* infections with decrease of performance losses and reduction of clinical signs and lesions (del Pozo Sacristán, 2014). However, in many cases, persistence of *M. hyopneumoniae* and shedding was evidenced during and after medication and clinical signs of EP may reappear after cessation of the antimicrobial treatment. Selection of resistant strains of *M. hyopneumoniae* during antimicrobial treatment has been described, but bacterial persistence is also observed without selection of resistance mechanisms (Le Carrou et al., 2006). Even if acquired resistance to macrolides/lincosamides (Felde et al., 2018a; Stakenborg et al., 2005a), tetracyclines (Inamoto et al., 1994) and fluoroquinolones (Felde et al., 2018a; Le Carrou et al., 2006; Vicca et al., 2007) has been evidenced, most strains of *M. hyopneumoniae* remain susceptible to the antimicrobial families used for treatment of EP. Therefore, antimicrobial resistance does currently not seem to constitute a major problem for treatment of *M. hyopneumoniae* infections, but studies should be conducted regularly to monitor the evolution of the susceptibility levels of field strains.

CHAPTER 11

VACCINES AND VACCINATION AGAINST *MYCOPLASMA HYOPNEUMONIAE*

Dominiek Maes[1]

Filip Boyen[1]

Odir Dellagostin[2]

Guoqing Shao[3]

Freddy Haesebrouck[1]

1 Faculty of Veterinary Medicine, Ghent University, Ghent, Belgium
2 Unit of Biotechnology, Federal University of Pelotas, Pelotas, Brazil
3 Institute of Veterinary Science, Jiangsu Academy of Agriculture Sciences, Nanjing, Jiangsu, China

11.1 Introduction

Control of *Mycoplasma hyopneumoniae (M. hyopneumoniae)* infections in pig herds can be accomplished in a number of different ways. First, management and biosecurity practices and housing conditions should be optimized within the herd. These relate to proper purchase policy and gilt acclimation strategies, all-in/all-out production, limiting factors that may destabilize herd immunity and/or disease dynamics, maintaining optimal stocking densities, prevention of other respiratory diseases, and optimal housing and climatic conditions (Maes et al., 2008). Improvements in these factors will decrease infection levels of *M. hyopneumoniae*. However, they will not eliminate the pathogen from the farm and will not always lead to the expected results. In addition, it might not always be possible to implement changes in management and housing because of financial, logistical and practical constraints.

Treatment and control can also be done using medication with antimicrobials active against *M. hyopneumoniae* and the major secondary respiratory bacteria. Antimicrobial medication can limit the consequences of the disease and decrease the infection load (Thacker et al., 2004), but it does not prevent pigs from becoming infected with *M. hyopneumoniae*. Strategic medication with antimicrobials during extended periods of time should also be discouraged because of both the increased risk of antimicrobial resistance development (Collignon and McEwen, 2019) and antimicrobial residues in the pig carcasses at slaughter.

Vaccination against *M. hyopneumoniae* has been shown to be a very useful tool to control *M. hyopneumoniae* infections. Different commercial *M. hyopneumoniae* vaccines are available and vaccination is frequently practiced worldwide. In infected herds, vaccination improves the health and performance of the animals and reduces the use of antimicrobials. This chapter reviews the current knowledge on vaccination against *M. hyopneumoniae*. The currently available vaccines and their effects will be discussed, the mechanisms of protection, the different vaccination strategies, and finally, different experimental vaccines that have been developed to further improve the efficacy of vaccination.

11.2 Commercial vaccines against *Mycoplasma hyopneumoniae*

Vaccination is used worldwide to control *M. hyopneumoniae* infections. Commercial vaccines mostly consist of inactivated, adjuvanted whole-cell preparations (Maes et al., 2018). Table 11.1 gives an overview of different commercially available *M. hyopneumoniae* bacterin vaccines. Most vaccines are based on the strain J, possibly because it is considered the reference strain of *M. hyopneumoniae*. This strain was isolated in 1963 from a field outbreak in sows experiencing a mild course of enzootic pneumonia (Goodwin and Whittlestone, 1963). The strain was originally a virulent strain but after continued *in vitro* passaging, the strain J lost its virulence (Zielinski and Ross, 1993). Commercial bacterin vaccines are licensed either for single or double vaccination (Table 11.1), and combinations with porcine circovirus type 2 (PCV-2) or *Glaesserella (Haemophilus) parasuis* are available. Some vaccines are also licensed for use in breeding animals. Most bacterin vaccines should be administered intramuscularly, but also bacterins for intradermal administration are licensed.

An inactivated vaccine based on soluble antigens of *M. hyopneumoniae* is also commercially available (USA: Fostera® PCV MH, Zoetis; in Europe: Suvaxyn Circo + MH RTU, Zoetis). It is a one-dose vaccine combined with PCV-2 that can be administered to piglets from three weeks of age onwards (Park et al., 2016a).

Attenuated vaccines against *M. hyopneumoniae* have been licensed in Mexico and in China (Feng et al., 2013). The vaccine in Mexico is a thermosensitive mutant of *M. hyopneumoniae* (strain LKR) (VaxSafe® MHYO, AviMex) that should be applied intranasally once from 3 days of age onwards. The attenuated Chinese vaccine strain is derived from a virulent parent strain 168 isolated in 1974, from an Er-hua-nian pig (local Chinese breed, very sensitive to *M. hyopneumoniae*) with typical signs and lesions of enzootic pneumonia (Feng et al., 2010a). This field strain was gradually attenuated by continuous alternate passage through KM2 cell-free medium (a modified Friis medium) and pigs. The field strain was completely attenuated by over 300 passages in KM2 media, and kept antigenicity by pig regression. The attenuated strain contains 60 insertions and 43 deletions. Mutations in genes related to metabolism and growth may contribute to the attenuated virulence in the 168 strain, in addition to variations previously described in *M. hyopneumoniae* adhesins (P97, P102, P146, P159, P216, and LppT), cell envelope proteins (P95), cell surface antigens (P36), and secreted proteins and chaperone protein (DnaK) (Liu et al., 2013). The attenuated vaccine strain is mostly used by intrapulmonary administration (Feng et al., 2010a).

Table 11.1. Most commonly used commercially available *M. hyopneumoniae* bacterin vaccines – Bacterin vaccines available in only one or a few countries are not included the table

Company	Vaccine	Antigen / Strain	Adjuvant	Licensed for	Dosage (ml)	Administration route	Vaccination schedule	Age of administration (days)	Onset of Immunity	Duration of Immunity
Boehringer Ingelheim	Ingelvac Mycoflex	J strain isolate B-3745	ImpranFLEX (carbomer)	pigs sows	1 1	IM IM	single single	≥21 d Semi-annually	2 w 2 w	26 w
Ceva	Hyogen	Ceva strain BA 2940-99	Imuvant (W/O J5 LPS)	pigs	2	IM	single	≥21 d	3 w	26 w
Elanco	Stellamune Mycoplasma	Strain NL 1042	Mineral oil and lecithin	pigs	2	IM	double	d2-14 + 2w later	2 w	22 w
	Stellamune One		Amphigen Base, and Drakeol 5 (mineral oil)	pigs	2	IM	single	Between d3 and d21	18d/3w	26 w/23 w
Hipra	Mypravac suis	J strain	Levamisole and carbomer	pigs	2	IM	double	1w + 3w later	2 w	26 w
MSD, Merck Animal Health	M+Pac	J strain	Mineral oil and $Al(OH)_3$	pigs	2 1	IM IM	single double	3w 1w + 2/4 w later	21 d	26 w
	Porcilis Mhyo	Strain 11	Tocopherol	pigs	2	IM	double	≥1w + 3w later	2 w	20 w
	Porcilis Myo ID ONCE	Strain 11	Paraffin oil and tocopherol	pigs	0.2	ID	single	≥2 w	3 w	22 w
	Porcilis PCV Mhyo[a]	J Strain	Mineral oil and $Al(OH)_3$	pigs	2	IM	single	≥3 w	4 w	21 w

Zoetis	Respisure	Strain NL 1042	Mineral oil	pigs dams PS boars	2 2 2 2	IM IM IM IM	double double double double	1 w + 2 w later 2 w BF 6 w + 2 w BF Semi-annually	3 w NA NA NA	23 w NA NA NA
	Respisure-ONE	Strain NL 1042	Mineral oil	pigs dams	2 2	IM	single single	> 1 d Semi-annually	18 d NA	25 w NA
	Suvaxyn Mhyo	Strain P-5722-3	Carbomer	pigs	2	IM	double	≥1 w + 2 w later	1 w	20 w
	Suvaxyn MH-one [b]		Mineral oil + carbomer	pigs	2	IM	single	≥1 w	2 w	6 m
	Suvaxyn Circo + MH[a]		Squalene + C/W emulsion	pigs	2	IM	single	≥3 w	3 w	23 w
	Suvaxyn MHYO – PARASUIS [c]		Carbomer	pigs	2	IM	double	≥1 w + 2 w later	1 w	6 m

(Based on Maes et al., 2017; Garza-Moreno et al., 2018).

a Combination vaccine with Porcine Circovirus type 2
b in the USA, the adjuvant is Amphigen
c Combination vaccine with *Glaesserella (Haemophilus) parasuis* - Named Suvaxyn RespiFend MH HPS in USA
IM intramuscular, ID intradermal, BF before farrowing, PS pregnant sows, NA no information available, $Al(OH)_3$ Aluminium hydroxide

11.3 Mechanisms of protection

The exact mechanisms of protection are not yet fully understood. In terms of innate immunity against *M. hyopneumoniae*, it has been shown that both Toll-like receptors 2 (TLR2) and TLR6 are important in the recognition of *M. hyopneumoniae* in porcine alveolar macrophages (Muneta et al., 2003). Blocking TLR2 and TLR6 receptors led to less TNF-α production by macrophages (Okusawa et al., 2004). Another study showed that TLR6 mRNA expression was higher in the Commercial White line than in the Piau breed after vaccination, and that there was differential expression before and after vaccination in the Commercial White line (Regia Silva et al., 2011).

In terms of adaptive immune responses, Marchioro et al. (2013) showed that *M. hyopneumoniae*-specific IgG, IgM and low concentrations of IgA were detected in BAL fluid of *M. hyopneumoniae*-free pigs after intramuscular administration of a commercial inactivated *M. hyopneumoniae* whole-cell vaccine. The importance of local mucosal antibodies in protection remains unclear. Antibody concentrations in respiratory tract washings did not correlate with protection in some studies (Djordjevic et al., 1997; Thacker et al., 1998), whereas opposite results were obtained in later studies (Thacker et al., 2000b; Sarradell et al., 2003).

Commercial vaccines also induce serum *M. hyopneumoniae*-specific antibodies. The percentage of animals seroconverting after vaccination may range from 30 to 100% (Sibila et al., 2004a), and serological responses differ among vaccines (Thacker et al., 1998). These varying serological responses after vaccination could be attributed to differences in adjuvants, administration routes, antigen dose, levels of antigen expression in bacterins, vaccination strategy (single or double vaccination) and/or varying individual animal responses (Fisch et al., 2016; Matthijs et al., 2019a). Serum antibodies are usually detected from two to four weeks after two-dose vaccination and they remain detectable for weeks to months. In the absence of infection that boosts the immune system, the serological response upon a single vaccination is generally lower than after double vaccination. In the absence of natural infections that boost the immune system, antibody titers decrease below detection limits one to three months after vaccination (Maes et al., 1999c). It is generally accepted that serum antibodies are not suited to evaluate protective immunity, since no direct correlation could be demonstrated between serum antibody concentration and protection against *M. hyopneumoniae* challenge (Djordjevic et al., 1997; Thacker et al., 1998).

A cell-mediated immune response upon vaccination or challenge is considered important for protection. Cell-mediated immune responses induced by *M. hyopneumoniae* vaccination have been measured mainly by determining the proliferative responses of lymphocytes following *in vitro* stimulation with *M. hyopneumoniae*-specific antigens (Thacker et al., 2000a; Marchioro et al., 2013, 2014a; Huang et al., 2014). Results of these studies indicate a considerable variation among individual pigs. Thacker et al. (2000a) reported that pigs vaccinated against and challenged with *M. hyopneumoniae* had the greatest number of IFN-γ secreting cells on day seven post-infection. They suggested that T helper (Th-1 response, type 1 immunity) and natural killer cells secreting IFN-γ may be important in mediating a protective immune response against *M. hyopneumoniae*-induced pneumonia through the activation of macrophages. Marchioro et al. (2013) observed significantly higher numbers of IL-12 secreting cells in the lung of vaccinated pigs compared to the non-vaccinated group. The number of IL-10 secreting cells from bronchial lymph nodes was also higher in the vaccinated group. After re-stimulation *in vitro*, lymphocytes from bronchial lymph nodes and lungs secreted significantly higher levels of IL-12 in the vaccinated group. These findings suggest that vaccination may modulate different immune and inflammatory responses in pigs.

Increases in concentrations of pro-inflammatory cytokines such as TNF-α, are thought to be associated with lesion development (Meyns et al., 2007) and to be one possible mechanism for the potentiation of porcine reproductive and respiratory syndrome (PRRSV)-induced pneumonia by *M. hyopneumoniae*. Vaccination against *M. hyopneumoniae* reduced the potentiation of PRRSV-induced pneumonia by *M. hyopneumoniae* (Thacker et al., 2000b). This may be because the vaccine prevents an increase in lung TNF-α concentrations in response to challenge with *M. hyopneumoniae*, thus impeding potentiation of PRRSV-induced pneumonia (Thacker et al., 2000b).

Studies suggest that cell mediated immunity may both help and hinder the development of mycoplasmal pneumonia (Tajima et al., 1984). Therefore, it is important that immune responses are tightly regulated. Vranckx et al. (2012b) reported a lower infiltration of macrophages in the lung tissue in vaccinated animals upon infection with *M. hyopneumoniae*, indicating that vaccination modulates the immune response following infection.

Recently, it has been shown that a type 3 immune response is important to protect mucosal surfaces, to promote epithelial cell regeneration, mucous and antimicrobial protein production, and the release of neutrophil recruitment (Abbas et al., 2016). A type 3 response is mainly induced by extracellular pathogens. These responses are associated with a strong Th-17 response with cytokines of the IL-17 family being the key T-cell cytokines (see chapter on Immunity).

11.4 Effects of vaccination

The major advantages of piglet vaccination relate to a decrease of the performance losses due to *M. hyopneumoniae* infections: improvement of daily weight gain (2-8%), feed conversion ratio (2-5%) and sometimes mortality rate. Additionally, shorter time to reach slaughter weight, less variation in slaughter weight (more homogeneous carcasses), reduced clinical signs (coughing), lower prevalence and severity of *Mycoplasma*-like lung lesions and lower treatment costs are observed (Maes et al., 1998, 1999c). Although *M. hyopneumoniae* infections do not cause pleurisy lesions, infections gained early during the rearing of fatteners have been shown to predispose for pleurisy recorded at slaughter (Holmgren et al., 1999). Correspondingly, studies showed that *M. hyopneumoniae* vaccination may lead to a decrease in pleurisy prevalence in slaughter pigs (Maes et al., 1999c; Del Pozo et al., 2014), probably due to a lower number of secondary bacterial infections such as those with *Pasteurella multocida* and/or *Actinobacillus pleuropneumoniae* (Marois et al., 2009). The currently used vaccines reduce the number of *M. hyopneumoniae* organisms in the respiratory tract (Meyns et al., 2006; Vranckx et al., 2012b) and decrease the infection level in a herd (Sibila et al., 2007).

However, the protection against clinical signs and *Mycoplasma*-like lung lesions is often incomplete and vaccination does not prevent colonization. Transmission models under experimental (Meyns et al., 2006) and field conditions (Pieters et al., 2010; Villarreal et al., 2011a) also showed that vaccination conferred only a limited and non-significant reduction of the transmission ratio of *M. hyopneumoniae*. Therefore, these authors concluded that vaccination with the current vaccines as the only control measure will not eliminate *M. hyopneumoniae* from infected pig herds.

A meta-analysis, based on 63 studies that had investigated the efficacy of *M. hyopneumoniae* vaccination, showed that vaccinated animals on average had 22 gram higher average daily weight gain (ADWG) than non-vaccinated animals (Jensen et al., 2002). Studies on the economic impact of vaccination are rare, but vaccination against *M. hyopneumoniae* in 14 Belgian pig herds with a moderate to low infection pressure of *M. hyopneumoniae* and practicing all-in/all-out production revealed a daily weight gain of +22 g/day, a feed conversion ratio of -0.07, a mortality rate -0.23%, and medication costs -€0.45 per vaccinated pig (Maes et al., 1999c; 2003). No changes in carcass quality following vaccination were observed. Partial budgeting analysis showed that the average additional net return to labor was +€0.72 per pig. Depending on the market conditions of a specific year, the additional net return to labor through vaccination ranged from -€0.28 per pig to €11.30 per pig (Maes et al., 2003). Thus, even in these herds with a low infection level of *M. hyopneumoniae*, vaccination was economically attractive, except in periods with extremely bad market conditions. In addition to information of production benefits resulting from vaccination, data of market conditions are necessary to precisely quantify the financial benefit of vaccination in a specific herd. In addition, to assess the full beneficial effects of vaccination, all piglets in the farm should be vaccinated and effects should be measured several months after the initiation of vaccination.

Pig producers sometimes wonder after having vaccinated for many years whether *M. hyopneumoniae* infections are still of importance in their herd and whether it is necessary to continue vaccination. To this end, the infection levels of *M. hyopneumoniae* could be assessed in the vaccinated animals or in a group of non-vaccinated sentinel animals in the farm. This could be accomplished using PCR testing on respiratory tract samples from pigs of different age groups. Serum antibodies against *M. hyopneumoniae* can also be tested for. In case of vaccinated animals however, the interpretation of the outcome of the serological testing is more difficult. A sIgA-ELISA method has been developed (Feng et al., 2010b) and used for differentiation of *M. hyopneumoniae* infected from vaccinated pigs (Bai et al., 2018). Monitoring other parameters such as the severity of coughing, *Mycoplasma*-like lung lesions and performance may also be helpful. It is up to the herd veterinarian after having received all data to decide whether vaccination should be continued or not.

11.5 Vaccination strategies

In *M. hyopneumoniae*-free herds or in herds with very low infection levels, vaccination may not be recommended since under these conditions, the benefits of vaccination may not be sufficient to outweigh the costs. In herds with higher infection levels without obvious clinical signs, or in herds with clinical disease, vaccination is often economically justified (Maes et al., 2003). Different vaccination strategies have been adopted, depending on the type of herd, production system and management practices, infection pattern and preferences of the pig producer.

11.5.1 Piglet vaccination

Since infections with *M. hyopneumoniae* may already occur during the first weeks of life (Villarreal et al., 2010), vaccination of piglets is most commonly used. Its efficacy has been demonstrated by means of numerous studies under experimental, as well as field conditions (Jensen et al., 2002). Vaccination of suckling piglets (early vaccination; < 4 weeks of age) is most common, especially in single-site herds, whereas vaccination of nursery/early fattening pigs (late vaccination; between 4 and 10 weeks) is sometimes practiced, especially in three-site systems where late infections are more common. Originally, double vaccination was the most frequent practice. Currently, one-dose vaccination is more frequently used, mainly because it requires less labor and it can be implemented more easily into routine management practices at the farm (Baccaro et al., 2006). With one-dose vaccines, the skill of the pig producer or employee to vaccinate properly might be more critical for vaccine compliance since only one injection is administered. Single vaccination at either 7 or 21 days of age was efficacious (performance, lung lesions) in a pig herd with clinical respiratory disease during the second half of the fattening period (Del Pozo Sacristan et al., 2014). Experimental (Arsenakis et al., 2016) and field studies (Arsenakis et al., 2017) have shown that vaccinating piglets three days prior to weaning conferred slightly better results (performance, lung lesions) than vaccination at weaning.

As a general rule, vaccination needs to be administered before animals become infected. Vaccination of suckling piglets has the advantage that immunity can be induced before pigs become infected with *M. hyopneumoniae*, and that less infections with other pathogens are present that can interfere with the immune response. The role of maternal immunity on vaccine responses will be discussed later (see further). Vaccination of nursery pigs has no or less interference with possible maternally derived antibodies. However, nursery pigs may already be infected with *M. hyopneumoniae* (Villarreal et al., 2010). In addition, the age of infection or the age-window in which the piglets become infected may vary between successive groups within a herd (Sibila et al., 2004b).Vaccination strategies tailored to specific farm infection patterns *e.g.* vaccination of piglets at least two weeks prior to seroconversion have been successful in decreasing the impact *of M. hyopneumoniae* infection (Wallgren et al., 2000). Some pathogenic infections *e.g.* with PRRSV, porcine circovirus type 2 (PCV-2) or *M. suis* mainly take place after weaning and may affect the general health status of the pigs, and consequently also interfere with proper immune responses after vaccination. Also, in case of intramuscular administration, iatrogenic transmission of these pathogens may take place during vaccination.

11.5.2 Breeding gilt vaccination

Vaccination of breeding gilts during gilt acclimation or quarantine unit is the most common gilt acclimation practice used in Europe and North America (Garza-Moreno et al., 2018). Vaccination is recommended in case negative gilts or gilts with unknown infection status will enter a herd that is infected with *M. hyopneumoniae*. The aim is to stimulate and homogenize the immunity of the replacement gilt population and to avoid destabilization of the breeding stock immunity.Vaccination of gilts twice (at 2 and 6 weeks post entry) in the gilt acclimation unit significantly reduced the proportion of PCR positive gilts at 14 weeks post entry in a farm that was clinically infected with *M. hyopneumoniae* (Garza-Moreno et al., 2019a). A vaccination protocol with four doses showed an even lower proportion of *M. hyopneumoniae* positive pigs, but based on the small and non-significant difference, more than two vaccine doses did not seem to be justified.

Vaccination also increased the antibody levels of the gilts and their offspring. Although vaccination did not provide full protection, the infection level within the gilt population of the studied herd was significantly reduced compared to a group of gilts that was not vaccinated. Because of the partial protection induced by vaccination, there is also a current trend, mainly in North-America, to deliberately expose gilts at a young age to infection with *M. hyopneumoniae* (Garza-Moreno et al., 2018; Robbins et al., 2019). This strategy allows to develop good immunity, decrease bacterial shedding by the time of farrowing, and having low colonized offspring at weaning (Pieters and Fano, 2016; see chapter 9). However, further research is needed.

11.5.3 Sow vaccination

Vaccination of sows at the end of gestation is less commonly practiced. It aims to both reduce the shedding of *M. hyopneumoniae* from the sow to the offspring and to confer protection to the piglets via maternally-derived immunity. Wallgren et al. (1998) showed that serum antibodies in the sow start declining during the last month of gestation and therefore, they recommended to vaccinate sows at least four weeks prior to expected farrowing. Maternally derived immunity in piglets provides partial protection against infection of *M. hyopneumoniae* and reduces the severity of clinical signs and *Mycoplasma*-like lung lesions upon challenge infection. The initial antibody titers in newborn piglets depend on the immune status of the sow and the amount of colostrum ingested by the piglets (Wallgren et al., 1998). As maternally derived immunity decrease with age of the piglets, also the protection may decrease with age. Morris et al. (1994) showed that *M. hyopneumoniae* colostral antibodies wane at 30, 45, and 63 days in piglets with low, medium, or high initial antibody concentrations, respectively. Field studies showed that vaccination of sows at the end of gestation resulted in a lower number of *M. hyopneumoniae* colonized piglets at and shortly after weaning, both in farrow-to-finish operations (Arsenakis et al., 2019) and in multisite production systems (Ruiz et al., 2003; Sibila et al., 2008). In addition, it was shown that pigs from vaccinated sows also had a lower number of *Mycoplasma*-like lung lesions at slaughter, compared to pigs from non-vaccinated sows. Since piglets from vaccinated sows can still be infected, additional measures to control *M. hyopneumoniae* during the nursery and finishing phases may be warranted. In farms with high levels of *M. hyopneumoniae* infections or in case of clinical symptoms in the sow population, vaccination of all breeding animals at the same moment may be practiced in order to stabilize and homogenize the breeding stock immunity.

11.5.4 Administration routes

Intramuscular injection is the most frequently used route of administration of *M. hyopneumoniae* vaccination.

Intradermal vaccination is also possible with some vaccines. This administration route directly targets epidermal Langerhans cells and dermal dendritic cells, which are essential for efficient T and B cell priming. In this sense, intradermal vaccination against *M. hyopneumoniae* can be an asset, as more of these specialized antigen presenting cells are present in the skin compared to muscle tissue (Fu et al., 1997). In addition, no needles are used as the vaccine is administered intradermal via pressure, which may reduce the risk for iatrogenic infections. The higher dispersion of the antigen at the site of injection may also reduce injection site reactions (Del Pozo Sacristán, 2014). Indeed, Beffort et al. (2017) observed less injection site reactions and better efficacy of intradermal compared to intramuscular vaccination in terms of the reduction of clinical signs and macroscopic *Mycoplasma*-like lung lesions. Martelli et al. (2014) showed that intradermal vaccination with a bacterin induced a systemic humoral and cell-mediated immune response as well as local humoral immunity, which was comparable to that obtained by the intramuscular administration of an inactivated, whole cell, adjuvanted vaccine. Good results (performance, lung lesions) with intradermal vaccination against *M. hyopneumoniae* have also been obtained in other studies (Jones et al., 2005; Ferrari et al., 2011; Tassis et al., 2012).

The attenuated vaccine used in Mexico is a thermosensitive mutant of *M. hyopneumoniae* that is applied intranasally. Intrapulmonary administration of the Chinese attenuated Mhp 168 strain was reported to induce a greater immunity than that of a bacterin administered intramuscularly (Feng et al., 2010a). In a subsequent study, Feng et al. (2013) showed that the attenuated vaccine could also be administrated as an aerosol. To this end, a nebulizer was used to produce an aerosol with vaccine particle sizes less than 5 μm, and a buffer with 5% glycerol to prevent inactivation of *M. hyopneumoniae* caused by aerosolization and evaporation. Before nebulization, the room temperature and relative humidity were standardized to 20 – 25°C and 70 – 75%, respectively. The aerosol vaccine was effectively delivered to the lower respiratory tract, and *M. hyopneumoniae*-specific sIgA secretion was detected in nasal swab samples at 14 days post-immunization. Li et al. (2015) showed that both intrapulmonary and intranasal administration of this vaccine induced local cellular and humoral immunity in the respiratory tract, but the intrapulmonary vaccination increased the levels of IFN-γ and sIgA more than the intranasal vaccination. The attenuated vaccine has also been shown to confer protection in experimentally infected pigs when administered intramuscularly together with an adjuvant such as an ISCOM-matrix (Xiong et al., 2014a) or a carbomer–astragalus polysaccharide mixture adjuvant (Xiong et al., 2014b).

11.6 Factors influencing efficacy of vaccination

Vaccination usually confers beneficial effects, but the results may be variable. This may be caused by non-compliance with the basic principles of good vaccination practices, *e.g.* improper storage conditions and administration of the vaccine, poor hygiene at vaccination, and not following the guidelines mentioned in the leaflet. However, also other factors such as stress at vaccination, infections with other pathogens at the moment of vaccination, important co-infections involved in PRDC, diversity of *M. hyopneumoniae* strains, and maternal immunity can have an effect of vaccination efficacy.

11.6.1 Stress factors

Vaccination of piglets is commonly practiced when piglets are handled anyway *e.g.* during ear notching, castration, weaning, re-grouping, etc. In this manner, vaccination can easily be included in the routine farm management practices without requiring much extra labor. However, such practices may cause stress to the animals and interfere with proper immune responses. This might be particularly the case at weaning as this is one of the most stressful events in the pig's life (Campbell et al., 2013). Wallgren et al. (1994) showed that severe and/or prolonged stress reactions mainly hamper the primary response of *M. hyopneumoniae* vaccination, but did not affect the immune responses following booster vaccination. Arsenakis et al. (2016; 2017) showed slightly decreased efficacy when pigs were vaccinated at the moment of weaning.

11.6.2 Infections with other pathogens at the moment of *Mycoplasma hyopneumoniae* vaccination

Under experimental conditions, PRRSV infection or administration of a modified live virus (MLV) PRRS vaccine (US genotype) at the time of *M. hyopneumoniae* vaccination significantly reduced the efficacy (lung lesions) of a *M. hyopneumoniae* vaccine (Thacker et al., 2000b; see chapter 7). However, this interference might depend on the timing of PRRV vaccination (Boettcher et al., 2002). Similarly, for the single dose vaccine, vaccination against *M. hyopneumoniae* at one week of age followed by vaccination with a PRRSV-2 MLV vaccine two weeks later had no negative impact on the efficacy of the bacterin (Park et al., 2014). In addition, PRRSV-1 (European genotype)

MLV vaccine administered simultaneously with the *M. hyopneumoniae* booster vaccination at four weeks of age did not affect the efficacy of the *M. hyopneumoniae* vaccine (Drexler et al., 2010; Bourry et al., 2015). Interestingly, in these studies, the combined vaccination had a cumulative positive effect on the efficacy of each respective vaccine (Bourry et al., 2015). However, under field conditions the beneficial effect of combined *M. hyopneumoniae*/PRRSV vaccination has been only evident in improved lung lesion score (Stricker et al., 2013). Therefore, administration of the *M. hyopneumoniae* vaccine in relation to the PRRSV vaccination should be carefully considered to avoid negative influence from this latter vaccine. Also, when combining *M. hyopneumoniae* vaccines with vaccines against other pathogens, administered either simultaneously or sequentially, it is important to strictly adhere to the guidelines described in the leaflet of the *M. hyopneumoniae* vaccine.

In an experimental trial, severe concurrent *Ascaris suum* infection of pigs has been shown to impair *M. hyopneumoniae* vaccination efficacy resulting in increased pneumonic lesions upon *M. hyopneumoniae* challenge (Steenhard et al., 2009). Possibly, also other co-infections or mycotoxin contamination of the feed at the moment of vaccination may interfere with the immune responses following vaccination. Nevertheless, further research in this area is warranted.

11.6.3 Co-infections with other pathogens involved in PRDC

It is also possible that co-circulation of other respiratory pathogens involved in PRDC *e.g.* swine influenza A virus, PRRS virus, PCV-2, *A. pleuropneumoniae* may explain why vaccination against *M. hyopneumoniae* does not confer the expected results in some farms. In such cases, *M. hyopneumoniae* might not be the main pathogen involved in the PRDC and other pathogens might contribute more to the clinical signs and lung lesions (Wallgren et al., 2016). In such cases, it might be more appropriate to focus on the control of these other major pathogens rather than on *M. hyopneumoniae*. This scenario also emphasizes the importance of establishing a correct diagnosis in case of PRDC.

11.6.4 Diversity of *Mycoplasma hyopneumoniae* strains

Variation across several isolates of *M. hyopneumoniae* has been described at antigenic, proteomic, transcriptomic, pathogenic, and genomic levels (Betlach et al., 2019). Antigenic differences between infection strain and vaccine strain have been suspected to be involved in the variable outcome of vaccination. However, solid scientific evidence is not available. Villarreal et al. (2011b) showed that vaccination conferred a more pronounced reduction of clinical signs and lung lesions after infection with a highly virulent strain compared to infection with a low virulent strain of *M. hyopneumoniae*. In a sense, this is logical as there is more room for improvement in a case of more severe clinical signs and lesions. Another study compared the protective effect of different *M. hyopneumoniae* bacterins and showed that a bacterin based on the challenge strain (homologous situation) did not perform better than bacterins based on the J-strain or other strains (Villarreal et al., 2012). Although differences between strains circulating in pig herds and strains included in commercial vaccines exist (Charlebois et al., 2014; Garza-Moreno et al., 2019b) and might result in variable vaccination effects, it is clear that this is not the only factor involved.

11.6.5 Maternally derived immunity

The influence of maternally derived antibodies on vaccine responses in piglets is not fully elucidated. Martelli et al. (2006) showed that passively acquired antibodies have little or no effect on either a vaccine induced priming or subsequent anamnestic response, even if pigs do not show a clear serological response after vaccination. Other studies showed that pigs with high maternally-derived antibody titers elicited lower (Maes et al., 1998; 1999c; Hodgins et al., 2004) serological responses after vaccination, but there was no significant decrease of clinical signs and lung lesions. Jayappa et al. (2001) showed that high titers of maternally-derived antibodies, induced by infection and vaccination of the sows, had negative effects on the efficacy of piglet vaccination. Lehner et al. (2008) applied sow vaccination at 6 and 3 weeks prior to farrowing in a pig herd. They reported that ADG and lung lesion scores in pigs born from these sows were numerically better in cases where piglets were vaccinated at six weeks instead of at three weeks of age. Also Wallgren et al. (1998) obtained better results in cases where piglets were vaccinated at an older age. Bandrick et al. (2014b) showed that piglets with *M. hyopneumoniae*-specific maternally derived immunity indeed failed to show a rise in vaccine induced *M. hyopneumoniae* antibody levels, but did show primary (antigen-specific lymphoproliferation) and secondary (delayed type hypersensitivity) *M. hyopneumoniae*-specific cell-mediated immunity responses following vaccination. Hodgins et al. (2004) showed that colostral antibody titers influenced the serological response following vaccination, but not the age of the pigs at vaccination (either 2, 3 or 4 weeks of age). It can be expected that the interaction of vaccine and maternally-derived antibodies vary with the exact formulation of the vaccine and also with the maternal antibodies titers present in the pig.

11.7 Experimental vaccines

Although the commercial vaccines are widely used and cost-efficient in many farms, only partial protection against clinical signs and lung lesions is induced, and there is no prevention of infection. Therefore, research on the development of new vaccines that may offer better protection against *M. hyopneumoniae* infections is ongoing. Table 11.2 provides an overview of the experimental vaccines against *M. hyopneumoniae* that have been tested in mice. These studies focused on the development of the vaccines and the evaluation of the immune responses, not the efficacy of the vaccines against challenge infection. As the immune responses in mice cannot be extrapolated as such to other species like the pig, the results of these studies should be considered as preliminary and warrant further confirmation in pigs. Table 11.3 shows studies that have tested experimental *M. hyopneumoniae* vaccines in pigs. These studies focused on mucosal and systemic immune responses, and many of them also assessed the efficacy (lung lesions) of the vaccines against experimental *M. hyopneumoniae* challenge infection. The tables summarize the antigens, adjuvants, vectors and routes of immunization that were used in these studies.

Different recombinant proteins of *M. hyopneumoniae*, used in combination with a vector or adjuvant and administered in various forms, have been tested, mainly in mice. The protein P97, which was identified as an important adhesion molecule of *M. hyopneumoniae*, has been included in most of the experimental vaccines, either as single antigen or in combination with other antigens. Bacteria (*Salmonella, Pseudomonas aeruginosa, Erysipelothrix rhusiopathiae, A. pleuropneumoniae*) as well as viruses (adenovirus) have been used as vectors.

Many different adjuvants have been tested. Aluminum hydroxide or phosphate have been commonly used and have a demonstrated safety profile. However, they have been associated with local reactions, such as erythema, subcutaneous nodules and contact hypersensitivity (Baylor et al., 2002). Oil emulsions of various compositions have also been widely used as adjuvant in *M. hyopneumoniae* vaccines. They are considered potent adjuvants, even though they have a tendency to induce granulomas at the injection sites (Aucouturier et al., 2001). Also, bacterial enterotoxins such as cholera toxin (CT) from *Vibrio cholera* or the related subunit B of heat-labile enterotoxin of *Escherichia coli* (LTB), a non-toxic subunit, have been used (Cox et al., 2006). They are considered potent mucosal and parenteral adjuvants, enhancing both mucosal IgA, as well as systemic antibody responses. Despite the advantage of mucosal vaccinations for the prevention of *M. hyopneumoniae,* the number of experimental vaccines evaluating this route of immunization is still very limited. This is mainly due to important difficulties that are encountered by using the mucosal route such as:
1. the antigen should be present at the mucosal barrier in sufficient amount,
2. mucosal tolerance mechanisms should be overcome,
3. protective immune mechanisms should be activated and
4. minimal/or no influence on mucosal functionality should occur.

Mucosal adjuvants are crucial in reaching these goals (Cox et al., 2006).

Only a few studies with DNA vaccines have been tested in mice. Significant immune responses were elicited in mice, based on the expression of a heat shock protein P42 gene (Chen et al., 2003) or a ribonucleotide reductase R2 subunit gene fragment of *M. hyopneumoniae* (Chen et al., 2006b). The studies suggest that these vaccines have potential for controlling *M. hyopneumoniae* infections, but they need to be validated in pigs. Most experimental vaccines were administered via the parenteral route (intramuscular, subcutaneous, intradermal), although some of them were applied via other routes (intranasal, intraperitoneal, oral). In general, most of the experimental vaccines developed so far have been less efficacious than the commercially available vaccines.

Matthijs et al. (2019a,b) investigated the safety, immune responses and also the efficacy of different *M. hyopneumoniae* bacterin vaccines in pigs. The bacterins were based on a more recently isolated *M. hyopneumoniae* strain compared to the J-strain, in combination with innovative adjuvants namely Toll-like receptor ligands, Mincle agonist trehalose 6,6-dibehenate (TDB), and STING ligand cyclic diadenylate monophosphate (c-di-AMP), and using different carriers namely liposomes, polylactic-co-glycolic acid (PLGA) microparticles or squalene-in-water emulsion. Some of these bacterins induced *M. hyopneumoniae*-specific antibody and also T-cell responses in pigs (Matthijs et al., 2019a), and showed promising results in terms of reduction of clinical signs, lung lesions and infection load in the lungs upon experimental challenge infection (Matthijs et al., 2019b).

Attenuated vaccines might have more potential to better stimulate a mucosal immune response at the respiratory tract compared to parenterally administered vaccines, but apart from the two currently commercially available attenuated *M. hyopneumoniae* vaccines, very few studies have focused on testing experimental attenuated *M. hyopneumoniae* vaccines. This might be due to the difficult isolation and cultivation characteristics of *M. hyopneumoniae*. One study (Villarreal et al., 2009) tested whether a low virulent strain could be used as a potential vaccine. However, infection with this strain did not protect piglets against infection with a highly virulent *M. hyopneumoniae* isolate one month later, suggesting that low virulent strains might not be suitable as such to be used as vaccines.

Development of new experimental vaccines and validation of the efficacy of promising experimental vaccines in pigs is needed. From an immunological point of view, a major challenge is to induce a protective immunity at the mucosal level. A comprehensive understanding of the pathobiology of *M. hyopneumoniae* infections and the molecular basis of the pathogenicity of this micro-organism are therefore required. Bacterial genes and antigens involved in survival of the bacterium in the host and/or that render the bacterium harmful to the host need to be identified. They could be included in subunit vaccines or could be targets for attenuation to develop safe attenuated vaccines. From a practical viewpoint, also aspects such as cost of the vaccine, ease of administration, differentiation of antibodies upon vaccination versus infection, and possible combination with other vaccines are important.

CHAPTER 12

ERADICATION OF *MYCOPLASMA HYOPNEUMONIAE* FROM PIG HERDS

Paul Yeske[1]

Sam Holst[1]

Alyssa Betlach[1, 2]

Maria Pieters[2]

1 Swine Vet Center, P.A., St. Peter, USA
2 College of Veterinary Medicine, University of Minnesota, St. Paul, US

12.1 Introduction

Mycoplasma hyopneumoniae (*M. hyopneumoniae*) is one of the most prevalent and economically significant respiratory pathogens in the swine industry (Pieters and Maes, 2019). This bacterium is the etiologic agent of enzootic pneumonia (EP), a chronic respiratory disease in swine characterized by non-productive cough (Goodwin et al., 1965; Mare and Switzer, 1965). Economic losses related to *M. hyopneumoniae* are associated with decreased feed efficiency, reduced average daily gain, and increased medication costs (Maes et al., 2018). Economic data gathered from a North American production system for over a decade suggested that the cost of *M. hyopneumoniae* positive pigs was almost $5 greater than that of *M. hyopneumoniae* negative pigs (Silva et al., 2019). In addition, *M. hyopneumoniae* plays a key role in the development of the porcine respiratory disease complex (PRDC; Dee, 1996), which can add up to even more important economic losses. Thus, these have become the primary reasons for herd owners to consider *M. hyopneumoniae* eradication programs.

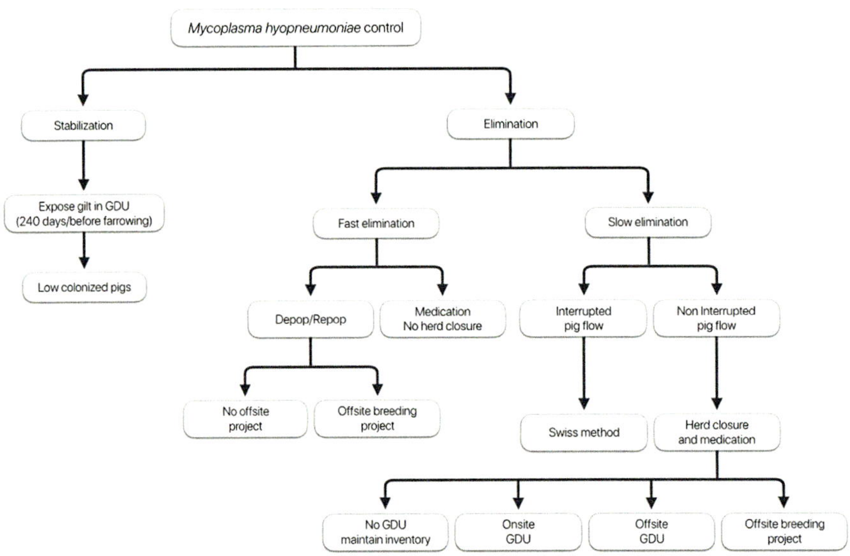

GDU: Gilt Developer Unit. Offsite: Taking place in external facilities. Stabilization: Pathogen under control regardless of positive disease status.

Figure 12.1. Decision tree for *M. hyopneumoniae* disease control options. (Source: P. Yeske and M. Pieters)

One of the best ways to guide the decision to eliminate *M. hyopneumoniae* is to evaluate performance and economic data from herds that are free from the pathogen and from endemically infected herds, in order to demonstrate the value of freedom of disease.

Nevertheless, the decision to eradicate *M. hyopneumoniae* is frequently made based on other activities, such as eradication programs for other pathogens, which may take place at the farm at the same time. The diagram shown in Figure 12.1 can aid the eradication decision-making process.

The ultimate control method for *M. hyopneumoniae* infections is the eradication of the microorganism from a herd. In the North American industry, *M. hyopneumoniae* eradication has been attempted predominately in farrow-to-wean farms. Eradication of *M. hyopneumoniae* is also possible from farrow-to-finish farms, as it has been performed historically in European countries.

Different eradication protocols that have been employed around the world, along with their intrinsic improvements in health, performance, and economics are discussed in the present chapter. As scientific information on *M. hyopneumoniae* eradication is rather scarce, some of the data presented in this chapter is based on clinical expertise and many years of experience, more than on peer-reviewed publications.

12.2 *Mycoplasma hyopneumoniae* eradication protocols

There are multiple methods that have been used to eliminate *M. hyopneumoniae* from a herd. The options for *M. hyopneumoniae* eradication protocols include: A) Depopulation and repopulation, B) Swiss method (partial depopulation), C) Herd closure and medication, D) Whole herd medication without herd closure and E) Other methods for eradication. A graphic representation of the main four methods used for *M. hyopneumoniae* eradication is shown in Figure 12.2.

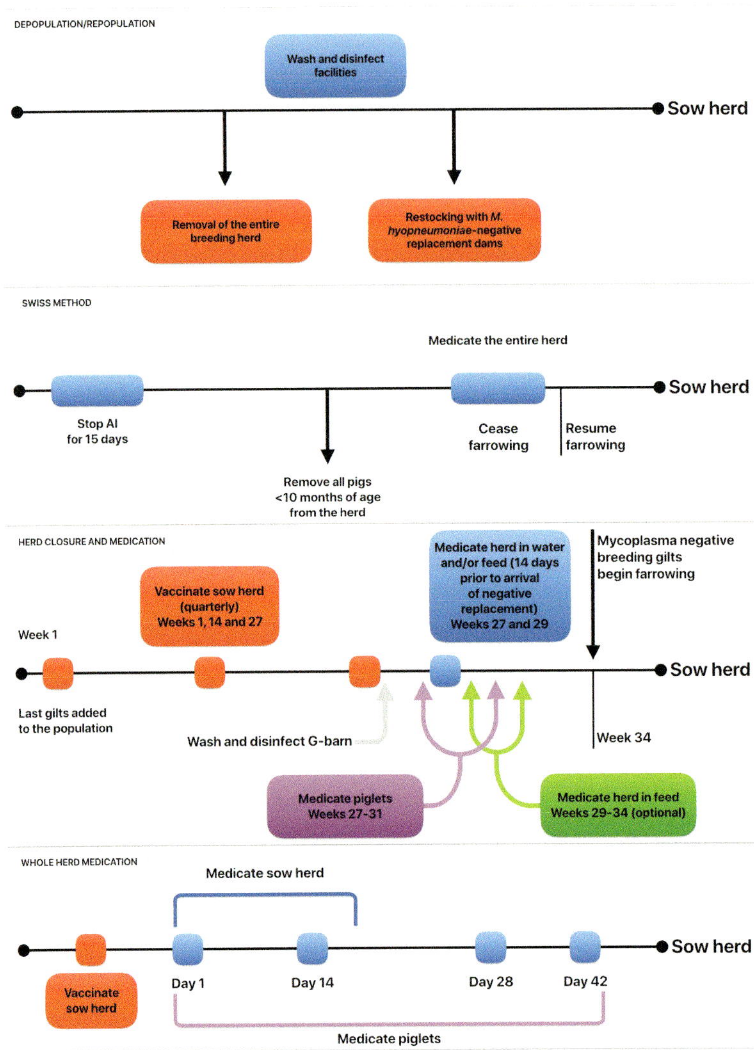

AI: Artificial insemination. G-barn: Gestation barn.

Figure 12.2. Graphical representation of various *M. hyopneumoniae* eradication protocols. Different colors represent different activities in the eradication protocol: Red: Vaccination. Blue: Sow medication. Purple: Piglet medication. Green: Feed medication. (Source: P. Yeske and M. Pieters)

12.2.1 Depopulation/Repopulation

Depopulation followed by repopulation is the most effective and direct approach for *M. hyopneumoniae* eradication as it involves removal of all animals and restocking with *M. hyopneumoniae* negative replacements (Yeske, 2007). The most important advantage of this method is that the overall success rate is extremely high, with virtually no failures. Other advantages of depopulation and repopulation include the ability to eliminate more than one disease at once and the opportunity to improve genetics or other production parameters in the herd, such as a poor parity distribution (Yeske et al., 2007). However, the significant disadvantage is a complete loss of production from the time the breeding herd is liquidated until replacement females begin farrowing. The production loss period can be shortened dramatically (to as little as 4 weeks), if an offsite breeding project is performed in conjunction with the depopulation. An offsite breeding project refers to the temporary use of a different site to house the clean population of pigs to repopulate the herd and breeding the sows, and housing during gestation up to the time dams are due to farrow. Even though there are costs related to the offsite breeding project, they are far outweighed by the cost of down time and lost production. Thus, these projects will greatly reduce the cost of depopulation and repopulation.

Another disadvantage of this method is that the total depopulation of the breeding herd may be undesirable on farms with pigs that have a high genetic potential (*i.e.*, genetic nucleus or multiplier farms).

In summary, depopulation and repopulation is the costliest approach for *M. hyopneumoniae* eradication and takes the longest time for economic break-even. The offsite breeding project can reduce the cost of production loss, but challenges of identifying a site to house the repopulating herd exist, such as assuring sufficient biosecurity to limit infection risks, and having the labor available to execute the program. Based on practical personal experiences, the time to repay the investment of a depopulation and repopulation of a sow herd without an offsite breeding project has been calculated to be 4.4 years, and with an offsite breeding project being 2.5 years. A calculation of the time to repay the investment of a depopulation and repopulation eradication program in a 1000-sow herd is shown in Table 12.1.

Table 12.1. Time to repay investment of depopulation and repopulation on a 1000-sow herd infected with *M. hyopneumoniae*.

	Cost of depopulation/ repopulation *with* offsite breeding project	Cost of depopulation/ repopulation *without* offsite breeding project
Total	$307,355	$533,249
Per Sow	$307.36	$533.25
Per Pig	$11.13	$19.30
Time to return investment	30.5 months	52.8 months

12.2.1.1 Complementary strategies to Depopulation and Repopulation

Specific Pathogen Free (SPF)/Cesarean-Derived pigs

Specific pathogen free derivation of piglets has been used to eliminate many diseases. This is achieved performing a cesarean section, where piglets are removed surgically in a sterile plastic bubble and moved to a clean location to be raised. This strategy is particularly suited for eliminating *M. hyopneumoniae,* and it is based on the premise that piglets are born *M. hyopneumoniae*-negative. This method is very costly due to the expenses associated with cesarean section and raising the pigs without a sow. This method implies a long-term investment based on the long time that is required for all replacements to reach production age. Nevertheless, it is a very effective and highly repeatable means to eliminate infection (Harris and Alexander, 1999). It is an option if genetics of the original herd need to be maintained. Once pigs are old enough and available, the original herd is depopulated and repopulated.

Medicated early weaning

The medicated early weaning method has been used to preserve genetics from the original herd (Harris and Alexander, 1999). The principles of this program include multiple vaccinations of sows pre-farrowing to stimulate the development of immunity, antimicrobial medication of sows pre-farrowing (usually with injectable compounds) and during the lactation, antimicrobial medication of piglets pre-weaning with injectable compounds, and weaning piglets at an early age of usually less than 2 weeks. Once adequate numbers of replacement gilts have been produced with the medicated early weaning, then the original herd is depopulated and repopulated with the new negative gilts.

12.2.2 Swiss method

The "Swiss" method, also known as the partial depopulation method, gained recognition in the 1990s, when Switzerland implemented a national program to eliminate *M. hyopneumoniae* and *Actinobacillus pleuropneumoniae* (Stark et al., 2007). The following items are the framework for the Swiss method (Baekbo, 1999; Bara, 2002; Stark et al., 2007):

1. Remove all animals younger than 10 months of age from the herd,
2. Cease farrowing for at least 2 weeks,
3. Medicate the remaining animals with an antimicrobial agent licensed for *M. hyopneumoniae* during the non-farrowing period.

Moreover, the method includes thorough cleaning and disinfection of all facilities where pigs are housed. This method is particularly well suited to farms that are farrowing in a batch system, since you can simply skip a batch of production and return to normal flow. Eradication projects in Norway, Denmark, and Finland were also successful with slight modifications to the Swiss partial depopulation protocol (Lium, 1992; Baekbo et al., 1994). The premise for this system is the same, in which all pigs are exposed at a young age and once animals are >10 months of age, they are no longer shedding the bacterium and being infectious to other pigs. If piglets are absent in the site during the non-farrowing period, this breaks the shedding-infection cycle. This protocol allows negative replacement gilts to enter the farm and the herd should remain negative.

12.2.3 Herd closure and whole herd medication

Herd closure and medication protocols for *M. hyopneumoniae* eradication are adaptations of the Swiss method. Herd closure refers to the process of stopping the entry of replacement gilts. The herd may be loaded with a high proportion of replacement gilts that can be housed in the facilities and then no additions are made from this point forward until the eradication program is completed. Modifications to the Swiss method include the allowance for farrowing to continue during the medication period, in order to minimize production losses during the eradication. The herd closure and medication approach utilizes the following key principles (Holtz, 2015):

1. Establishing exposure of all females, including replacement gilts, to *M. hyopneumoniae*,
2. Closure of the herd for at least 8 months (*i.e.* 240 days). The start of the herd closure follows the confirmed exposure to *M. hyopneumoniae* (based on clinical and diagnostic testing) to allow for immunity to develop and for dams to stop shedding the organism,
3. Entire herd vaccination (quarterly) with a *M. hyopneumoniae* bacterin to boost herd immunity,
4. Antimicrobial medication of the entire sow herd and piglets prior to introduction of *M. hyopneumoniae*-negative replacement gilts.

For this method to be successful, it is critical that all replacement gilts are exposed to *M. hyopneumoniae* and effectively colonized prior to the beginning of the herd closure (establishment of Day 0; Pieters, 2018). Replacement gilts can be exposed to *M. hyopneumoniae* in a number of ways (see Chapter 9), including natural contact with infected pigs and/or purposeful exposure to infective lung material (Robbins et al., 2019). Herd closure of at least 8 months (240 days) is based on published research indicating that pigs can shed *M. hyopneumoniae* up to 200 days post-infection (Pieters et al., 2009). Whole herd vaccination of all sows and gilts with a *M. hyopneumoniae* bacterin is usually performed to increase herd immunity during the process, and potentially reducing shedding (Woolley et al., 2013; Garza-Moreno et al., 2018). Finally, all sows and piglets on-site are medicated with an approved antimicrobial effective against *M. hyopneumoniae* (Holst et al., 2015). There have been many diverse protocols using different medications successfully, and each plan needs to be adapted to meet the unique conditions of each farm.

The herd closure and medication protocol has been the most widely adopted method of eradication in North America. This protocol can be performed regardless of the *M. hyopneumoniae* health status of replacement gilts, but once eradication is completed, there needs to be a negative source of replacements available. If an internal replacement program is used, there will be an extended time without gilts available, since no gilts can be "saved" during the closure. This is why most farms that go through the closure choose to purchase gilts from a confirmed *M. hyopneumoniae*-negative source, to supplement replacements until internal replacement gilts are available again. Whether positive or negative replacements are used during the elimination process, it is critical that all gilts are exposed to *M. hyopneumoniae* at the start of the closure, in the original sow herd, in order to establish Day 0.

The rationale for stopping new introductions to the herd is to decrease the number of susceptible animals for the pathogen to replicate in (Dee et al., 1995; 1996), eventually reducing the number of susceptible animals to zero (leading to infection die out). The herd remains closed to new animals until enough time has passed for the pathogen to have infected all animals on the farm and infected animals have had time to mount an immune response, clear the pathogen, and no longer be infectious (Pieters and Fano, 2016).

Following successful eradications of porcine reproductive and respiratory syndrome virus (PRRSV) using the herd closure technique, veterinarians in the USA have adapted the strategy for utilization in *M. hyopneumoniae* eradication projects, and very often incorporating both pathogens in the program to gain more benefit from the closure.

To maintain a practical replacement rate and to continue utilizing gilts in weekly or batch breeding groups, a 6 to 9 month supply of gilts (equal to the desired length of the closure period) are stocked in an on-site isolation, gilt developer unit (GDU) or secure off-site location. As previously stated, the recommended length of herd closure for *M. hyopneumoniae* eradication is at least 8 months (240 days from Day 0). Therefore, an 8 month supply of gilts would be required to avoid gaps in production. Generally, there is at least 1 - 2 months of acclimation time to make sure the herd has been properly exposed (set Day 0). Additionally, it is recommended for the entire adult pig population to be over 10 months of age (to be consistent with the Swiss method) at the time that negative replacements are introduced (Yeske, 2007), in order to increase the likelihood that no animals will be actively shedding.

One of the key elements for success of this approach is that gilts have been exposed to *M. hyopneumoniae* before the beginning of the closure. Thus, to satisfy the 10 month age recommendation, gilts should be a minimum of 2 months of age when stocked into isolation or the GDU. Furthermore, gilts of various ages and weights should be included to avoid a surplus of gilts that are too old or too big at the end of the *M. hyopneumoniae* eradication project (Yeske, 2007). The culling routine may need to be modified in the procedure to ensure that adequate herd inventory is maintained in the process. Once the eradication is completed, replacements can enter the herd at an accelerated rate in order to get the population and parity distribution back under control and at the targeted level.

A potential obstacle that may present itself for the application of this eradication method is that not all sow farms have an on-site isolation or GDU facilities. However, the eradication protocol can be accomplished even if there is no on-site isolation or GDU, or if this space is limited, and adequate numbers of gilts cannot be brought into the herd. If this is the case, an off-site breeding project could be considered as an alternative plan to alleviate the problems with losses in production. An off-site breeding project allows for an extended closure time without having to sacrifice inventory and pig flow. Site selection is critical to ensure that the herd is not at additional risk for disease introduction and that there is adequate labor available to accomplish the project. In addition, an off-site breeding project allows for gilts to be bred at a distinct location and then added back to the herd at the time of farrowing (Yeske, 2002). Gilts should be bred in weekly groups, or batches, that match the herd schedule, and breeding should be timed so that the first group of gilts is due to farrow shortly after the herd closure period is completed. This allows breeding and farrowing targets to be met and for pig flow to be maintained as consistently as possible once the herd closure is lifted, with only minimal impact on parity distribution, as well as production numbers (Yeske, 2002).

Additionally, it is recommended to expose the surplus of gilts if there is not good evidence of exposure based on herd diagnostic testing. Recent application of exposure to lung homogenate (Robbins et al., 2019) has been added to the protocol as an effective means to establish herd exposure and Day 0 for the closure (see Chapter 9 for a detailed explanation of gilt exposure and confirmation of *M. hyopneumoniae* infection).

Commercial *M. hyopneumoniae* bacterins are commonly used in swine production worldwide (Haesebrouck et al., 2004). While vaccination does have several advantages in regards to increased performance, it does not prevent *M. hyopneumoniae* colonization (Meyns et al., 2006; Thacker et al., 1998; Pieters et al., 2010). Other studies have shown that *M. hyopneumoniae* vaccination is associated with a reduction in the number of organisms in the respiratory tract (Meyns et al., 2004) as well as a decreased infection level within a herd (Sibila et al., 2007). Based on the previously stated arguments, *M. hyopneumoniae* vaccination is incorporated into eradication programs. Vaccination of sows has been utilized to reduce the pathogen spread from sow to offspring and to confer immunity to piglets via colostrum (Ruiz et al., 2003; Sibila et al., 2008; Bandrick et al., 2008). Sow vaccination during *M. hyopneumoniae* eradication projects is aimed at increasing herd immunity (Yeske, 2007) and has been implemented on a quarterly basis, prior to whole herd antimicrobial medication, or on a pre-farrow schedule. Although many different vaccination protocols can be applied, there are no published studies that would indicate an advantage of utilizing one protocol over another for eradication purposes (Holst et al., 2015). Nevertheless, a recent study has suggested that multiple vaccination in gilts during acclimation can help decreasing bacterial shedding (Garza-Moreno et al., 2019a).

There have been many successful programs with different antimicrobial treatments, which are determined working with the herd veterinarian, while complying with judicious antimicrobial usage. Since there is limited information and antimicrobial sensitivities are not routinely performed for *M. hyopneumoniae*, knowledge of the herd, flow of pigs, and their historical response to antimicrobials has been the best guide on compounds to use.

The following is an example timeline of action items associated with a herd closure and medication eradication protocol for *M. hyopneumoniae* eradication (Schneider, 2006; Yeske, 2007; Table 12.2). It is important to note that normal herd gilt acclimation procedures should be incorporated into the framework of the eradication protocol.

Table 12.2. Timeline of a herd closure and medication protocol for *M. hyopneumoniae* eradication. This example timeline uses the hypothetical start day of January 2 for practicality.

Day	Week	Activity
January 2	1	Herd closure
January 4	1	Gestation barn exposed
February 4	6	Tracheal swabs and blood samples of sows in herd and GDU to confirm exposure
February 10	7	Exposure complete (when confirmed)
February 11	7	Whole herd vaccination (sow unit & GDU)
May 12	20	Whole herd vaccination (sow unit & GDU)
July 7	28	Laryngeal swabs gilts in GDU. Test all pens and keep track of pens for retest if needed
August 4	32	Wash gestation barns
August 18	34	Whole sow herd medication (antibiotic in water)
August 18	34	Piglet treatment all piglets on site (larger pigs will need to adjust dose by weight)
August 19	34	Piglet treatment at birth
September 1	36	Stop whole herd medication in water. Begin antibiotic in feed during lactation and gestation
September 1	36	Begin piglet treatment at 14 days of age (treat all piglets 14 days of age and older)
September 15	38	Stop antibiotic in gestation and lactation
September 15	38	Last piglets treated at birth
September 29	40	Last piglet treatment at 14 days of age
October 8	41	Elimination completed
October 9	41	Negative gilts enter breeding herd (replacement gilts need to be confirmed free of pathogen)
October 9	41	Date pigs are expected negative

GDU: Gilt development unit.

12

12.2.4 Whole herd medication without closure

The whole herd medication without herd closure protocol is the most recent *M. hyopneumoniae* eradication strategy used in the field (Yeske, 2008; Yeske, 2010; Holst et al., 2015). This protocol involves medicating the entire herd (gilts, sows, boars, and piglets) with a long acting antimicrobial (typically administered via injection) with activity against *M. hyopneumoniae*. An example timeline for *M. hyopneumoniae* elimination using whole herd medication without herd closure is shown in Table 12.3. Antimicrobial sensitivity testing is not routinely performed (due to the difficulty of isolating the bacterium) and because of this, antibiotics are generally selected based on clinical response in the herd. Vaccination is used in this method, just as in the herd closure method, for a reduction in the number of organisms in the respiratory tract (Woolley et al., 2012), as well as a decreased infection pressure within the herd (Sibila et al., 2007). Vaccination of sows has been utilized in an attempt to reduce vertical spread of the pathogen and to confer immunity to piglets (Ruiz et al., 2003; Sibila et al., 2008; Bandrick et al., 2008). Herds have been administered a booster dose of vaccine prior to antibiotic treatment. The entire herd is treated via antimicrobial injection on day 1 of the eradication project, followed by another treatment 2 weeks later.

Table 12.3. Timeline of a whole herd medication protocol without herd closure for eradication of *M. hyopneumoniae*.

Day	Week	Activity
February 4	6	Whole herd vaccination (sow unit & GDU)
February 11	7	Wash gestation barn
February 16	7	Measure sows for whole herd injection with antibiotic
February 16	7	1st whole herd injection with antibiotic (all pigs on site sows, boars, gilts, growing gilts & suckling piglets)
February 17	8	Start injecting piglets day 1 with antibiotic
March 2	9	2nd whole herd injection with antibiotic (all pigs on site sows, boars, gilts, growing gilts & suckling piglets)
March 3	10	Start injecting piglets day 14 with antibiotic
March 16	11	Stop day 1 injections to piglets with antibiotics
March 30	13	Stop day 14 injections to piglets with antibiotics
April 6	14	Whole injection elimination completed
April 7	15	Negative gilts enter breeding herd
April 7	15	Date pigs are expected negative

GDU: Gilt development unit

Additionally, piglets born up to 4 weeks following the initial whole herd injection are treated at birth and at 14 days of age with an anti-*Mycoplasma* antibiotic. Replacement gilt flow is maintained per normal farm protocol (however in many cases one gilt shipment has been skipped to facilitate the process) while the farm remains open to new animal introductions. The advantage of whole herd medication without herd closure is that the herd has a faster return to *M. hyopneumoniae* negative status when successful. However, this protocol has been less effective at eliminating *M. hyopneumoniae* than the herd closure and medication protocol (Yeske et al., 2020). It has the highest risk of not being successful in eliminating the *M. hyopneumoniae* from the herd, but the fastest return to negative pigs.

12.2.5 Other protocols

Simple herd closure (stopping the entry of purchased replacement animals and/or stopping the rearing of self-replacement gilts) with no medication or vaccination, and using an extended closure period (over 300 days), has been successfully applied for *M. hyopneumoniae* eradication. In most cases, this type of program is unexpectedly occurring after some other event has caused the herd closure to be extended. Examples of these cases are prolonged PRRSV closures (herd not turning negative to PRRSV at the expected time and the herd remains closed until negativity is achieved). Post-closure diagnostics have shown that negativity to *M. hyopneumoniae* has been also obtained.

There have been hybrid protocols for *M. hyopneumoniae* eradication of short herd closure with medication, using injection of all animals on the herd at the end of the program (Yeske, field experience). In addition, short closure times similar to the whole herd medication without closure program, but at 150 days or as long as the farm could operate without gilt introduction. This later combined program appears to have a greater success rate than whole herd medication with no herd closure protocols.

Without a doubt, there are other methods that have not been mentioned in this chapter but have been employed for *M. hyopneumoniae* eradication. In the future, there are likely to be new protocols developed, as more and improved technologies are designed and applied at herd level. New strategies might allow to facilitate procedures, shorten herds closures or reduce the use and cost of treatments.

12.3 The value of *Mycoplasma hyopneumoniae* eradication

Why do farms undergo *M. hyopneumoniae* eradication? The answer to that question may have been best summarized by Dr. Carlos Pijoan's pragmatic words: "The benefits of eradication are obvious and justify all of our continued efforts in this arena" (Pijoan, 2001). Generally, *M. hyopneumoniae* eradication methods are employed because of the relatively quick return of the cost of the investment and the long-term internal rate of return. Other factors for eradicating *M. hyopneumoniae* include decreased exacerbation of secondary pathogens in PRDC (such as PRRSV and Influenza A virus (IAV-S)), improved animal welfare, increased production, decreased cost of production, reduction in preventative or therapeutic antimicrobial usage, and increased caretaker morale (Pijoan, 2001). These are all convincing arguments for performing eradication. One continued question in segregated production systems, that has been recently explored is: What is the likelihood (even after successful *M. hyopneumoniae* eradication in the sow herd, but growing pigs in pig-dense areas) that growing pigs will become infected once in finishing and the eradication benefit will be lost? In a field study, Yeske et al. (2017b) followed 200 groups (sites) of conventional *M. hyopneumoniae* negative pigs placed in pig dense areas (southern Minnesota and northern Iowa) and monitored them at the end of the finishing period, prior to marketing, to evaluate their *Mycoplasma*-free status. Under the conditions of the investigation, only 6% of *M. hyopneumoniae*-free groups became laterally infected as evaluated during summer/fall and winter/spring seasons. Those results suggest that the likelihood of lateral *M. hyopneumoniae* infections occurring in single batches of finishing pigs was a relatively rare event (Yeske et al., 2017b). However, it remains to be investigated whether these results are similar in regions with a high pig density, as well as a high pig herd density, with more single site farrow-to-finish pig herds, and with different meteorological conditions.

Swine producers and veterinarians have had success in eradicating *M. hyopneumoniae* using the protocols, principles, and techniques described in this paper. However, comprehensive comparisons of the success of the different protocols for *M. hyopneumoniae* eradication are lacking in the literature. A comparison of the success of the herd closure with medication and whole herd medication with no herd closure methods, over time, from farms that have tracked their status (the longest herds for over 16 years and a minimum of one year) is shown in Table 12.4. Herd statuses are determined by different means, from observation of clinical signs and lung lesions in mortality, to the testing of seroconversion and detection of *M. hyopneumoniae* genetic material in sentinel animals introduced into the herd (or sentinel pigs, downstream at the end of finishing phase), to having negative wean to finish production, being able to capture the benefits of eradication.

Table 12.4. Comparing the success of two *M. hyopneumoniae* eradication protocols (Yeske, 2018).

	Herd closure	No herd closure
Number of herds	52	20
Number of sows	145038	47450
Percent (%) negative to date	76	53
Number of negative herds	45	10
Average herd size	2789	2373
Average months negative	58	52
Years negative	4.9	4.4

12.4 Eradication economics

M. hyopneumoniae is a significant cause of economic loss to swine producers. In 2012, Haden et al. quantified the economic impact of IAV, PRRSV, and *M. hyopneumoniae* on a large US production system over a 4 year period. The cost of uncomplicated *M. hyopneumoniae* was determined to be \$0.63 per head placed in grow to finish. Unfortunately, respiratory disease in growing pigs is usually not limited to one uncomplicated pathogen, but is rather a mixed infection of *M. hyopneumoniae,* viruses, and other bacteria (Thacker et al., 2001; Holtkamp et al., 2007). Haden et al. (2012) also calculated the cost per head in situations in which *M. hyopneumoniae* was complicated by co-infection with PRRSV or IAV. The combination of PRRSV and *M. hyopneumoniae* resulted in a loss of \$9.69 per head, while co-infections of IAV and *M. hyopneumoniae* inflicted a loss of \$10.12 per head. The loss incurred due to the combination of IAV and *M. hyopneumoniae* was only surpassed by IAV and PRRSV co-infections (\$10.41 per head). While these loss calculations are specific to one particular production system, similar losses are likely incurred by other swine producers as well. A 2005 and 2006 survey completed by Holtkamp et al. (2007) attempted to estimate the impact of major swine health challenges in the USA. The participants in the survey comprised companies that produced approximately 50% of the total number of pigs marketed annually in the USA. The results of the survey indicated that PRRSV, IAV, and *M. hyopneumoniae* were the top three health challenges experienced in finishing herds (Holtkamp et al., 2007). Although recent data regarding this topic is not available, it is suggested that a similar trend is observed nowadays.

An economic model has been developed using performance data of a production system involving herds that were both *M. hyopneumoniae* positive and negative. Performance was tracked by health status of the wean to finish flow from 2007-2016 (Silva et al., 2019). Results from this model are shown in Table 12.5.

Table 12.5. Effects on production of *M. hyopneumoniae* infection in a production system (Silva et al., 2019).

Production parameter	Value in *M. hyopneumoniae* positive pigs	Value in *M. hyopneumoniae* negative pigs	Difference
Average daily gain (kg)	0.818	0.854	0.036
Feed conversion rate (feed:gain)	2.840	2.820	-0.018
Mortality (%)	5.37	4.11	-1.26

The cost of eradication (per sow) using the herd closure method was $22.14 and using the whole herd medication and no herd closure method was $37.14. Results showed a payback period of 3.1 months for the herd closure protocol and 11.4 months for the whole herd medication program (Table 12.6; Silva et al., 2019). Both methods showed less than one year to return the investment, based on data from large sow farms in North American production systems. A comprehensive review of the potential factors affecting the economic model are available in Silva et al. (2019).

Table 12.6. Costs and returns of two *M. hyopneumoniae* eradication programs (Silva et al., 2019).

	Herd closure & medication (cost per sow)	No herd closure & medication (cost per sow)
Investment	$22.14	$37.14
Net benefit per year	$124.71	$124.71
Probability of success	86%	58%
Projected value after 1 year	$85.11	$39.24
Months to payback cost	3.12	11.36

12.5 *Mycoplasma hyopneumoniae* eradication trends

Official figures of herds undergoing *M. hyopneumoniae* eradication are not available, as efforts are not required to be reported, except in Switzerland, where EP is a reportable disease. Annual confirmed cases of EP in Switzerland are available on the Swiss Federal Food Safety and Veterinary Office website (https://www.infosm.blv.admin.ch/public/?lang=en).

Historically, *M. hyopneumoniae* eradications were initially applied in European countries, with Switzerland achieving eradication, or a very low prevalence, at the national level (Stark et al., 2007; Overesch and Kuhnert, 2017). Other countries, such as Finland, have reported to have accomplished a similar goal (Rautianen et al., 2001). Renewed efforts for *M. hyopneumoniae* eradication have been reported in North America, where eradication strategies and technologies have been adapted to large herds using multisite production (Holst et al., 2015). The combination of available technologies and improved production costs and benefits (Schwartz et al., 2015) for *M. hyopneumoniae*-free herds have generated data that has been shared in the swine community, aiding the decision making process of veterinarians and producers. Whether total *M. hyopneumoniae* eradication or a very low level of prevalence is achieved, the success rate of the protocols and the financial benefit outweigh doubts and have generally led the industry to attempt eradication, not only in single farms, but in entire swine production systems (Schwartz et al., 2018). Numerous successes for *M. hyopneumoniae* eradication have made the practice a frequent one in North America, as producers realize the economic results of eradication (Yeske et al., 2020).

12

CHAPTER 13

MYCOPLASMA HYORHINIS AND *MYCOPLASMA HYOSYNOVIAE* IN PIG HERDS

Andreas Palzer[1]

Mathias Ritzmann[2]

Joachim Spergser[3]

1 Veterinary Pig Practice, Scheidegg, Germany
2 Faculty of Veterinary Medicine, Ludwig-Maximilians-Universität München, Germany
3 University of Veterinary Medicine, Vienna, Austria

13.1 *Mycoplasma hyorhinis*

Mycoplasma hyorhinis (*M. hyorhinis*) is ubiquitous in the swine population worldwide. Although frequently present as commensal in the upper respiratory tract, it can cause polyserositis, arthritis, abortion, otitis and conjunctivitis. Furthermore, it may be involved in the porcine respiratory disease complex (PRDC).

13.1.1 Etiology

M. hyorhinis is a frequent inhabitant of the porcine upper respiratory tract (Friis and Feenstra, 1994; Kobisch and Friis, 1996) and the first *Mycoplasma* species ever described in swine (Switzer, 1955). *M. hyorhinis* is also a common contaminant of cell and tissue cultures (Tang et al., 2000) and persistent exposure of cells has been associated with malignant transformation of human prostate cells (Namiki et al., 2009).

Like all mollicutes, *M. hyorhinis* is small, pleomorphic and lacks a cell wall. It ferments glucose but does not hydrolyze arginine or urea (Kobisch and Friis, 1996). A broad range of carbohydrates is metabolized by this microorganism, possibly reflecting its ability to aggressively grow in diverse host sites and to cross host species barriers (Siqueira et al., 2013; Ferrarini et al., 2016). Also, *M. hyorhinis* grows well in and on Friis medium (Kobisch and Friis, 1996; Cook et al., 2016) which provides all nutrients required for its *in vitro* growth, such as amino acids, nucleic acid precursors, cholesterol and serum. Based on glucose fermentation, an acidification of the culture medium is observed. Colonies of *M. hyorhinis* develop within 3 - 5 days on agar plates, exhibit fried egg morphology and are variable in size (0.2 - 1mm in diameter) and in opacity (Figure 13.1).

To date, complete genome sequences of three porcine *M. hyorhinis* stains including the type strain BTS-7T (LS991950), the arthritogenic strain SK76 (CP003914; Goodison et al., 2013), and strain HUB-1 isolated from the respiratory tract (CP002170; Liu et al., 2010) have been published and/or deposited at the NCBI genome database (www.ncbi.nlm.nih.gov/genome/genomes/1922, on September 2019). Each genome of the three porcine strains is composed of a single circular chromosome and genome sizes range from 837 to 840 kb, which is on the small side for mycoplasmas, with an overall G+C content of 25.9%.

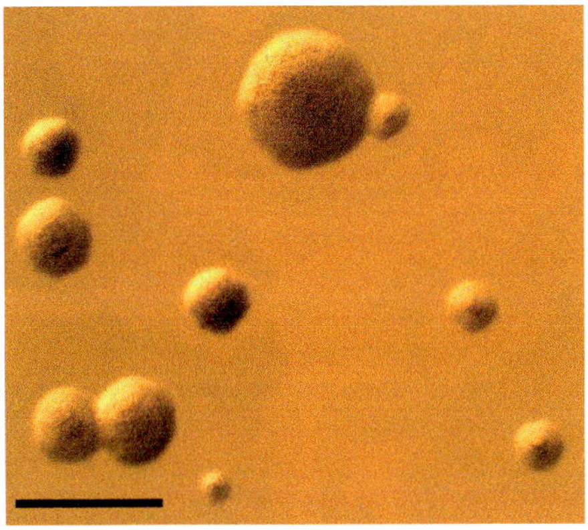

Figure 13.1. Colonies of *M. hyorhinis* BTS-7T on modified Hayflick's agar after 5 days of incubation exhibiting fried egg morphology and variability in size and in opacity. Bar, 1 mm. (Source: J. Spergser)

The virulence factors of *M. hyorhinis* and mechanisms of pathogenicity are still largely unknown. However, *M. hyorhinis* typically possesses a family of up to seven genes encoding variable surface lipoproteins (VLPs) that undergo high-frequency phase and size variation (Rosengarten and Wise, 1991). While phase-variable expression of each Vlp is governed by mutations in a characteristic promoter region (poly(A) tract), size variation results from insertion/deletion of tandemly repeated intragenic sequences within the 3' region of individual *vlp* genes (Yogev et al., 1995; Citti et al., 2000). Altering the antigenic make-up of surface associated VLPs may enhance colonization and adherence (Xiong et al., 2016a, b) and facilitate evasion of the host's immune response (Citti et al., 1997). Recently, the capability of *M. hyorhinis* to produce cytotoxic hydrogen peroxide *via* glycerol conversion to dihydroxyacetone-phosphate has been shown *in silico* and *in vitro* (Ferrarini et al., 2016).

Phylogenetically, *M. hyorhinis* is located in the *Neurolyticum-Hyopneumoniae* clade and thus closely related to the two porcine *Mycoplasma* species, *M. hyopneumoniae* and *M. flocculare* (Volokhov et al., 2012; Siqueira et al., 2013; Figure 13.2).

13

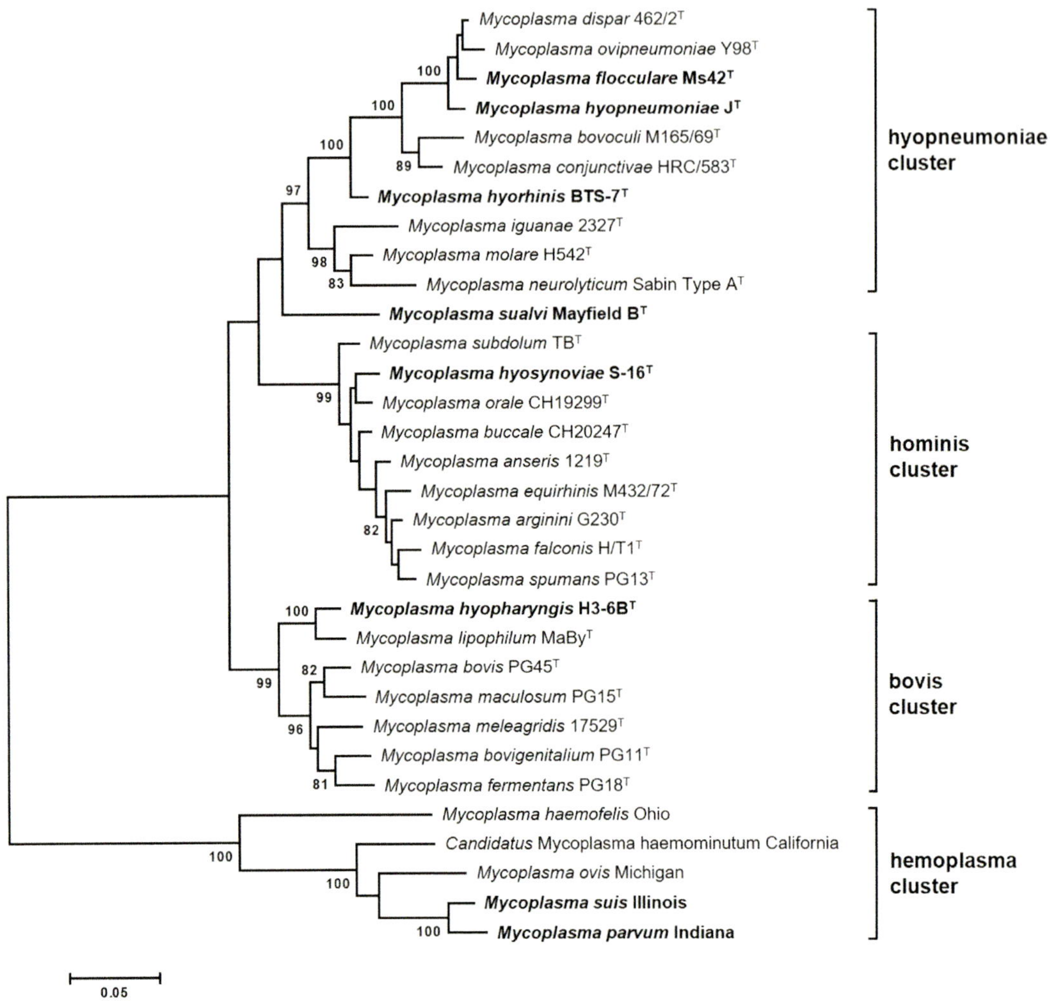

Figure 13.2. Phylogenetic tree based on 16S rRNA gene sequences of porcine mycoplasmas (in bold) and closely related species. Phylogenetic analysis was performed using the maximum likelihood method based on Kimura two-parameter model. Numbers at nodes are bootstrap confidence values (1000 replications, only values greater than 75% are shown). Bar, 0.05 substitution per nucleotide position. (Source: J. Spergser)

13.1.2 Epidemiology

Piglets are colonized *via* their mother, sows or older pigs in the herd. Different studies revealed that *M. hyorhinis* could often be detected in older animals like fattening pigs (Clavijo et al., 2017; Luehrs et al., 2017; Roos et al., 2019). The number of positive pigs differ between countries and herds. In Switzerland, 10% of lungs of animals with pneumonia were tested positive, while in Germany the prevalence was 18.5% (Luehrs et al., 2017). In the United States, *M. hyorhinis* has been detected in a variable number of piglets and dams. The prevalence in suckling piglets was low (8%) while up to 98% of animals post-weaning were tested positive by PCR (Clavijo et al., 2017). Roos et al. (2019) reported high detection rates in three-week-old piglets, in comparison to week one post-farrowing. These findings are comparable to those of other studies showing that positive results in sows may indicate a higher infection dynamic during the fattening period (Palzer et al., 2015c). In contrast, genotyping methods such as multilocus sequence tying (MLST) or multilocus variable number tandem repeats analysis (MLVA), which enable differentiating strains, showed that similar genotypes can be detected in different farms (Tocqueville et al., 2014; Dos Santos et al., 2015b), and that different genotypes can be identified within a farm (Trüeb et al., 2014). Rosales et al. (2008) investigated 40 British and Spanish field isolates by molecular typing methods revealing that they were highly diverse. An association between isolates and geographic origin could not be demonstrated (Rosales et al., 2008; Dos Santos et al., 2015b).

13.1.3 Pathogenesis

M. hyorhinis has frequently been isolated from nasal discharges or bronchoalveolar lavage (BAL; Schulman et al., 1970; Palzer et al., 2007b). The microorganism adheres to the ciliary epithelium of the upper and lower respiratory tract of pigs. Mechanisms leading to systemic spread resulting in polyserositis and arthritis are yet unknown. It can be assumed, however, that other pathogens (Friis and Feenstra, 1994; Palzer et al., 2008) and stress facilitate systemic dissemination of *M. hyorhinis*. Variable expression of certain surface lipoproteins (VLPs; Citti et al., 2000; Rosengarten et al., 2000;Yogey et al., 1995) may contribute to escape host immune recognition. After infection, pigs show a detectable immune response (Gomes-Neto et al., 2014). Simultaneous infections with *M. hyorhinis* and *Glaesserella (Haemophilus) parasuis* (*G. parasuis*) may occur since both bacteria have frequently been detected in field samples, such as swabs from the serosal surfaces like pericardium, pleura and peritoneum (Palzer et al., 2015b). It has not yet been determined whether these two pathogens promote each other's virulence or whether concurrent detection is merely based on their similar route of infection (Palzer et al., 2015b). Dual PCR detection of *M. hyorhinis* and *M. hyosynoviae* in tonsils of sows prior to weaning has been reported (Roos et al., 2019). However, an apparent colonization relationship between these two microorganisms has not been established.

13.1.4 Clinical-pathological presentation

In 1975, Friis already observed that *M. hyorhinis* could cause serofibrinous inflammation of serous body cavities and joints of piglets under 10 weeks of age. Approximately 20 years later, Friis and Feenstra (1994) induced the typical serofibrinous inflammation of serous body cavities in young piglets by experimental infection with *M. hyorhinis*. An earlier study had shown experimentally induced pleurisy and pleuropneumonia by inoculating gnotobiotic pigs with *M. hyorhinis* (Poland et al., 1971). Recently, the disease was reproduced by experimental infections of seven-week-old cesarean-derived, colostrum-deprived (CDCD) piglets (Martinson et al., 2018a) and Specific-Pathogen-Free (SPF) piglets at 7 weeks of age (Fourour et al., 2019). The infection can be induced intravenously, intraperitoneally, intratracheally or intranasally, though the intravenous inoculation caused the most severe infection (Martinson et al., 2018a). Fibrinous polyserositis caused by *M. hyorhinis* is one of the most important differential diagnosis of Glässer's disease due to the very similar, hardly distinguishable clinical and pathological-anatomical presentation.

Table 13.1 Clinical-pathological presentation of *M. hyorhinis*, study and type of study.

Clinical-pathological presentation	Study	Type of study
Polyserositis	Roberts et al., 1963 Martinson et al., 2018 Giménez-Lirola et al., 2019	Experimental Experimental Experimental
Arthritis	Martinson et al., 2018 Giménez-Lirola et al., 2019	Experimental Experimental
Eustachitis	Morita et al., 1999	Experimental
Otitis	Morita et al., 1998	Experimental
Conjunctivitis	Resende et al., 2019	Case report
Meningoencephalitis	Friis, 1975	Case report
Pneumonia	Lin et al., 2006 Palzer et al., 2007b Pereira et al., 2017	Experimental Case report Case report

The clinical presentation is clearly age-dependent, as older piglets are less susceptible to infection when challenged (Martinson et al., 2017). These findings are comparable to those observed in the field. Furthermore, *M. hyorhinis* has been shown to cause arthritis (Barden and Decker, 1971; Ennis et al., 1971), a clinical and pathological presentation that has been reproduced experimentally (Jannson et al., 1983; Martinson et al., 2018b;

Giménez-Lirola et al., 2019). However, a singular study indicated that *M. hyorhinis*-associated arthritis did not occur very often (Hariharan et al., 1992). In addition to polyserositis and arthritis, *M. hyorhinis* has been described as a cause of abortion, otitis or conjunctivitis (Shin et al., 2003; Morita et al., 1995; Resende et al., 2019). The question remains whether *M. hyorhinis* is involved in the development of pneumonia. Thus far, *M. hyorhinis* has been considered as secondary pathogen in enzootic pneumonia (EP) that may manifest only in previously damaged lung tissue (Zielinski and Ross 1993). In literature, studies conjecturing its causative involvement are available (Friis, 1971b; Gois et al., 1971; Lin et al., 2006; Palzer et al., 2007b; Fourour et al., 2018). Nevertheless, investigations questioning its involvement have been reported as well (Luehrs et al., 2017). In a study by Falk et al. (1991), *M. hyorhinis* was detected in 37% of pneumonic lesions in slaughter lungs and was associated with pleurisy. Palzer et al. (2007a) detected *M. hyorhinis* significantly more often in the bronchoalveolar lavage (BAL) fluid of pneumonic animals by PCR than in BAL fluid of animals without clinical signs. In other studies, the microorganism has been identified as the sole causative agent of EP in several cases (Lin et al., 2006; Pereira et al., 2017), and its relevance to the PRDC has been described (Lee et al., 2016). Furthermore, Lee et al. (2018) were able to reproduce pneumonic lesions in the lung and trachea of pigs by an experimental infection with *M. hyorhinis.*

13.1.4.1 Clinical signs

Most pigs colonized with *M. hyorhinis* do not show clinical signs (Rapp-Gabrielson et al., 2006). However, the microorganism can induce polyserositis, arthritis, eustachitis, otitis, and conjunctivitis with the corresponding clinical signs (Jannson et al., 1983; Morita et al., 1998; Thacker, 2006; Martinson et al., 2018b; Giménez-Lirola et al., 2019; Resende et al., 2019). *M. hyorhinis*-associated polyserositis (pericarditis, pleuritic or peritonitis) generally occurs in three to ten-week-old pigs, and occasionally also in older animals (Heinritzi, 2006). After experimental infections, pigs display clinical signs like rough hair coat, loss of condition and mild front and hind limb lameness (Martinson et al., 2018b; Giménez-Lirola et al., 2019). Younger pigs can waste after infection and the daily weight gain can be reduced (Martinson et al., 2018). Mortality in experimental infections was not very high (Martinson et al., 2018b; Giménez-Lirola et al., 2019). The induced polyserositis typically presents itself 3 to 10 days after exposure to the bacterium or after the induction of a systemic reaction due to factors diminishing body resistance or due to a previous EP (Heinritzi, 2006). Pigs exhibit an elevated body temperature (up to 40.5°C), apathy, loss of appetite, lameness, swollen joints and labored breathing (Heinritzi, 2006; Thacker and Minion, 2012). Clinical signs usually subside after two weeks. However, some animals do not show any improvement and suddenly perish. In case of arthritis, lameness and swollen joints will persist for 2 to 3 months, and sometimes up to 6 months. The joints are usually bulging with synovia. Conjunctivitis can be observed as well in connection with infection with *M. hyorhinis* (Friis, 1976; Resende et al., 2019).

Meningoencephalitis only rarely develops in the case of *M. hyorhinis* infection (Friis, 1975; Friis, 1976; Heinritzi, 2006). Morita et al. (1998) achieved to induce an otitis media in SPF piglets by experimentally inoculating *M. hyorhinis* in the tympanic cavity. Nevertheless, lesions were only temporary and subsided 25 days after inoculation. Intranasal inoculation with *M. hyorhinis* can result in eustachitis, but only seldom in otitis media (Morita et al., 1999). In case of otitis media, pigs typically show tilting of the head and in special cases in which both ears are affected, clinical signs resembling those of meningitis, *i.e.* the animals lie on the side and are unable to get up. *M. hyorhinis* can be detected significantly more often in pigs with pneumonia, which indicates its importance as a potential pathogen triggering pneumonia (Palzer et al., 2007b). However, if *M. hyorhinis* leads to clinical signs, arthritis, polyserositis, eustachitis and otitis are possible, in addition to pneumonia (Thacker and Minion, 2012).

13.1.4.2 Pathological changes

The polyserositis in the acute stage of the disease is characterized by fibrinopurulent pericarditis, pleurisy and sometimes also peritonitis (Roberts et al., 1963; Poland et al., 1971; Martinson et al., 2018b; Giménez-Lirola et al., 2019; Figures 13.3, 13.4, 13.5). Due to synovitis, the synovial membrane is swollen and hyperemic. After some days, this inflammation can lead to adhesions. Capsular fibrosis can develop and erosions on articular surfaces may occur (Thacker and Minion, 2012). Even though the involvement of *M. hyorhinis* in the PRDC is debated, some investigations indicate that *M. hyorhinis* can cause typical lesions otherwise known to be caused by infection with *M. hyopneumoniae* (Gois et al., 1968; Gois et al., 1971; Poland et al., 1971; Lin et al., 2006; Pereira et al., 2017). Nevertheless, it is very difficult to investigate the involvement of *M. hyorhinis* in cases of pneumonia as the microorganism can be detected in the lungs of clinical healthy pigs, without causing any disease (Palzer et al., 2007b), and as normally many different viruses and bacteria are involved in cases of pneumonia in pigs (Palzer et al, 2007b).

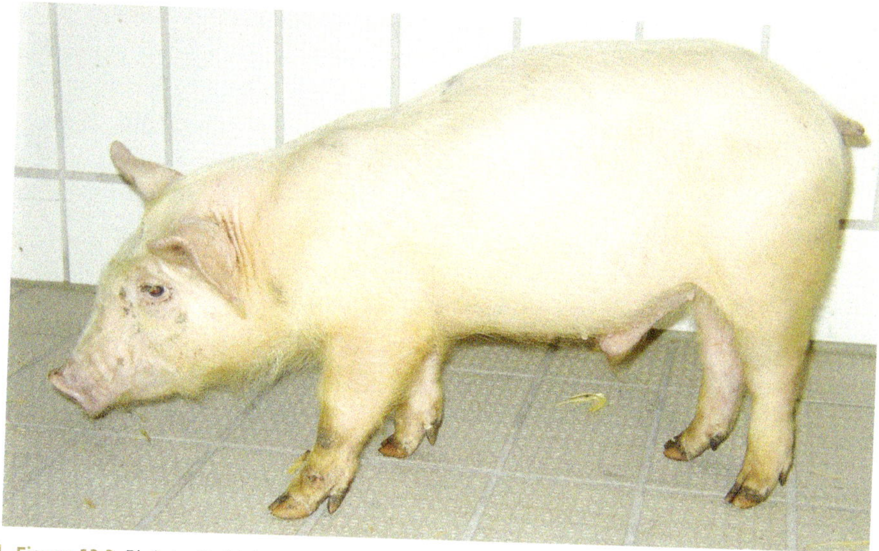

Figure 13.3. Piglet with *M. hyorhinis* infection and diagnosed polyserositis. (Source: J. Spergser)

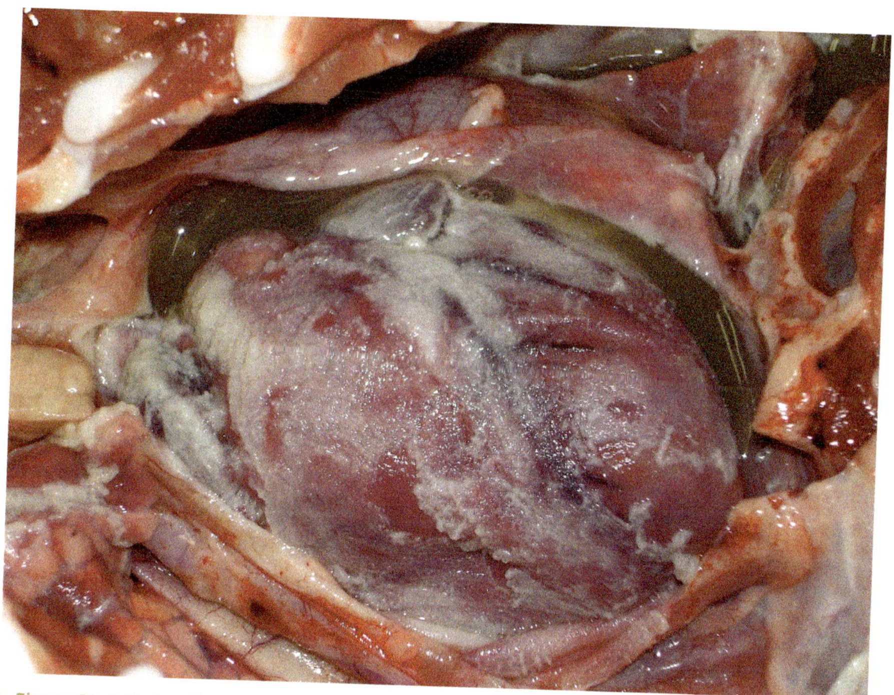

Figure 13.4. Pericarditis and pleurisy in *M. hyorhinis* infection. (Source: A. Palzer)

13

In general, bacterial culture and isolation of *M. hyosynoviae* from affected joints is challenging since the agent has shown to be often present at concentrations below the detection limit, or even absent during subacute and chronic phases of infection. Nevertheless, isolation of *M. hyosynoviae* may be achieved by cultivating ten-fold dilutions of joint samples collected from acutely infected pigs for approximately one week in and on modified Hayflick's medium supplemented with porcine mucin and contamination-inhibiting antibiotics (Kobisch and Friis, 1996). The same cultivation procedure is recommended for the isolation of *M. hyosynoviae* from tonsil scrapings. Growth of *M. hyosynoviae* is indicated by a color change of the phenol red indicator in the medium (alkaline pH resulting from arginine hydrolysis) and colonies produced on agar plates that uniformly exhibit characteristic fried-egg morphology (Figure 13.3). However, further tests that allow definite species identification are required including antigenic or genetic identification (Johansson et al., 1998; Poveda and Nicholas, 1998; McAuliffe et al., 2005; Stakenborg et al., 2006a; Nathues et al., 2011; Volokhov et al., 2012, Gomes-Neto et al., 2012), or MALDI-ToF mass spectrometry (Spergser et al., 2019).

For direct detection of *M. hyosynoviae* in clinical samples, real-time PCR has shown to be the most sensitive assay available (Gomes-Neto et al., 2012; Gomes-Neto et al., 2015; Roos et al., 2019; Pillman et al., 2019). However, molecular diagnosis of *M. hyosynoviae*-associated arthritis can be influenced by the strain involved since duration of joint infection has shown to differ between strains. In consequence, varying molecular detection results can be partially attributed to the variability in timing of the necropsy relative to the onset of clinical disease, and in addition, to the low or absent bacterial load in joints at necropsy (Gomes-Neto et al., 2016). Complete diagnosis of *M. hyosynoviae*-associated arthritis requires the presence of characteristic histologic lesions such as hyperplasia of the synovial membrane, synovial villous hypertrophy, edema, mononuclear cell infiltration in the synovial sub-intima, and mild fibrosis of the capsule (Gomes-Neto et al., 2012; Macedo et al., 2019). Serodiagnosis of *M. hyosynoviae* infection and carriership has been used in field and experimental studies, some revealing promising test results (Zimmerman and Ross, 1982; Nielsen et al., 2005; Gomes-Neto et al., 2014). However, validated immunoassays are currently not available for routine diagnostic purposes.

13.2.6 Therapy and control

M. hyosynoviae isolates have been shown to be sensitive to tetracyclin, tiamulin and enrofloxacin (Nielsen et al., 2001). An antimicrobial therapy commonly applied was lincomycin *via* the feed or water (Schmitt, 2014). There is a range of sensitivity, with some isolates being overall more resistant while others appear more susceptible (Schultz et al., 2012). Recent studies have shown susceptibility of *M. hyosynoviae* to tiamulin, valnemulin, gentamicin, enrofloxacin, and danofloxacin (Gautier-Bouchardon, 2018).

No commercial vaccine for the prevention of disease caused by *M. hyosynoviae* has been developed. Manufacturers and producers have used autogenous vaccines, but field trial data are very limited (Schmitt, 2014). In one field trial, an autogenous *M. hyosynoviae* vaccine could not successfully prevent the development of lameness in finishing pigs presumably caused by *M. hyosynoviae*-associated arthritis (Lauritsen et al., 2013).

CHAPTER 14

MYCOPLASMA SUIS INFECTIONS IN PIGS

Ludwig E. Hoelzle[1]

Katharina Hoelzle[1]

1 Department behavioral physiology of livestock, University of Hohenheim, Stuttgart, Germany

14.1 Pathogen history

In 1931, a novel disease in eight-week old pigs was recognized in the Midwest of the United States of America (USA). Clinical signs of this disease comprised icterus, anemia, and hemorrhages in the lungs, heart, and kidney, a swollen dark spleen as well as a yellowish-discolored enlarged liver. *Anaplasma*-like particles were observed in Giemsa-stained blood smears of the diseased pigs (Kinsley, 1932; Kinsley and Ray, 1934). Doyle (1932) confirmed these observations identifying *Anaplasma*-like bodies on red blood cells as well as floating free in the blood plasma of pigs suffering from anemia and icterus. In the following years, the clinical significance of this anaplasmosis-like disease was proved in several pig herds (Campbell, 1945; Dicke, 1934; Quin, 1938; Robb, 1943; Spencer, 1940).

Due to its morphological similarity with *Eperythrozoon wenyonii* and *Eperythrozoon ovis*, Splitter initially named the disease swine icteroanemia (Splitter, 1950; Splitter and Williamson, 1950). For a long time, the genus agent *Eperythrozoon suis* was classified within the family *Anaplasmataceae*, order *Rickettsiales*. Phylogenetic analyses of the 16S rRNA genes of members of the genus *Eperythrozoon* led to the reclassification of *Eperythrozoon suis* as *Mycoplasma suis* (*M. suis*) within the class *Mollicutes*, order *Mycoplasmatales* family *Mycoplasmataceae*, and the genus *Mycoplasma* (Neimark et al., 2001, 2002). However, the unique biological characteristics of the former members of the genus *Eperythrozoon* (mainly the erythrocytes tropism), in association with the lack of an *in vitro* culture system and the insufficient level of sequence similarity of 77 - 83% when compared to the closest related *Mycoplasma* spp. (Uilenberg et al., 2004) have raised uncertainties in their classification. Currently, the Bergey's manual classifies *M. suis* within *Mycoplasmatales* family, *incertae sedis*, and the taxonomic position will possibly remain uncertain until *M. suis* can be cultivated *in vitro*.

14.2 Pathogen characteristics

M. suis belongs to the highly specialized and unique group of hemotrophic mycoplasmas, which target red blood cells of a wide range of vertebrate animals (Hoelzle, 2008; Messick, 2004; Neimark et al., 2001). The resulting disease is known as infectious anemia of pigs (IAP) or, historically, as porcine eperythrozoonosis. *M. suis* is a spherical, small bacterium that is remarkably pleomorphic in size and shape (375 to 600 nm in diameter; coccoid, discoid, ring-shaped, rod-shaped; Messick, 2004; Pospischil and Hoffmann, 1982; Zachary and Basgall, 1985). Like all hemotrophic organisms, *M. suis* can be observed individually or in chains on the erythrocyte surface (Figure 14.1). Some isolates are capable of invading erythrocytes (Groebel et al., 2009) or to form micro-colonies on the vessel endothelium (Sokoli et al., 2011).

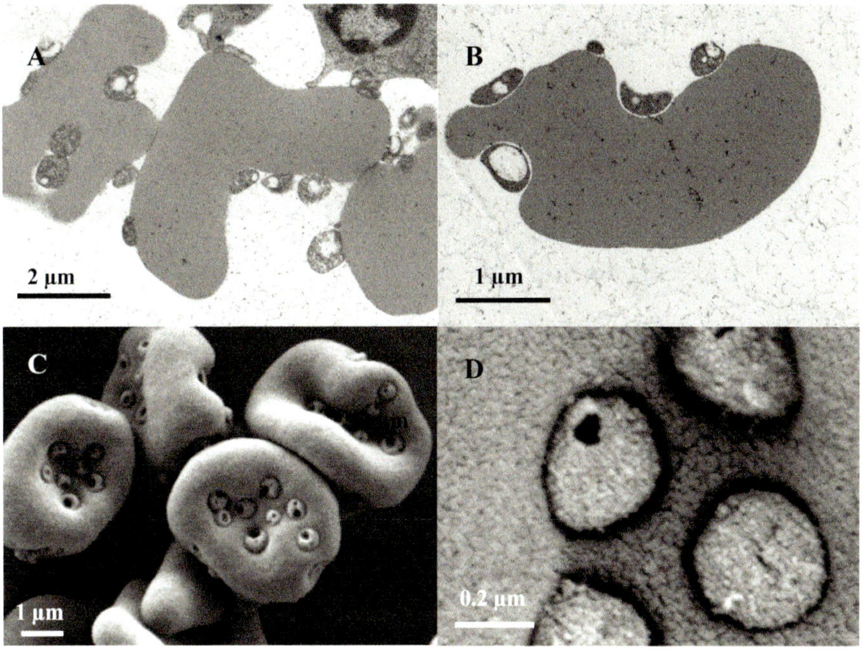

Figure 14.1. Transmission electron microscope (A, B) and scanning electron microscope (C, D) images of *M. suis* infected erythrocytes. During the acute phase of infection, numerous *M. suis* cells are visible on the surface of erythrocytes. Erythrocytes show the formation of marked invaginations upon the adhesion of *M. suis*. The contact between *M. suis* and the erythrocyte surface is intimate and seems to be mediated by fibrils. (Source: K. Hoelzle)

Presently, *M. suis* cannot be cultivated *in vitro* - neither in cell-free media nor in cell cultures (Schreiner et al., 2012a). Therefore, the only way to propagate this bacterium is by serially inoculating splenectomized pigs with blood derived from infected pigs at the peak of bacteremia (Nonaka et al., 1996; Oberst et al., 1990).

Up to date, sequencing and genetic characterization of *M. suis* was performed for two strains (strain Illinois and strain KI3806). A phenotypic difference between these strains was the cell invasiveness of KI_3806 (Guimaraes et al., 2011a; Oehlerking et al., 2011). The single, circular, double-stranded genomes of both strains differ considerably in length (709 kb for strain

KI3806 to 742 kb for strain Illinois) and in the number of annotated encoded protein sequences (795 proteins in KI3806 to 845 in strain Illinois). Comparison of both genomes reveals that they are very similar, *i.e.*, most genes that are present in both genomes show high similarity (98%). However, both contain a high number of hypothetical proteins that are not present in the other genome.

14.3 Epidemiology: prevalence and transmission

Infections with *M. suis* are widespread in domestic pigs of all ages and production levels. *M. suis* infections have been reported in Europe (Germany, Switzerland, Austria, Belgium; Hoelzle, 2008; Ritzmann et al., 2009), North and South America (Guimaraes et al., 2007; Messick, 2004) and Asia, especially China (Yuan et al., 2009; Zhou et al., 2009). PCR-based prevalence studies in Germany in fattening pig herds revealed a pig-level prevalence of 13.9% and a herd prevalence of 43.7% (Ritzmann et al., 2009). In sow herds, the pig-level prevalence was 26.9% and the herd prevalence 76.2% (Stadler et al., 2019). In the USA, sera from about 6300 pigs were tested for *M. suis* by hemagglutination test. Approximately 16% of the pigs were serologically positive (Sisk et al., 1980). In Brazil, a PCR study detected a prevalence of 18.2% in 186 pigs (121 sows, 61 piglets, and four boars) from four farms (Guimaraes et al., 2007). In China, two recent prevalence studies used a recombinant blocking ELISA and determined seroprevalences of 31.9% in 4004 pigs (Song et al., 2014) as well as 33.3% in 3458 pigs and 95.65% at herd level (Zhongyang et al., 2017). Both studies investigated pigs of different age groups, including piglets, nursery pigs, fattening pigs, sows, and boars. On the other hand, a prevalence study in Korea detected strikingly lower prevalences: only 0.2% of 1867 pigs were *M. suis*-positive by PCR (Seo et al., 2019). *M. suis* has also been detected in wild boars by PCR in Germany (Hoelzle et al., 2010) and Brazil (Dias et al., 2019). The pathogenic significance of *M. suis* in wild boars is not yet known, but it is likely that wild boars act as a reservoir for *M. suis*.

The transmission of *M. suis* from pig to pig occurs via blood, urine, saliva as well as via nasal and vaginal secretions. Transmission is often iatrogenic in the context of interventions (*e.g.* tail-docking, iron application and other injections) or occurs because of skin lesions (*e.g.* during fighting) or smaller injuries (Dietz et al., 2014; Plank and Heinritzi, 1990). Other potential transmission routes include vertical transmission (intra uterine) and transmission by blood-suckling arthropods such as stable flies, pig lice and mosquitoes (Hoelzle, 2008; Prullage et al., 1993; Stadler et al., 2019). The presence of *M. suis* in lice (*Haematopinus suis*) was detected by PCR (Acosta et al., 2019) and *M. suis* transmission through stable flies and mosquitoes was experimentally demonstrated in splenectomized pigs (Prullage et al., 1993). For the vertical transmission route strong indications are based on the detection of *M. suis* in neonatal pre-suckling piglets (Stadler et al., 2019).

A crucial aspect in the *M. suis* epidemiology is the long-term persistence. Despite an intense immune response and/or antibiotic treatment, *M. suis* is known to persist in subclinical infected (carrier) pigs for years without any detectable clinical signs (Hoelzle, 2008; Hoelzle et al., 2014). The determining factors for the long persistence are poorly understood so far. In these pigs, the reoccurrence of the disease can be triggered by immunosuppressive events (*e.g.* stress, transport, and infections with other pathogens; Hoelzle et al., 2014). Carrier pigs are the major reservoir of *M. suis* and key players in the epidemiology of *M. suis* infections due to the undetected persistence of the pathogen in pig herds. During the reactivation and recurring *M. suis* bacteremia, the infection can spread through the herd.

14.4 Pathogenesis

Our knowledge on *M. suis* pathogenesis is still limited due to the inability to cultivate the pathogen *in vitro*. Most of the studies have been performed in pig models. Nevertheless, the introduction of highly sensitive molecular methods has helped in clarifying the *M. suis* pathogenesis. However, sequencing and annotation of the *M. suis* genomes provided no clear evidence of virulence factors *e.g.* adhesins, invasins, hemolysins, haemagglutinins, toxins or known persistence factors (Guimaraes et al., 2011a; Oehlerking et al., 2011). Therefore, pathogenicity mechanisms of *M. suis* seem to be complex and genetically unapparent or "hidden" (Hoelzle et al., 2014).

Two different forms of anemia formation are described. In the first form, anemia is observed in association with massive bacteremia. The second anemia form is associated with minimal bacteremia (Heinritzi et al., 1990a; 1990b; Zachary and Basgall, 1985). Therefore, several mechanisms might account for the destruction and elimination of *M. suis* infected erythrocytes from blood.

The direct interaction between *M. suis* and the porcine erythrocytes (adhesion, invasion) leads to destruction, eryptosis, phagocytosis or splenic sequestration of the infected, often deformed or modified erythrocytes. In addition, *M. suis* is capable of transiently inducing the expression of anti-erythrocytic IgM and IgG auto-antibodies. These induce immune mediated lysis of erythrocytes, causing hemolytic anemia (Felder et al., 2010, 2011; Groebel et al., 2009; Hoelzle et al., 2006; Zachary and Smith, 1985).

14.4.1 Adhesion and invasion

Pathogenesis models for *M. suis* induced anemia have identified mechanical and/or osmotic damage of erythrocytes due to the attachment of *M. suis* as one major trigger for extravascular hemolysis in the spleen, liver, lungs, and bone marrow (Guimaraes et al., 2011a; Heinritzi and Plank, 1992; Messick, 2004). Indeed, electron microscopic analyses demonstrated that *M. suis* has a highly intimate contact with its host erythrocytes resulting in severe membrane deformations and damages (Figure 14.1). Fibrillary structures, which are responsible for the attaching of the bacteria to the host cell membrane, could be observed in a 30-nm electron lucent zone that separates *M. suis* from the erythrocyte membrane (Zachary and Basgall, 1985). The *in vivo* interaction is accompanied by severe and irreversible depressions and cup-like invaginations of the erythrocytes (Groebel et al., 2009). However, bacterial adhesins that mediate the attachment of *M. suis* to the porcine erythrocytes and the molecular basis of adherence have not been fully elucidated. To date, only few proteins involved in adherence have been identified. Two adhesion proteins, *i.e.* MSG1 and alpha-enolase are so called moonlighting proteins, with at least more than one function. The primary function is enzymatic catalysis in the carbohydrate metabolism. In addition, the secondary and non-enzymatic role is adhesion (Hoelzle et al., 2007b; Schreiner et al., 2012b). MSG1 was the first identified adhesin of *M. suis*. It is a glyceraldehyde-3-phosphate dehydrogenase (GAPDH) which catalyzes the conversion of glyceraldehyde 3-phosphate to D-glycerate 1, 3-bisphosphate within the glycolysis, but in its surface localized form, MSG1 is also involved in the adhesion of the pathogen to the erythrocyte membranes (Hoelzle et al., 2007b). The second moonlighting protein, alpha-enolase, catalyzes the reaction of phosphoglycerate to phosphoenolpyruvate but was also shown to be surface localized (a prerequisite of adhesion involvement) and adhesion-related (Schreiner et al., 2012b). MSG1 and alpha-enolase are expressed during acute infection in experimentally infected pigs as demonstrated by a proteome study (Felder et al., 2012) and elicit strong immunological response in infected pigs. These anti-MSG1 and anti-alpha-enolase antibodies block the adhesion, indicating an important role of these proteins in pathogenesis (Hoelzle et al., 2007b; Schreiner et al., 2012b). Interestingly, an alternative enzyme for the MSG1 *i.e.* the NADP-dependent glyceraldehyde-3-phosphate dehydrogenase (*i.e.* GAPN) is also expressed during acute IAP. The use of this enzyme during acute infection adds an additional energy-providing step that is of special interest, since GAPN may ensure glycolysis when GAPDH serves as an adhesion protein (Felder et al., 2012). Other putative *M. suis* adhesion proteins include the heat shock protein DnaK that was observed in *M. suis* (*i.e.* MSA1) to be surface localized, immunogenic, and expressed during acute IAP (Felder et al., 2012; Hoelzle et al., 2007a).

Cell adhesion complexes or tip organelles that are typical in many other mycoplasmas were not observed in *M. suis* by electron microscopic investigations (Groebel et al., 2009; Zachary and Basgall, 1985). In addition, sequence data of two strains of *M. suis* did not provide further information on such adhesion mechanisms. Only two additional putative adhesion proteins, *i.e.* orthologues of proteins involved in adhesion that are MgpA orthologues (DHH family phosphoesterase proteins) could be detected in the *M. suis* genomes. In *M. suis*, one MgpA orthologous protein has a nearly full-length match to other mycoplasmal MgpA proteins; the other MgpA was C-terminally extended for 140 amino acids. One of the MgpA proteins was observed to be expressed during acute phase indicating an important role in acute *M. suis* infection (Dietz et al., 2016; Guimaraes et al., 2011a; Oehlerking et al., 2011).

For a long time, it was believed that *M. suis* remained only attached to the surface of mature erythrocytes or could be identified free in plasma. However, Groebel et al. (2009) evidenced that (i) *M. suis* parasitizes also normoblasts and reticulocytes, and (ii) *M. suis* invades into the erythrocyte, a process which contributes to erythrocyte damage and destruction. The observed invasion process of *M. suis* is similar to the mechanisms used by other hemoparasites (*e.g. Plasmodium falciparum*, *Bartonella bacilliformis*). *M. suis* uses an endocytosis-like invasion mechanism which begins with invagination of the erythrocyte membrane, followed by deepened depressions. Upon entry, the aperture of the invaginated erythrocyte apparently fuses, and as a result, *M. suis* could be observed in intracellular vacuoles (Groebel et al., 2009). Factors which are used by *M. suis* to initiate and perform invasion are unknown so far and the genome sequencing did not identify any of these factors (Guimaraes et al., 2011a). However, the shift from epi- to intraerythrocytic life-style provides a possible strategy of preventing bacterial damage by antibiotics and host immune response, and, thus, of life-long persistence (Rottem, 2003). Erythrocytes carrying only intracellular *M. suis* cells show no surface modification and are not recognized by spleen macrophages. The hypothesis is supported by the fact that the shift from epi- to intracellular life-style is accompanied by a significant increase of virulence of *M. suis* (Groebel et al., 2009).

14.4.2 Nutrient scavenging

As genetically proved, *M. suis* competes for and scavenges glucose, inosine, hypoxanthine, amino acids, NADH/NADPH, and ribose from the host cell (Guimaraes et al., 2011a). This competition might lead to reduced energy production and, thus, to oxidative stress and a reduced life span of the porcine erythrocytes and their premature removal from the circulation. The substantial decrease of glucose in the blood of naturally and experimentally infected pigs during acute IAP supports this hypothesis (Heinritzi et al., 1990a; 1990b; Hoelzle et al., 2014; Nonaka et al., 1996; Smith et al., 1990). However, it is still unclear how *M. suis* acquires metabolites from the host erythrocytes. Only a few metabolite transporters are encoded by *M. suis* which lead to the assumption that these transport systems might have a broad substrate specificity or other, so far not assigned CDS functions as biomolecule transporters through the bacterial membrane (Dietz et al., 2016). The high portion of predicted hypothetical proteins (64.6%) further supports the latter hypothesis. Furthermore, invasion into host cells might help in nutrient acquisition (Groebel et al., 2009; Guimaraes et al., 2011a).

14.4.3 Eryptosis

M. suis induced alterations of the membrane and cytoskeleton of the porcine erythrocytes are thought to be responsible for a special pathogenesis mechanism in *M. suis* infections, *i.e.* up-regulation of eryptosis, the suicidal programmed cell death of erythrocytes (Zachary and Basgall, 1985; Felder et al., 2011). As a function of different virulent strains, infected erythrocytes undergo eryptosis. This is a process of conformational changes of erythrocytes known from apoptosis in nucleated cells and as pathogenesis factor of malaria (Felder et al., 2011; Lang et al., 2005, 2006). Specific characteristics are cell shrinkage, blebbing, microvesiculation, activation of proteases, and phosphatidyl-serine exposure on the outer membrane leading to recognition by macrophages and, therefore, to phagocytosis (Felder et al., 2011; Lang et al., 2006). Figure 14.2 shows the typical cell shrinkage of eryptotic erythrocytes of a *M. suis* infected pig.

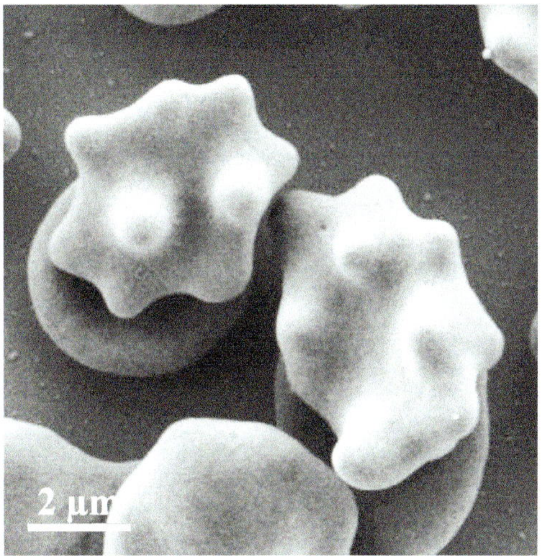

Figure 14.2. Scanning electron microscopic image of *M. suis* infected erythrocytes showing eryptotic cell shrinkage. (Source: K. Hoelzle)

Up to now, the trigger and signals for the induction of eryptosis in *M. suis* infections are not clear. *In vitro* assays implied that *M. suis* cells or soluble *M. suis* substances are responsible for the induction of eryptosis (Felder et al., 2011). *M. suis* induced massive alterations of the cell membranes, and particularly in the cytoskeleton may lead to channel formation and subsequently to an up-regulation of eryptosis. In addition, adhesion of *M. suis* may lead to oxidative stress and therefore, triggers eryptosis (Felder et al., 2011; Hoelzle, 2008).

In pigs infected with highly virulent strains, eryptosis is up-regulated but fails to control the infection. In contrast, pigs infected with low virulent strains develop a high eryptosis rate that successfully prevents a fatal course of infection indicated by mild anemia (Felder et al., 2011). Eryptosis seems to be responsible for phagocytosis of *M. suis* infected erythrocytes in early and mild anemia since phosphatidylserine-exposing erythrocytes are identified by and degraded by macrophages. Therefore, eryptosis might prevent massive *M. suis* proliferation in the blood so that the bacteremia remains at a low level (Felder et al., 2011).

14.4.4 Immunopathology

It has been known for a long time that autoimmune processes play an important role in the development of IAP associated clinical signs, *i.e.* anemia through immune mediated destruction of porcine erythrocytes as well as the acrocyanosis/acronecrosis (Felder et al., 2010; Hofmann et al., 1982; Zachary and Smith, 1985). Both, IgM and IgG types of autoreactive antibodies have been described.

Cold autoreactive antibodies of the IgM class (so-called cold agglutinins; CA) can mainly be detected during the chronic phase of IAP (Hoelzle, 2008). These *M. suis* induced CAs target sialo-glycosylated regions of proteins on the surface of the erythrocyte membrane (Hofmann et al., 1981; Jungling et al., 1994; Zachary and Smith, 1985). They occur during low *M. suis* bacteremia and are responsible for agglutination and therefore, for cyanosis and necrosis at the acra (extremities of peripheral body parts), where the body temperature is typically lower than in the body core. This IgM binding and peripheral agglutination leads, via the classical activation pathway of the complement system, to intravascular hemolysis and, subsequently, to anemia. However, the exact mechanism of CA induction is not clear yet (Gwaltney and Oberst, 1994; Zachary and Smith, 1985). Hoelzle et al. (2006) detected an up-regulation of so-called warm-autoreactive IgG antibodies directed against porcine actin (Felder et al., 2010; Hoelzle et al., 2006). In contrast to the CAs, the up-regulation of autoreactive IgGs occurs mainly during the acute phase in association with the peak of the *M. suis* bacteremia. IgG-targeted, *M. suis* infected erythrocytes are cleared from the circulation mainly by the spleen and other parts of the reticuloendothelial system leading to extravascular hemolysis. This process is often accompanied by a massive indirect hyperbilirubinemia because pre-hepatic bilirubin concentration increased significantly due to overloaded capacities of the hepatocytes. Accordingly, warm-reactive IgGs are assumed to be responsible for the destruction of the erythrocytes in the early acute infection leading to early initial anemia (Hoelzle et al., 2006; Zachary and Smith, 1985). The coincidence of high bacteremia and autoreactive IgGs indicates a direct effect of *M. suis* in terms of misguided up-regulation of naturally-occurring B cells specific for self-antigens, the occurrence of altered self-antigens, and loss of tolerance to self-antigens due to molecular mimicry (Felder et al., 2010). The adhesion protein MSG1 seems to be involved in inducing autoimmunity by molecular mimicry since cross-reactivity between MSG1 and porcine actin has been evidenced (Felder et al., 2010).

In addition, *M. suis* is suspected of suppressing the host's immune response that leads to an increased susceptibility to other infectious agents causing respiratory and enteric diseases (Zachary and Smith, 1985). This assumption was based on empirical data (Hoelzle, 2008) as well as on data from experimental infections where a suppression of the T-lymphocyte blastogenic response was observed (Zachary and Smith, 1985). Recently, analyses of the porcine transcriptome during subclinical infection under experimental conditions identified significant changes in the blood mRNA levels including a significant down-regulation of immune response associated genes suggesting a general immune suppression in infected animals (do Nascimento et al., 2018).

14.4.5 Endothelial targeting

M. suis is observed in close connection with endothelial cells. This endothelial targeting represents a novel pathway in the pathogenesis of *M. suis* that contributes to the clinical signs beyond anemia, *i.e.* coagulopathy, vascular thrombosis, and hemorrhagic diathesis (Sokoli et al., 2013). Electron microscopic investigations based on experimental infections revealed that *M. suis* and *M. suis*-infected erythrocytes establish intimate interactions with endothelial cells, triggering activation responses and coagulation processes. *M. suis* was attached to the endothelium of the vasculature of different organs either as a single cell or as biofilm-like micro-colonies (Sokoli et al., 2013).

Consequently, the endothelium is massively and widespread damaged and denudated. In addition, the endothelial activation due to *M. suis* is characterized by microvilli formation which contributes to a higher adhesion frequency of *M. suis*-infected erythrocytes (Sokoli et al., 2013).

Both, endothelial adherence and formation of biofilm-like micro-colonies on the endothelium contribute to immune evasion (particularly by prohibiting the filtration by the spleen), protection from antimicrobial agents, and persistence of *M. suis* infections. Circulation dysfunction and the necrosis of the acra in *M. suis* infected pigs can also be associated with this endothelial targeting. Furthermore, dysfunction of affected organs, circular disturbance, blood clotting and flow impairment, and local hypoxia may occur consequently. Actin condensation and up-regulation of endothelial activation markers in *in vitro* studies further support the interaction of *M. suis* with endothelial cells (Sokoli et al., 2013).

14.5 Incubation period and clinical signs

In experimental infections where *M. suis* negative animals are challenged with a high infectious dose of *M. suis* (10^8 to 10^{10} bacteria), clinical signs (fever, apathy, anemia) occur as early as 5 days post infection (dpi) in splenectomized pigs, and after 6 dpi in non-splenectomized pigs (Stadler et al., 2014). Splenectomized pigs reached the maximum severity of clinical signs 7 dpi and non-splenectomized pigs 9 - 10 dpi.

Under natural conditions, the incubation period is difficult to predict and appears to vary largely as it depends on several factors including the individual susceptibility and general health status of the animals, the infection dose and the virulence of the *M. suis* strain, the production and management system and the presence of other secondary infections (Hoelzle, 2008). Infected animals can remain apparently normal for months before developing clinical signs, and it is probable that many infected animals will never show any clinical signs at all.

The clinical spectrum of *M. suis* infections (acute, chronic or latent form) can vary from life-threatening acute disease to clinically unapparent chronic infections. The manifestation of the disease is highly dependent on the susceptibility of the host, endogenous and exogenous stress factors, but also on the pathogen virulence, the infection dose and the route of infection (Heinritzi, 1989). Predisposing endogenous stress factors include impairments, other infections (*e.g.* PRRSV, PCV), reduced immune status *e.g.* during weaning, birth and peri-parturient period. Examples of predisposing exogenous stress factors are several common farm practices (*e.g.* regrouping, castration, transport), crowding, inappropriate feeding, and poor barn air quality (Gresham et al., 1994; Heinritzi, 1990; McLaughlin et al., 1991). Little is known about differences in virulence of different *M. suis* strains. So far, it is only known that there are *M. suis* isolates that can invade erythrocytes, which is thought to lead to an increase in virulence (Groebel et al., 2009).

Acute *M. suis* infection is associated with massive bacteremia of erythrocytes, causing severe and sometimes lethal hemolytic anemia in suckling piglets, pregnant sows shortly before farrowing and during the suckling period, as well as in fattening pigs under stress (Messick, 2004; Zachary and Basgall, 1985). *M. suis* provokes changes in the erythrocyte surface structure (*i.e.* densifications and cytoskeletal rearrangements) leading to the induction of IgM cold agglutinins and IgG autoreactive antibodies. Moreover, there is a subsequent immune-mediated hemolytic anemia in the affected pigs that is concurrent with a drop in the number of bacteria in the blood (Felder et al., 2010; Groebel et al., 2009; Zachary and Basgall, 1985; Zachary and Smith, 1985). Typical clinical signs of acute IAP are fever, anemia, icterus, and cyanosis at the acra. In addition, skin alterations with petechiae and hemorrhagic diathesis as well as a higher bleeding tendency could be observed (Plank and Heinritzi, 1990; Stadler et al., 2014). Coincidentally, a severe decrease of the blood glucose concentration leading to a life-threatening hypoglycaemia can be observed (Gwaltney and Oberst, 1994; Heinritzi et al., 1984; Smith et al., 1990). In sows, infection may lead to pyrexia, anorexia, depression, decreased milk production, and poor maternal behavior (Brömel and Zettl, 1985; Heinritzi et al., 1984; Strait et al., 2012).

Laboratory diagnostic analyses reveal a normochromic normocytic anemia with a parallel decline in erythrocyte counts, hemoglobin concentration, and hematocrit. In addition, bilirubinemia, leucocytosis, and severe hypoglycemia, increased lactate levels, and negative baseline blood acidosis, as well as spontaneous agglutination of erythrocytes at 25°C and 4°C are described (Heinritzi et al., 1984, 1990a, 1990b). In addition, intravascular coagulation (DIC) and subsequent coagulopathy are accompanied by thrombocytopenia and prolonged partial thromboplastin and prothrombin time (Plank and Heinritzi, 1990; Stadler et al., 2014).

Reports on morbidity, mortality, and case fatality are, however, rare. The morbidity rate of weaning piglets infected with *M. suis* is considered to be 15 to 20% (Müller and Neddenriep, 1979; Preston and Greve, 1965). Case fatality in one-week old weaning piglets was determined to be one piglet per litter (Berrier and Gouge, 1954). Quin (1938) and Robb (1943) reported a mortality rate of 10 to 20%. Kloster et al. (1987) described a morbidity rate of 29% and a case fatality of 5% and 24%, respectively.

Most often, chronic *M. suis* infections are present in pig herds. These infections vary from subclinical courses to a range of clinical conditions including (i) anemia, mild icterus, and general unthriftiness in newborns and suckling pigs, (ii) growth retardation in feeder pigs, and (iii) poor reproductive performance in sows, including failure to return to estrus, increased abortions, stillbirths, weak piglets, as well as dysgalactia (Brownback, 1981; De Busser et al., 2008; Gresham et al., 1984; Henry, 1979; Schweighardt et al., 1986; Strait et al., 2012; Zinn et al., 1983).

14.6 Socio-economic impact

M. suis is a pig-specific pathogen. However, several reports especially from China raise the question of a zoonotic potential for *M. suis*. (Hu et al., 2009;Wu et al., 2006;Yang et al., 2000;Yuan et al., 2009). High prevalences of 49% (Yuan et al., 2009) or even 97.29% (Huang et al., 2012) were reported using either microscopic evaluation of stained blood smears or the amplification of the 16S rDNA as diagnostic methods. However, both methods do not allow a bacterial species diagnosis clearly differentiating *M. suis* from other hemotrophic mycoplasmas. In addition, outside China, no human *M. suis* infection has been detected so far. Nevertheless, the possibility of zoonosis should not be negated and should be a goal of future studies.

In the pig industry, infections with *M. suis* have major economic impact, mainly because of the reduced performance, reduced daily weight gain and the reproductive disorders (Hoelzle et al., 2014). Moreover, the immune modulatory and suppressive conditions in subclinical infected pigs enhance the susceptibility to other viral and bacterial infections (Hoelzle, 2008; Zachary and Smith, 1985). The impact of the subclinical infections should be further evaluated as basis for future eradication programs.

14.7 Diagnostics

Clinical signs are highly variable and not pathognomonic. Therefore, laboratory testing is necessary for a conclusive diagnosis. Isolation of *M. suis* as diagnostic method is not possible due to the lack of an *in vitro* cultivation system.

For a long time, microscopic detection in Giemsa or acridine orange stained blood smears was the method of choice. Under the microscope, stained *M. suis* organisms are seen individually, in pairs or in chains on the surface of erythrocytes. The disadvantage of this inexpensive and fast method is its low sensitivity and specificity. A reliable diagnosis succeeds only from a pathogen concentration of 10^5 bacteria per mL blood (Ritzmann et al., 2009). However, such high numbers of the pathogen are only evidenced in the blood during the short phase of an acute IAP with high bacteremia. In addition, there is quite a high number of false-positive microscopic findings due to misinterpretations by *e.g.* Howell Jolly bodies and staining artefacts. The correct assessment of a blood smear depends very much on the experience of the examiner (Ritzmann et al., 2009).

Significant improvements in diagnostic capabilities were achieved by the establishment of PCR methods (conventional PCR and real-time PCR) that enables the examination of different materials including blood, organs or abortion material. The published conventional PCR assays target either the 16S rRNA gene (Messick et al., 1999) or the ß-subunit of the DNA-directed RNA polymerase (Gwaltney et al., 1994). Conventional PCR assays showed high specificity and sensitivity. Specificity was significantly improved by the combination of the PCR assays with subsequent hybridization assays (Oberst et al., 1993). Messick et al. (1999) determined the lower detection limit of their PCR assay as 57 to 800 organisms per reaction. Due to this sensitivity, it is possible to diagnose chronically infected animals as well as pigs with low *M. suis* blood concentrations at the beginning of acute infections (Gwaltney et al., 1994; Messick et al., 1999).

As technology has improved, real-time PCR was developed and utilized in *M. suis* detection. Guimaraes et al. (2011b) developed a fluorescent reporter probe based real-time Taqman PCR assay, which targets the 16S rRNA gene of *M. suis*. In addition, a second quantitative PCR assay was established which targets the gene encoding the adhesion protein MSG1 (syn. GAPDH; Hoelzle et al., 2007c). This assay is based on the LightCycler technology also using one specific fluorescent reporter probe. Both qPCR assays detect as few as ten copies of *M. suis* and are 10 to 100-fold more sensitive than the conventional PCR assays.

The MSG1-based qPCR was also adopted to SYBR green dye chemistry (Stadler et al., 2019) with the advantage that the reactions are significantly less expensive, and the assay can be performed on many qPCR devices in routine diagnosis, independently of the LightCycler or Taqman technology. This SYBR green assay is also highly sensitive and specific (Stadler et al., 2019).

Due to the high sensitivity of the PCR assays, early *M. suis* infection stages (3 to 4 days post infection; Stadler et al., 2014) can be diagnosed. In addition, the qPCR assays enable the detection of sub-clinically infected pigs or pig herds. However, further research is warranted to investigate the significance of these subclinical infections. Other detection methods, *e.g.* immunohistochemistry or colloidal gold-based immunochromatographic assay (Meng et al., 2014) can only be used to a limited extent because *M. suis*-specific antibodies are not commercially available.

Circulating anti-*M. suis* antibodies can be detected using different serological assays. From experimental infections, it is well known that *M. suis* infected animals seroconvert between days 8 to 35 post infection (Hoelzle et al., 2006; Stadler et al., 2014). Antigen production for the serological assays is difficult due to the lack of a laboratory cultivation system. Whole cell antigens must be purified from the blood of experimentally infected pigs. Despite intense purification steps residues from the host are still present in the antigen preparations which might interfere with the serum antibodies as well as with the conjugated secondary anti-pig antibodies (Hoelzle, 2008; Hoelzle et al., 2006). Therefore, interpretation of serological results is difficult. Several whole antigen tests used for the diagnosis of *M. suis* infection are described in the literature including complement binding reaction (CBR), indirect hemagglutination test and ELISA (Schuller et al., 1990; Smith and Rahn, 1975; Splitter, 1958).

The introduction of defined recombinant *M. suis* proteins (MSG1, HspA1, and inorganic pyrophosphatase) as diagnostic ELISA antigens, significantly increased the specificity and sensitivity of serological tests and would enable *M. suis* serology in routine diagnostic laboratories (Hoelzle et al., 2007d; Liu et al, 2012; Zhang et al., 2012). However, these antigens are only available in a few laboratories and no commercial test is available so far. Other serological diagnostic approaches include a multiplex micro-bead immunoassay (Guimaraes et al., 2014). Further studies are necessary to compare serological parameters and clinical impact, and to improve serological diagnostics.

14.8 Treatment, general control measures and vaccination

Treatment with tetracyclines or gyrase inhibitors (*e.g.* enrofloxacin) is possible and cures diseased animals from clinical signs. However, it is well known from experimental infections that treatment does not eliminate *M. suis* from the pig (Stadler et al., 2014). Infected pigs remain *M. suis* carriers lifelong and are, therefore, crucial for epidemiology (Hoelzle, 2008). However, mechanisms and bacterial determinants of persistence are poorly understood. Important persistence factors could be the highly evolved host-pathogen interaction, the invasion of host cells and the high metabolic adaption of *M. suis* that might facilitate survival in specific niches (Groebel et al., 2009; Hoelzle et al., 2014). Avoiding entry of *M. suis* into negative herds is crucial to remain free from infection. There are no *M. suis*-specific biosecurity measures, but the application of general measures is recommended (controlled entry of persons, regular disinfection and decontamination). The main risk for *M. suis* introduction to pig herds involves infected incoming pigs. Screening of incoming pigs for *M. suis* can minimize this risk.

After pathogen entry, transmission mainly occurs through direct contact or contaminated instruments and devices. Thus, all measures improving the hygiene status, such as disinfection of instruments and equipment or use of disposable supplies are recommended. In addition, susceptibility to *M. suis* infection depends on the health and immune status of the animals. Therefore, optimal feeding and housing conditions, consistent endoparasite control - and blood suckling ectoparasite treatment (as potential *M. suis* vectors), and avoidance of general stress factors are important control measures.

Elimination protocols for *M. suis* are not established and sanitation of herds must be decided in the individual case. In sow herds *M. suis* carriers might be identified and eliminated using PCR-based methods.

Vaccines against *M. suis* are not yet available. The main reasons for this are the impossibility to produce classical whole bacteria vaccines and the complex immunity induced by the infection including autoreactive processes. A recombinant *M. suis* vaccine based on the highly antigenic adhesion protein MSG1 induced a strong humoral and cellular immune response in the experimental model of splenectomized pigs but showed no protective effect (Hoelzle et al., 2009).

REFERENCES

Aalund, O., Willeberg, P., Mandrup, M., & Riemann, H. (1976). Lung lesions at slaughter: Associations to factors in the pig herd. *Nordic Veterinary Medicine, 28*, 487-495.

Abbas, A. K., Lichtman, A. H., & Pillai, S. (2016). *Basic immunology: Functions and disorders of the immune system* (5th ed.). St. Louis, Missouri, USA: Elsevier.

Acosta, D.B., Ruiz, M., & Sanchez J.P. (2019). First molecular detection of *Mycoplasma suis* in the pig louse *Haematopinus suis* (Phthiraptera: Anoplura) from Argentina. *Acta Tropica, 194*, 165-168.

Adams, C., Pitzer, J., & Minion, F.C. (2005). In vivo expression analysis of the P97 and P102 paralog families of *Mycoplasma hyopneumoniae*. *Infection and Immunity, 73*, 7784-7787.

Advisory Group on Integrated Surveillance of Antimicrobial Resistance (AGISAR). (2016). WHO list of critically important antimicrobials for human medicine (5th ed.). https://www.who.int/foodsafety/publications/cia2017.pdf

Ahn, K. K., Kwon, D., Jung, K., Ha, Y., Seo, M. J., Kim, S. H., . . . Chae, C. (2009). Identification of interleukin-1, tumor necrosis factor-alpha, and interleukin-6 expression in lungs from pigs naturally infected with *Mycoplasma hyopneumoniae* by in situ hybridization. *Journal of Veterinary Medical Science, 71*(4), 441-445. doi:10.1292/jvms.71.441

Ahn, K. K., Lee, Y. H., Ha, Y., Kim, D., Chae, S., Kim, C. H., Lee, J. H., Kim, S. H., & Chae, C. (2008). Detection by in-situ hybridization of *Pasteurella multocida toxin* (toxA) gene in the lungs of naturally infected pigs. *Journal of Comparative Pathology, 139*(1), 51-53.

Aitken, I. A., Morgan, J. H., Dalziel, R., Burch, D. G. S., & Ripley, P. H. (1999). Comparitive *in vitro* activity of valnemulin against porcine bacterial pathogens. *Veterinary Record, 144*, 128.

Alabdullah, H. A., Fox, L. K., Gay, J. M., Barrington, G. M., & Mealey, R. H. (2015). Effects of dexamethasone and *Mycoplasma bovis* on bovine neutrophil function in vitro. *Veterinary Immunology and Immunopathology, 164*(1-2), 67-73. doi:10.1016/j.vetimm.2014.12.010

Alikhan, N.-F., Petty, N. K., Ben Zakour, N. L., Beatson, S. A. (2011). BLAST Ring Image Generator (BRIG): Simple prokaryote genome comparisons. *BMC Genomics, 12*, 402. https://doi.org/10.1186/1471-2164-12-402

Amass, S. F., Clark, L. K., van Alstine, W. G., Bowersock, T. L., Murphy, D. A., Knox, K. E., & Albregts, S. R. (1994). Interaction of *Mycoplasma hyopneumoniae* and *Pasteurella multocida* infections in swine. *Journal of American Veterinary Medical Association, 204*(1), 102-107.

Anderson, A., Sponheim, A., Dalquist, L., Fano, E. & Pieters, M. (2017). Use of molecular characterization tools to investigate *Mycoplasma hyopneumoniae* outbreaks. Oral presentation. American Association of Swine Veterinarians Annual Meeting. Denver, Colorado, February 2017.

Andrada, M., Quesada-Canales, O., Suarez-Bonnet, A., Paz-Sanchez, Y., Espinosa de Los Monteros, A., & Rodriguez, F. (2014). Cyclooxygenase-2 expression in pigs infected experimentally with *Mycoplasma hyopneumoniae*. *Journal of Comparative Pathology, 151*(2-3), 271-276. doi:10.1016/j.jcpa.2014.04.005

Andreasen, M., Baekbo, P., & Nielsen, J. P. (2000). Lack of effect of aerial ammonia on atrophic rhinitis and pneumonia induced by *Mycoplasma hyopneumoniae* and toxigenic *Pasteurella multocida*. *Journal of Veterinary Medicine B, 47*(3), 161-171.

Andreasen, M., Mousing, J., & Thomsen, L. K. (2001). No overall relationship between average daily weight gain and the serological response to *Mycoplasma hyopneumoniae* and *Actinobacillus pleuropneumoniae* in eight chronically infected Danish swine herds. *Preventive Veterinary Medicine, 49*(1-2), 19-28.

Aragon, V., Segalés, J., & Oliveira, S. (2012). Glässer's Disease. In J. J. Zimmerman, L. A. Karriker, A. Ramirez, K. J. Schwartz, & G. Stevenson (Eds.), *Diseases of swine* (10th ed.; pp. 760-769). Oxford: Wiley Blackwell.

Arfi, Y., Minder, L., Di Primo, C., Le Roy, A., Ebel, C., Coquet, L., Claverol, S., Vashee, S., Jores, J., Blanchard, A., & Sirand-Pugnet, P. (2016). MIB-MIP is a mycoplasma system that captures and cleaves immunoglobulin G. Proceedings National Academy of Sciences, USA, 113, 5406-5411.

Armstrong, C. H. (1994). Porcine mycoplasmas. In H. W. Whitford, R. F. Rosenbusch, & L. H. Lauerman (Eds.), *Mycoplasmosis in animals: Laboratory diagnosis* (pp. 68-83). Ames, Iowa: State University Press.

Armstrong, C. H., Freeman, M. J., Sands-Freeman, L., Lopez-Osuna, M., Young, T., & Runnel, L. J. (1983). Comparison of the enzyme-linked immunosorbent assay and the indirect haemagglutination and complement fixation tests for detecting antibody to *Mycoplasma hyopneumoniae. Canadian Journal of Comparative Medicine, 47*, 464-470.

Arsenakis, I., Michiels, A., del Pozo Sacristán, R., Boyen, F., Haesebrouck, F., & Maes, D. (2017). *Mycoplasma hyopneumoniae* vaccination at or shortly before weaning under field conditions: A randomised efficacy trial. *Veterinary Record, 181*, 19.

Arsenakis, I., Michiels, A., Schagemann, G., Gomez-Duran C. O., Boyen, F., Haesebrouck, F., & Maes, D. G. D. (2019). Effect of pre-farrowing sow vaccination against *Mycoplasma hyopneumoniae* on offspring colonization and lung lesion. *Veterinary Record, 184*(7), 578-583.

Arsenakis, I., Panzavolta, L., Michiels, A., Del Pozo Sacristan, R., Boyen, F., Haesebrouck, F., & Maes, D. (2016). Efficacy of *Mycoplasma hyopneumoniae* vaccination before and after weaning against experimental challenge infection in pigs. *BMC Veterinary Research, 12*, 63. doi:10.1186/s12917-016-0685-9

Artiushin, S., & Minion, F. C. (1996). Arbitrarily primed PCR analysis of *Mycoplasma hyopneumoniae* field isolates demonstrates genetic heterogeneity. *International Journal of Systematic and Evolutionary Microbiology, 46*, 324-328. https://doi.org/10.1099/00207713-46-1-324

Asai, T., Okada, M., Ono, M., Irisawa, T., Mori, Y., Yokomizo, Y., & Sato, S. (1993). Increased levels of tumor necrosis factor and interleukin 1 in bronchoalveolar lavage fluids from pigs infected with *Mycoplasma hyopneumoniae. Veterinary Immunology and Immunopathology, 38*(3-4), 253-260.

Asai, T., Okada, M., Ono, M., Mori, Y., Yokomizo, Y., & Sato, S. (1994). Detection of interleukin-6 and prostaglandin E2 in bronchoalveolar lavage fluids of pigs experimentally infected with *Mycoplasma hyponeumoniae. Veterinary Immunology and Immunopathology, 44*(1), 97-102.

Assunção, P., Antunes, N. T., Rosales, R. S., Poveda, C., de la Fe, C., Poveda, J. B., & Davey, H. M. (2007). Application of flow cytometry for the determination of minimal inhibitory concentration of several antibacterial agents on *Mycoplasma hyopneumoniae. Journal of Applied Microbiology, 102*(4), 1132-1137.

Assunção, P., Diaz, R., Comas, J., de Galarreta, C. M., González-Llamazares, O. R., & Poveda, J. B. (2005a). Evaluation of *Mycoplasma hyopneumoniae* growth by flow cytometry. *Journal of Applied Microbiology, 98*(5), 1047-1054. https://doi.org/10.1111/j.1365-2672.2005.02536.x

Assunção, P., Fe, C. D. la, Ramírez, A. S., Llamazares, O. G., Poveda, J. B. (2005b). Protein and antigenic variability among *Mycoplasma hyopneumoniae* strains by SDS-PAGE and immunoblot. *Veterinary Research Communications, 29*, 563-574. https://doi.org/10.1007/s11259-005-3083-5

Aucouturier, J., Dupuis, L., & Ganne, V. (2001). Adjuvants designed for veterinary and human vaccines. *Vaccine, 19*, 2666-2672.

Babot, D., Sancho, V., García, E., Blanco, G., Sobrevia, L., & Dolz, N. (Grup de Gestió Porcina, UdL) (2017). *Informe Anual del Sector Porcí 2017*. Observatori del Porcí. Departament Agricultura, Ramaderia, Pesca i Alimentació. Generalitat de Catalunya, 51.

Baccaro, M. R., Hirose, F., Umehara, O., Gonçalves, L. C. B., Doto, D. S., Paixão, R., Shinya, L. T., & Moreno, A. M. (2006). Comparative efficacy of two single-dose bacterins in the control of *Mycoplasma hyopneumoniae* in swine raised under commercial conditions in Brazil. *Brazilian Veterinary Journal, 172*, 526-531.

Baekbo, P. (1990). Pollution in fattening units. Proceedings 11th International Pig Veterinary Society Congress, Lausanne, Switzerland, 395.

Baekbo, P. (1999). Procedures to eliminate *M. hyo* and produce *M. hyopneumoniae* free pigs: An update. Proceedings AASP, Saint Louis, Missouri, 479-481.

Baekbo, P., Madsen, K. S., Aagard, M., & Szancer, J. (1994). Eradication of *Mycoplasma hyopneumoniae* from infected herds without restocking. Proceedings 13th IPVS, 26-30 June 1994, Bangkok, Thailand, 135.

Bai, F., Ni, B., Liu, M., Feng, Z., Xiong, Q., & Shao, G. (2013) *Mycoplasma hyopneumoniae*-derived lipid-associated membrane proteins induce apoptosis in porcine alveolar macrophage via increasing nitric oxide production, oxidative stress, and caspase-3 activation. *Veterinary Immunology and Immunopathology, 155*, 155-161.

Bai, F., Ni, B., Liu, M., Feng, Z., Xiong, Q., & Shao, G. (2015). *Mycoplasma hyopneumoniae*-derived lipid-associated membrane proteins induce inflammation and apoptosis in porcine peripheral blood mononuclear cells in vitro. *Veterinary Microbiology, 175*, 58-67.

Bai, Y., Gan, Y., Hua, L.-Z., Nathues, H., Yang, H., Wei, Y.-N., Meng, W., Shao, G.-Q. & Feng, Z.-X. (2018). Application of a sIgA-ELISA method for differentiation of *Mycoplasma hyopneumoniae* infected from vaccinated pigs. *Veterinary Microbiology, 223*, 86-92.

Baietto, L., Corcione, S., Pacini, G., Perri, G. D., D'Avolio, A., & De Rosa, F. G. (2014). A 30-years review on pharmacokinetics of antibiotics: Is the right time for pharmacogenetics? *Current Drug Metabolism, 15*(6), 581-598.

Bandrick, M., Ariza-Nieto, C., Baidoo, S. K., & Molitor, T. W. (2014a). Colostral antibody-mediated and cell-mediated immunity contributes to innate and antigen-specific immunity in piglets. *Developmental and Comparative Immunology, 43*(1), 114-120. doi:10.1016/j.dci.2013.11.005

Bandrick, M., Pieters, M., Pijoan, C., Baidoo, S. K., & Molitor, T. W. (2011). Effect of cross-fostering on transfer of maternal immunity to *Mycoplasma hyopneumoniae* to piglets. *Veterinary Record, 168*, 100.

Bandrick, M., Pieters, M., Pijoan, C., & Molitor T. W. (2008). Passive transfer of maternal *Mycoplasma hyopneumoniae*-specific cellular immunity to piglets. *Clinical and Vaccine Immunology, 15*(3), 540-543.

Bandrick, M., Theis, K., & Molitor, T. W. (2014b). Maternal immunity enhances *Mycoplasma hyopneumoniae* vaccination induced cell-mediated immune responses in piglets. *BMC Veterinary Research, 10*, 1-11.

Banhazi, T. (2007). Improving air quality in piggery buildings. XIIIth International Congress in Animal Hygiene, Tartu, Estonia, 237-241.

Banhazi, T. M., Seedorf, J., Rutley, D. L., Cargill, C., & Hartung, J. (2004). Introduction - effects of airborne pollutants and factors affecting concentrations in livestock buildings. International Society for Animal Hygiene - Saint-Malo, 187-191.

Bara, M.R. (2002). Eradication of *Mycoplasma pneumonia*: First reported Swiss depopulation in Australia. Proceedings 17th IPVS, 2-5 June 2002, Ames, Iowa, 107.

Barate, A., Cho, Y., Truong, Q., & Hahn, T. (2014). Immunogenicity of IMS 1113 plus soluble subunit and chimeric proteins containing *Mycoplasma hyopneumoniae* P97 C-terminal repeat regions. *FEMS Microbiology Letters, 352*, 213-220.

Barden, J.A., & Decker, J.L. (1971). *Mycoplasma hyorhinis* swine arthritis. I. Clinical and microbiologic features. *Arthritis & Rheumatology, 14*, 193-201.

Batista, L., Pijoan, C., Ruiz, A., Utrera, V., & Dee, S. (2004). Assessment of transmission of *Mycoplasma hyopneumoniae* by personnel. *Journal of Swine Health and Production, 12*(2), 75-77.

Baumgartner, A., & Nicolet, J. (1984). Detection of antibodies against membrane-glycoproteins of swine erythrocytes after experimental infection with *Mycoplasma hyopneumoniae*. *Veterinary Immunology and Immunopathology,* 7(1), 73-79.

Cargill, C., & Banhazi, T. (1998). The importance of cleaning in all-in/all-out management systems. Proceedings 15th International Pig Veterinary Society Congress, Birmingham, England, 15.

Cargill, C., Murphy, T., & Banhazi, T. (2002). Hygiene and air quality in intensive housing facilities in Australia. *Animal Production in Australia, 24*, 387-393.

Carlson, S. A., Barnhill, A. E., & Griffith, R. W. (2012). Salmonellosis. In J. J. Zimmerman, L. A. Karriker, A. Ramirez, K. J. Schwartz, & G. Stevenson (Eds.), *Diseases of swine* (10th ed.; pp. 821-833). Oxford: Wiley Blackwell.

Caron, J., Ouardani, M., & Dea, S. (2000). Diagnosis and differentiation of *Mycoplasma hyopneumoniae* and *Mycoplasma hyorhinis* infections in pigs by PCR amplification of the p36 and p46 genes. *Journal of Clinical Microbiology, 38*, 1390-1396.

Cartner, S. C., Lindsey, J. R., Gibbs-Erwin, J., Cassell, G. H., & Simecka, J. W. (1998). Roles of innate and adaptive immunity in respiratory mycoplasmosis. *Infection and Immunity, 66*(8), 3485-3491.

Caruso, J. P., & Ross, R. F. (1990). Effects of *Mycoplasma hyopneumoniae* and *Actinobacillus* (*Haemophilus*) *pleuropneumoniae* infections on alveolar macrophage functions in swine. *American Journal of Veterinary Research, 51*(2), 227-231.

Castro, L. A. de, Pedroso, T. R., Kuchiishi, S. S., Ramenzoni, M., Kich, J. D., Zaha, A., Vainstein, M. H., & Ferreira, H. B. (2006). Variable number of tandem aminoacid repeats in adhesion-related CDS products in *Mycoplasma hyopneumoniae* strains. *Veterinary Microbiology,* 116, 258-269. https://doi.org/10.1016/j.vetmic.2006.04.022

Chae, C. (2016). Porcine respiratory disease complex: Interaction of vaccination and porcine circovirus type 2, porcine reproductive and respiratory syndrome virus, and *Mycoplasma hyopneumoniae*. *Veterinary Journal, 212*, 1-6. https://doi.org/10.1016/j.tvjl.2015.10.030

Charlebois, A., Marois-Créhan, C., Hélie, P., Gagnon, C. A., Gottschalk, M., & Archambault, M. (2014). Genetic diversity of *Mycoplasma hyopneumoniae* isolates of abattoir pigs. *Veterinary Microbiology, 168*(2-4), 348-356. https://doi.org/10.1016/j.vetmic.2013.11.006

Chauvin, C., Beloeil, P. A., Orand, J. P., Sanders, P., & Madec, F. (2002). A survey of group-level antibiotic prescriptions in pig production in France. *Preventive Veterinary Medicine, 55*(2), 109-120.

Chen, A. Y., Fry, S., Daggard, G., & Mukkur, K. (2008). Evaluation of immune response to recombinant potential protective antigens of *Mycoplasma hyopneumoniae* delivered as cocktail DNA and/or recombinant protein vaccines in mice. *Vaccine, 26*, 372-378.

Chen, A. Y., Fry, S., Forbes-Faulkner, J., Daggard, G., & Mukkur, T. (2006a). Evaluation of the immunogenicity of the P97R1 adhesin of *Mycoplasma hyopneumoniae* as a mucosal vaccine in mice. *Journal of Medical Microbiology, 55*, 923-929.

Chen, A. Y., Fry, S., Forbes-Faulkner, J., Daggard, G., & Mukkur, T. (2006b). Comparative immunogenicity of *M. hyopneumoniae* NrdF encoded in different expression systems delivered orally via attenuated *S.* Typhimurium aroA in mice. *Veterinary Microbiology, 114*, 252-259.

Chen, J., Liao, C., Mao, S., & Weng, C. (2001). A recombinant chimera composed of repeat region RR1 of *Mycoplasma hyopneumoniae* adhesin with Pseudomonas exotoxin: In vivo evaluation of specific IgG response in mice and pigs. *Veterinary Microbiology, 80*, 347-357.

Chen, Y., Wang, S., Yang, W., Chen, Y., Lin, H., & Shiuan, D. (2003). Expression and immunogenicity of *Mycoplasma hyopneumoniae* heat shock protein antigen p42 by DNA vaccination. *Infection and Immunity, 71*, 1155-1160.

Choi, C., Kwon, D., Jung, K., Ha, Y., Lee, Y. H., Kim, O., Park, H. K., Kim, S. H., Hwang, K. K., & Chae, C. (2006). Expression of inflammatory cytokines in pigs experimentally infected with *Mycoplasma hyopneumoniae*. *Journal of Comparative Pathology, 134*(1), 40-46. doi:10.1016/j.jcpa.2005.06.009

Choi, Y. K., Goyal, S. M., & Joo, H. S. (2003). Retrospective analysis of etiologic agents associated with respiratory diseases in pigs. *Canadian Veterinary Journal, 44*(9), 735-737.

Chopra-Dewasthaly, R., Baumgartner, M., Gamper, E., Innerebner, C., Zimmermann, M., Schilcher, F., Tichy, A., Winter, P., Jechlinger, W., Rosengarten, R., & Spergser, J. (2012). Role of Vpma phase variation in *Mycoplasma agalactiae* pathogenesis. *FEMS Immunology and Medical Microbiology, 66*, 307-322. https://doi.org/10.1111/j.1574-695X.2012.01010.x

Chopra-Dewasthaly, R., Citti, C., Glew, M. D., Zimmermann, M., Rosengarten, R., & Jechlinger, W. (2008). Phase-locked mutants of *Mycoplasma agalactiae*: Defining the molecular switch of high-frequency Vpma antigenic variation. *Molecular Microbiology, 67*, 1196-1210. https://doi.org/10.1111/j.1365-2958.2007.06103.x

Chopra, I., & Roberts, M. (2001). Tetracycline antibiotics: mode of action, applications, molecular biology, and epidemiology of bacterial resistance. *Microbiology and Molecular Biology Reviews, 65*(2), 232-260.

Chu, H. W., Jeyaseelan, S., Rino, J. G., Voelker, D. R., Wexler, R. B., Campbell, K., Harbeck, R. J., & Martin, R. J. (2005). TLR2 signaling is critical for *Mycoplasma pneumoniae*-induced airway mucin expression. *Journal of Immunology, 174*(9), 5713-5719.

Ciprián, A., Cruz, T. A., & de la Garza, M. (1994). *Mycoplasma hyopneumoniae*: Interaction with other agents in pigs, and evaluation of immunogens. *Archives in Medical Research, 25*, 235-239.

Ciprián, A., Palacios, J. M., Quintanar, D., Batista, L., Colmenares, G., Cruz, T., Romero, A., Schnitzlein, W., & Mendoza, S. (2012). Florfenicol feed supplemented decrease the clinical effects of *Mycoplasma hyopneumoniae* experimental infection in swine in Mexico. *Research in Veterinary Science, 92*(2), 191-196.

Ciprián, A., Pijoan, C., Cruz, T., Camacho, J., Tórtora, J., Colmenares, G., López-Revilla, R., & de la Garza, M. (1988). *Mycoplasma hyopneumoniae* increases the susceptibility of pigs to experimental *Pasteurella multocida* pneumonia. *Canadian Journal of Veterinary Research, 52*(4), 434-438.

Citti, C., Dordet-Frisoni, E., Nouvel, L. X., Kuo, C. H., & Baranowski, E. (2018). Horizontal gene transfers in *Mycoplasmas* (*Mollicutes*). *Current Issues in Molecular Biology, 29*, 3-22.

Citti, C., Kim, M. F., & Wise, K. S. (1997). Elongated versions of Vlp surface lipoproteins protect *Mycoplasma hyorhinis* escape variants from growth-inhibiting host antibodies. *Infection and Immunity, 65*, 1773-1785.

Citti, C., Nouvel, L.-X., & Baranowski, E. (2010). Phase and antigenic variation in mycoplasmas. *Future Microbiology, 5*, 1073-1085. https://doi.org/10.2217/fmb.10.71

Citti, C., Watson-McKnown, R., Droesse, M., & Wise, K. S. (2000). Gene families encoding phase- and size-variable surface lipoproteins of *Mycoplasma hyorhinis*. *Journal of Bacteriology, 182*, 1356-1363.

Clark, K., Wu, C. C., Van Alstine, W. G., & Knox, K. E. (1998). Evaluation of the effectiveness of a macrolide antibiotic on reduction of respiratory pathogens in 12 and 21 day weaned pigs. *Journal of Swine Health and Production, 6*, 257-262.

Clavijo, M. H., Murray, D., Oliveira, S., & Rovira, A. (2017). Infection dynamics of *Mycoplasma hyorhinis* in three commercial pig populations. *Veterinary Record, 181*, 68.

Clavijo, M. J., Oliveira, S., Zimmerman, J., Rendahl, A., & Rovira, A. (2014). Field evaluation of a quantitative polymerase chain reaction assay for *Mycoplasma hyorhinis*. *Journal of Veterinary Diagnostic Investigation, 26*, 755-760.

Clawson, M. L., Murray, R. W., Sweeney, M. T., Apley, M. D., DeDonder, K. D., Capik, S. F., Larson, R. L., Lubbers, B. V., White, B. J., Kalbfleisch, T. S., Schuller, G., Dickey, A. M., Harhay, G. P., Heaton, M. P., Chitko-McKown, C. G., Brichta-Harhay, D. M., Bono, J. L. & Smith, T. P. L. (2016). Genomic signatures of *Mannheimia haemolytica* that associate with the lungs of cattle with respiratory disease, an integrative conjugative element, and antibiotic resistance genes. *BMC Genomics, 17*, 982. https://doi.org/10.1186/s12864-016-3316-8

CLSI. (1999). Methods for determining bactericidal activity of antibacterial agents; approved guideline. In *CLSI document M26-A*. Wayne, PA, USA: Clinical and Laboratory Standards Institute.

CLSI. (2011). Methods for antimicrobial susceptibility testing for human mycoplasmas; approved guideline. In *CLSI document M43-A*. Wayne, PA, USA: Clinical and Laboratory Standards Institute.

CLSI. (2013). Performance standards for antimicrobial disk and dilution susceptibility tests for bacteria isolated from animals, 4th Edition. In *CLSI document VET01-A4*. Wayne, PA, USA: Clinical and Laboratory Standards Institute.

CLSI. (2015). Performance standards for antimicrobial disk and dilution susceptibility tests for bacteria isolated from animals, 3rd Edition. In *CLSI document VET01S*. Wayne, PA, USA: Clinical and Laboratory Standards Institute.

Collignon, P. J., & McEwen, S. A. (2019). One health. Its importance in helping to better control antimicrobial resistance. *Tropical Medicine and Infectious Disease, 4*(1), tropicalmed4010022.

Comité de l'Antibiogramme de la Société Française de Microbiologie. (2019). Recommandations vétérinaires 2019. Available on https://www.sfm-microbiologie.org/wp-content/uploads/2019/07/CASFM_VET2019.pdf.

Conceição, F., Moreira, A., &. Dellagostin, O. (2006). A recombinant chimera composed of R1 repeat region of *Mycoplasma hyopneumoniae* P97 adhesin with Escherichia coli heat-labile enterotoxin B subunit elicits immune response in mice. *Vaccine, 24*, 5734-5743.

Cook, B. S., Beddow, J. G., Manso-Silván, L., Maglennon, G. A., & Rycroft, A. N. (2016). Selective medium for culture of *Mycoplasma hyopneumoniae. Veterinary Microbiology, 195*, 158-164.

Correa-Fiz, F., Gonçalves dos Santos, J. M., Illas, F., & Aragon, V. (2019). Antimicrobial removal on piglets promotes health and higher bacterial diversity in the nasal microbiota. *www.nature.com/scientificreports, 9*, 6545.

Cox, E., Verdonck, F., Vanrompay, D., & Goddeeris, B. (2006). Adjuvants modulating mucosal immune responses or directing systemic responses towards the mucosa. *Veterinary Research, 37*, 511-539.

Crosbie, P. J. A., & Woodhead, M. A. (2009). Long-term macrolide therapy in chronic inflammatory airway diseases. *European Respiratory Journal, 33,* 171-181.

Cullen, J. M., & Rycroft, A. N. (1994). Phagocytosis by pig alveolar macrophages of *Actinobacillus pleuropneumoniae* serotype 2 mutant strains defective in haemolysin II (ApxII) and pleurotoxin (ApxIII). *Microbiology, 140* (Pt 2), 237-244.

Cury, J., Touchon, M., & Rocha, E. P. C. (2017). Integrative and conjugative elements and their hosts: Composition, distribution and organization. *Nucleic Acids Research, 45*, 8943-8956. https://doi.org/10.1093/nar/gkx607

Czaja, T., Kanci, A., Lloyd, L. C., Markham, P. F., Whithear, K. G., & Browning, G. F. (2002). Induction of enzootic pneumonia in pigs by the administration of an aerosol of in vitro-cultured *Mycoplasma hyopneumoniae. Veterinary Record, 150*, 9-11.

Darling, A. E., Jospin, G., Lowe, E., Iv, F. A. M., Bik, H. M., Eisen, J. A. (2014). PhyloSift: Phylogenetic analysis of genomes and metagenomes. *PeerJ,* 2, e243. https://doi.org/10.7717/*peerj*.243

Darling, A. C. E., Mau, B., Blattner, F. R., & Perna, N. T. (2004). Mauve: Multiple alignment of conserved genomic sequence with rearrangements. *Genome Research, 14*, 1394-1403. https://doi.org/10.1101/gr.2289704

Darling, A. E., Miklós, I., & Ragan, M. A. (2008). Dynamics of genome rearrangement in bacterial populations. *PLoS Genetics, 4*. https://doi.org/10.1371/journal.pgen.1000128

De Busser, E., Mateusen, B., Vicca, J., Hoelzle, L., Haesebrouck, F., & Maes, D. (2008). *Mycoplasma suis* infection in suckling pigs on a Belgian farm. *Vlaams Diergeneeskundig Tjidschrift,* 77, 182-186.

de Castro, L. A., Pedroso, T. R., Kuchiishi, S. S., Ramenzoni, M., Kich, J. D., Zaha, A., Vainstein, M. H., & Ferreira, H. B. (2006). Variable number of tandem aminoacid repeats in adhesion-related CDS products in *Mycoplasma hyopneumoniae* strains. *Veterinary Microbiology, 116*, 258-269.

De Oliveira, N. R., Jorge, S., Gomez, C. K., Rizzi, C., Dias Pacce, V., Farias Collares, F., Garcia Monte, L., & Dellagostin, O. (2017). A novel chimeric protein composed of recombinant *Mycoplasma hyopneumoniae* antigens as a vaccine candidate evaluated in mice. *Veterinary Microbiology, 201*, 146-153.

DeBey, M. C., & Ross, R. F. (1994). Ciliostasis and loss of cilia induced by *Mycoplasma hyopneumoniae* in porcine tracheal organ cultures. *Infection and Immunity, 62*(12), 5312-5318.

DeBey, M. C., Roth, J. A., & Ross, R. F. (1993). Enhancement of the increase in intracellular calcium concentration in stimulated neutrophils by *Mycoplasma hyopneumoniae*. *Veterinary Research Communications,* 17, 249-257.

Deblanc, C., Delgado-Ortega, M., Gorin, S., Berri, M., Paboeuf, F., Berthon, P., Herrler, G., Meurens, F., & Simon, G. (2016). *Mycoplasma hyopneumoniae* does not affect the interferon-related anti-viral response but predisposes the pig to a higher level of inflammation following swine influenza virus infection. *Journal of General Virology, 97*(10), 2501-2515.

Deblanc, C., Gorin, S., Queguiner, S., Gautier-Bouchardon, A.V., Ferre, S., Amenna, N., Cariolet, R., & Simon, G. (2012). Pre-infection of pigs with *Mycoplasma hyopneumoniae* modifies outcomes of infection with European swine influenza virus of H1N1, but not H1N2, subtype. *Veterinary Microbiology, 157*(1-2), 96-105.

Deblanc, C., Robert, F., Pinard, T., Gorin, S., Queguiner, S., Gautier-Bouchardon, A. V., Ferre, S., Garraud, J. M., Cariolet, R., Brack, M., & Simon, G. (2013). Pre-infection of pigs with *Mycoplasma hyopneumoniae* induces oxidative stress that influences outcomes of a subsequent infection with a swine influenza virus of H1N1 subtype. *Veterinary Microbiology, 162*(2-4), 643-651.

Declerck, I., Dewulf, J., Sarrazin, S., & Maes, D. (2016). Long-term effects of colostrum intake on piglet mortality and performance. *Journal of Animal Science, 94*(4), 1633-1643.

Dee, S. A. (1996). The porcine respiratory disease complex: Are subpopulations important? (commentary). *Journal of Swine Health and Production, 4*(3), 147-149.

Dee, S. A., Joo, H. S., Henry, S., Tokach, L., Park, B. K., Molitor, T., & Pijoan, C. (1996). Detecting subpopulations after PRRS virus infection in large breeding herds using multiple serologic tests. *Journal of Swine Health and Production, 4*(4), 181-184.

Dee, S. A., Joo, H. S., & Pijoan, C. (1995). Controlling the spread of PRRS virus in the breeding herd through management of the gilt pool. *Journal of Swine Health and Production, 3*(2), 64-69.

Deeney, A. S., Maglennon, G. A., Chapat, L., Crussard, S., Jolivet, E., & Rycroft, A. N. (2019). *Mycoplasma hyopneumoniae* evades phagocytic uptake by porcine alveolar macrophages in vitro. *Veterinary Research, 50*(1), 51. doi:10.1186/s13567-019-0667-6

Del Pozo Sacristán, R. (2014). *Treatment and control of Mycoplasma hyopneumoniae infections*. PhD thesis, Faculty ofVeterinary Medicine, Ghent University.

Del Pozo Sacristán, R., Rodriguez, A. L., Sierens, A., Vranckx, K., Boyen, F., Dereu, A., Haesebrouck, F., & Maes, D. G. (2012a). Efficacy of in-feed medication with chlortetracycline in a farrow-to-finish herd against a clinical outbreak of respiratory disease in fattening pigs. *Veterinary Record, 171*(25), 645.

Del Pozo Sacristán, R., Sierens, A., Marchioro, S., Vangroenweghe, F., Jourquin, J., Labarque, G., Haesebrouck, F., & Maes D. (2014). Efficacy of early *Mycoplasma hyopneumoniae* vaccination against mixed respiratory disease in older fattening pigs. *Veterinary Record, 174*, 197.

Del Pozo Sacristán, R., Thiry, J., Vranckx, K., Lopez Rodriguez, A., Chiers, K., Haesebrouck, F., Thomas, E., & Maes, D. (2012b). Efficacy of florfenicol injection in the treatment of *Mycoplasma hyopneumoniae* induced respiratory disease in pigs. *Veterinary Journal, 194*(3), 420-422.

Deutscher, A. T., Jenkins, C., Minion, F. C., Seymour, L. M., Padula, M. P., Dixon, N. E., Walker, M. J., & Djordjevic, S. P. (2010). Repeat regions R1 and R2 in the P97 paralogue Mhp271 of *Mycoplasma hyopneumoniae* bind heparin, fibronectin and porcine cilia. *Molecular. Microbiology, 78*, 444-458. https://doi.org/10.1111/j.1365-2958.2010.07345.x

Deutscher, A. T., Tacchi, J. L., Minion, F. C., Padula, M. P., Crossett, B., Bogema, D. R., Jenkins, C., Kuit, T. A., Walker, M. J., & Djordjevic, S. P. (2012). *Mycoplasma hyopneumoniae* surface proteins Mhp385 and Mhp384 bind host cilia and glycosaminoglycans and are endoproteolytically processed by proteases that recognize different cleavage motifs. *Journal of Proteome Research, 11*, 1924-1936. https://doi.org/10.1021/pr201115v

Devillers, N., Farmer, C., Le Dividich, J., & Prunier, A. (2007). Variability of colostrum yield and colostrum intake in pigs. *Animal, 1*(7), 1033-1041.

Di Teodoro, G., Marruchella, G., Mosca, F., Di Provvido, A., Sacchini, F., Tiscar, P. G., & Scacchia, M. (2018). Polymorphonuclear cells and reactive oxygen species in contagious bovine pleuropneumonia: New insight from in vitro investigations. *Veterinary Immunology and Immunopathology, 201*, 16-19. doi:10.1016/j.vetimm.2018.04.011

Dias, G. B., do Amaral, R. B., Gatto, I. R. H., Lapera, I. M., de Oliveira, L. G., Lux Hoppe, E. G., Machado, R. Z., & André, M. R. (2019). Molecular detection of *Mycoplasma suis* in captivewhite-lipped peccaries (Tayassu pecari) and wild boars (Sus scrofa) in Brazil. *Comparative Immunology, Microbiology & Infectious Diseases, 63*, 94-96.

Dicke, W. E. (1934). Anaplasmosis-like disease in swine. *Veterinary Medicine, 29*, 288.

Dietz, S., Lassek, C., Mack, S. L., Ritzmann, M., Stadler, J., Becher, D., Hoelzle, K., Riedel, K., & Hoelzle, L. E. (2016). Updating the proteome of the uncultivable hemotrophic *Mycoplasma suis* in experimentally infected pigs. *Proteomics, 16*, 609-613.

Dietz, S., Mack, S. L., Hoelzle, K., Becker, K., Jannasch, C., Stadler, J., Ritzmann, M., & Hoelzle, L. E. (2014). Quantitative PCR analysis of *Mycoplasma suis* shedding patterns during experimental infection. *Veterinary Microbiology, 172*, 581-585.

Djordjevic, S. P., Cordwell, S. J., Djordjevic, M. A., Wilton, J., & Minion, F. C. (2004). Proteolytic processing of the *Mycoplasma hyopneumoniae* cilium adhesin. *Infection and Immunity,* 72, 2791-2802.

Djordjevic, S. P., Eamens, G. J., Romalis, L. F., Nicholls, P. J., Taylor, V., & Chin, J. (1997). Serum and mucosal antibody responses and protection in pigs vaccinated against *Mycoplasma hyopneumoniae* with vaccines containing a denatured membrane antigen pool and adjuvant. *Australian Veterinary Journal, 75*, 504-511. https://doi.org/10.1111/j.1751-0813.1997.tb14383.x

Djordjevic, S. R., Forbes, W. A., Forbes-Faulkner, J., Kuhnert, P., Hum, S., Hornitzky, M. A., Vilei, E. M., & Frey, J. (2001). Genetic diversity among *Mycoplasma* species bovine group 7: Clonal isolates from an outbreak of polyarthritis, mastitis, and abortion in dairy cattle. *Electrophoresis,* 22, 3551-3561. https://doi.org/10.1002/1522-2683(200109)22:16<3551::AID-ELPS3551>3.0.CO;2-#

do Nascimento, N. C., Guimaraes, A. M. S., Dos Santos, A. P., Chu, Y., Marques, L. M., & Messick, J. B. (2018). RNA-Seq based transcriptome of whole blood from immunocompetent pigs (*Sus scrofa*) experimentally infected with *Mycoplasma suis* strain Illinois. *Veterinary Research, 49*, 49.

Dobbs, N. A., Odeh, A. N., Sun, X., & Simecka, J. W. (2009). The multifaceted role of T cell-mediated immunity in pathogenesis and resistance to *Mycoplasma* respiratory disease. *Current Trends in Immunology, 10*, 1-19.

Dom, P., Haesebrouck, F., Kamp, E. M., & Smits, M. A. (1992). Influence of *Actinobacillus pleuropneumoniae* serotype 2 and its cytolysins on porcine neutrophil chemiluminescence. *Infection and Immunity, 60*(10), 4328-4334.

Done, S. H., Paton, D. J., & White, M. E. (1996). Porcine reproductive and respiratory syndrome (PRRS): A review, with emphasis on pathological, virological and diagnostic aspects. *British Veterinary Journal, 152*(2), 153-174.

Donham, K.J. (1991). Association of environmental air contaminants with disease and productivity in swine. *American Journal of Veterinary Research, 52*, 1723-1730.

Dorr, P. M., Baker, R. B., Almond, G. W., Wayne, S. R., & Gebreyes, W. A. (2007). Epidemiologic assessment of porcine circovirus type 2 coinfection with other pathogens in swine. *Journal of American Veterinary Medical Association, 230*(2), 244-250.

Dos Santos, L., Clavijo, M., Sreevatsan, S., Rovira, A., Moreira, M., & Pieters, M. (2015b). Genotyping of *Mycoplasma hyorhinis* using multiple-locus variable number tandem repeat analysis. *Journal of Microbiological Methods, 111*, 87-92.

Dos Santos, L. F., Sreevatsan, S., Torremorell, M., Moreira, M. A. S., Sibila, M., & Pieters, M. (2015a). Genotype distribution of *Mycoplasma hyopneumoniae* in swine herds from different geographical regions. *Veterinary Microbiology, 175*, 374-381. https://doi.org/10.1016/j.vetmic.2014.11.018

Doster, A. R., & Lin, B. C. (1988). Identification of *Mycoplasma hyopneumoniae* in formalin-fixed porcine lung, using an indirect immunoperoxidase method. *American Journal of Veterinary Research, 49*, 1719-1721.

Doyle, L. P. (1932). A rickettsia-like or anaplasmosis-like disease in swine. *Journal of the American Veterinary Medical Association, 81*, 668-671.

Drexler, C. S., Witvliet, M. H., Raes, M., van de Laar, M., Eggen, A. A., & Thacker, E. L. (2010). Efficacy of combined porcine reproductive and respiratory syndrome virus and *Mycoplasma hyopneumoniae* vaccination in piglets. *Veterinary Record, 166*(3), 70-74.

Drummond, J., Curtis, S., & Simon, J. (1978). Effects of atmospheric ammonia on pulmonary bacterial clearance in the young pig. *American Journal of Veterinary Research, 39*, 211-212.

Drummond, J., Curtis, S., Simon, J., & Norton, H. (1980). Effects of aerial ammonia on growth and health of young pigs. *Journal of Animal Science, 50*, 1085-1091.

Dubosson, C. R., Conzelmann, C., Miserez, R., Boerlin, P., Frey, J., Zimmermann, W., Häni, H., & Kuhnert, P. (2004). Development of two real-time PCR assays for the detection of *Mycoplasma hyopneumoniae* in clinical samples. *Veterinary Microbiology, 102*, 55-65.

Dybvig, K., Tu, A.-H., & Clapper, B. (2005). Mycoplasma Phages. *Phages*, 223-237. https://doi.org/10.1128/9781555816506.ch11

Dykhuis-Haden, C., Painter, T., Fangman, T., & Holtkamp, D. (2012). Assessing production parameters and economic impact of swine influenza, PRRS and Mycoplasma hyopneumoniae on finishing pigs in a large production system. Paper presented at the 43rd American Association of Swine Veterinarians Annual Meeting, Denver, Colorado.

Eamens, G. J., Gonsalves, J. R., Whittington, A. M., & Turner, B. (2012). Evaluation of serovar-independent ELISA antigens of *Actinobacillus pleuropneumoniae* in pigs, following experimental challenge with *A. pleuropneumoniae*, *Mycoplasma hyopneumoniae* and *Pasteurella multocida*. *Australian Veterinary Journal, 90*, 225-234.

Elbers, R., Tielen, J., Snijders, J., Cromwijk, W., & Hunneman, W. (1992). Epidemiological studies on lesions in finishing pigs in the Netherlands. I. Prevalence, seasonality and interrelationship. *Preventive Veterinary Medicine, 14*(3-4), 217-231.

Ennis, R. S., Dalgard, D., Willerson, J. T., Barden, J. A., & Decker, J. L. (1971). *Mycoplasma hyorhinis* swine arthritis. II. Morphologic features. *Arthritis & Rheumatology, 14*, 202-211.

Erickson, B. Z., Ross, R. F., Rose, D. L., Tully, J. G., & Bové, J. M. (1986). *Mycoplasma hyopharyngis*, a new species from swine. *International Journal of Systematic Bacteriology, 36*, 55-59.

Erlandson, K. R., Evans, R. B., Thacker, B. J., Wegner, M. W., & Thacker, E.L. (2005). Evaluation of three serum antibody enzyme-linked immunosorbent assays for *Mycoplasma hyopneumoniae*. *Journal of Swine Health and Production, 13*(4), 198-203.

European Committee on Antimicrobial Susceptibility Testing. Data from the EUCAST MIC distribution website, last accessed 27/09/2019. http://www.eucast.org.

Fablet, C., Marois, C., Dorenlor, V., Eono, F., Eveno, E., Jolly, J. P., Le Devendec, L., Kobisch, M., Madec, F., & Rose, N. (2012a). Bacterial pathogens associated with lung lesions in slaughter pigs from 125 herds. *Research in Veterinary Science, 93*(2), 627-630.

Fablet, C., Marois, C., Kobisch, M., Madec, F., & Rose, N. (2010). Estimation of the sensitivity of four sampling methods for *Mycoplasma hyopneumoniae* detection in live pigs using a Bayesian approach. *Veterinary Microbiology, 143*(2-4), 238-245.

Fablet, C., Marois-Créhan, C., Simon, G., Grasland, B., Jestin, A., Kobisch, M., Madec, F., & Rose, N. (2012b). Infectious agents associated with respiratory diseases in 125 farrow-to-finish pig herds: A cross-sectional study. *Veterinary Microbiology*, *157*, 152-163.

Fagan, P., Djordjevic, S., Chin, J., Eamens, G., &. Walker, M. (1997). Oral immunization of mice with attenuated *Salmonella typhimurium* aroA expressing a recombinant *Mycoplasma hyopneumoniae* antigen (NrdF). *Infection and Immunity*, *65*, 2502-2507.

Fagan, P., Walker, M., Chin, J., Eamens, G., &. Djordjevic, S. (2001). Oral immunization of swine with attenuated *Salmonella typhimurium* aroA SL3261 expressing a recombinant antigen of *Mycoplasma hyopneumoniae* (NrdF) primes the immune system for a NrdF specific secretory IgA response in the lungs. *Microbial Pathogenesis, 30*, 101-110.

Fagan, P.K., Djordjevic, S. P., Eamens, G. J., Chin, J., & Walker, M. J. (1996). Molecular characterization of a ribonucleotide reductase (nrdF) gene fragment of *Mycoplasma hyopneumoniae* and assessment of the recombinant product as an experimental vaccine for enzootic pneumonia. *Infection and Immunity*, 64, 1060-1064.

Falk, K., Hoie, S., & Lium, B. M. (1991). An abattoir survey of pneumonia and pleurisy in slaughter weight swine from 9 selected herds. II. Enzootic pneumonia of pigs: Microbiological findings and their relationship to pathomorphology. *Acta Veterinaria Scandinavica, 32*, 67-77.

Fano, E., Pijoan, C., & Dee, S. (2005). Dynamics and persistence of *Mycoplasma hyopneumoniae* infection in pigs. *Canadian Journal Veterinary Research, 69*, 223-228.

Fano, E., Pijoan, C., & Dee, S. (2006b). Assessing the M. hyopneumoniae infection pattern in a sow herd following an increase in the replacement rate. Proceedings 19th IPVS Congress, Copenhagen, Denmark, Volume 1, 135.

Fano, E., Pijoan, C., & Dee, S. (2007). Effect of *Mycoplasma hyopneumoniae* colonization at weaning on disease severity in growing pigs. *Canadian Journal of Veterinary Research*, *71*, 195-200.

Fano, E., Pijoan, C., Dee, S., & Deen, J. (2012). Longitudinal assessment of two *Mycoplasma hyopneumoniae* enzyme-linked immunosorbent assays in challenged and contact-exposed pigs. *Journal of Veterinary Diagnostic Investigation, 24*, 383-387.

Fano, E., Pijoan, C., Dee, S., & Torremorell, M. (2006a). Assessment of the effect of sow parity on the prevalence of *Mycoplasma hyopneumoniae* in piglets at weaning. Proceedings 19th IPVS Congress, Copenhagen, Denmark, Volume 1, 96.

Feld, N. C., Qvist, P., Ahrens, P., Friis, N. F., & Meyling, A. (1992). A monoclonal blocking ELISA detecting serum antibodies to *Mycoplasma hyopneumoniae*. *Veterinary Microbiology*, *30*, 35-46.

Felde, O., Kreizinger, Z., Sulyok, K. M., Hrivnak, V., Kiss, K., Jerzsele, A., Biksi, I., & Gyuranecz, M. (2018a). Antibiotic susceptibility testing of *Mycoplasma hyopneumoniae* field isolates from Central Europe for fifteen antibiotics by microbroth dilution method. *PLoS One, 13*(12), e0209030.

Felde, O., Kreizinger, Z., Sulyok, K. M., Marton, S., Bányai, K., Korbuly, K., Kiss, K., Biksi, I., & Gyuranecz, M. (2018b). Genotyping *Mycoplasma hyopneumoniae* isolates based on multi-locus sequence typing, multiple-locus variable-number tandem repeat analysis and analysing gene p146. *Veterinary Microbiology*, *222*, 85-90. https://doi.org/10.1016/j.vetmic.2018.07.004

Felder, K. M., Carranza, P. M., Gehrig, P. M., Roschitzki, B., Barkow-Oesterreicher, S., Hoelzle, K., Riedel, K., Kube, M., & Hoelzle, L. E. (2012). Insights into the gene expression profile of uncultivable hemotrophic *Mycoplasma suis* during acute infection, obtained using proteome analysis. *Journal of Bacteriology*, *194*, 1505-1514.

Felder, K. M., Hoelzle, K., Heinritzi, K., Ritzmann, M., & Hoelzle, L. E. (2010). Antibodies to actin in autoimmune haemolytic anaemia. *BMC Veterinary Research*, *6*, 18.

Felder, K. M., Hoelzle, K., Ritzmann, M., Kilchling, T., Schiele, D., Heinritzi, K., Groebel, K., & Hoelzle, L. E. (2011). Hemotrophic mycoplasmas induce programmed cell death in red blood cells. *Cellular Physiology and Biochemistry*, *27*, 557-564.

Feng, Z., Shao, G., Liu, M., Wang, H., Gan, Y., & Wu, X. (2010b). Development and validation of a SIgA-ELISA for the detection of *Mycoplasma hyopneumoniae* infection. *Veterinary Microbiology, 143*, 410-416.

Feng, Z., Shao, G., Liu, M., Wu, X., Zhou, Y., & Gan, Y. (2010a). Immune responses to the attenuated *Mycoplasma hyopneumoniae* 168 strain vaccine by intrapulmonic immunization in piglets. *Agricultural Sciences in China, 9*, 423-431.

Feng, Z., Wei, Y., Li, G., Lu, X., Wan, X., Pharr, G., Wang, Z., Kong, M., Gan, Y., Bai, F., Liu, M., Xiong, Q., Wu, X., & Shao, G. (2013). Development and validation of an attenuated *Mycoplasma hyopneumoniae* aerosol vaccine. *Veterinary Microbiology, 167*, 417-424.

Ferrari, L., Borghetti, P., Gozio, S., De Angelis, E., Ballotta, L., Smeets, J., Blanchaert, A., & Martelli, P. (2011). Evaluation of the immune response induced by intradermal vaccination by using a needle-less system in comparison with the intramuscular route in conventional pigs. *Research in Veterinary Science, 90*, 64-71.

Ferrarini, M. G., Mucha, S. G., Parrot, D., Meiffrein, G., Ruggiero Bachega, J. F., Comte, G., Zaha, A., & Sagot, M. F. (2018). Hydrogen peroxide production and myo-inositol metabolism as important traits for virulence of *Mycoplasma hyopneumoniae*. *Molecular Microbiology, 108*, 683-696.

Ferrarini, M. G., Siqueira, F. M., Mucha, S. G., Palama, T. L., Jobard, É., Elena-Hermann, B., Vasconcelos, A. T. R., Tardy, F., Schrank, I. S., & Zaha, A. (2016). Insights on the virulence of swine respiratory tract mycoplasmas through genome-scale metabolic modeling. *BMC Genomics,* 17, 353.

Fisch, A., Beutinger Marchioro, S., Klazer Gomes, C., Galli, V., Rodriguez de Oliveira, N., Simionatto, S., Dellagostin, O. A., Mendonça, M., Nunes Moreira, A., & Conceição, F. R. (2016). Commercial bacterins did not induce detectable levels of antibodies in mice against *Mycoplasma hyopneumoniae* antigens strongly recognized by swine immune system. *Trials in Vaccinology, 5*, 32-37.

Flesja, K. I., & Ulvesaeter, H. O. (1980). Pathological lesions in swine at slaughter. III. Inter-relationship between pathological lesions, and between pathological lesions and 1) carcass quality and 2) carcass weight. *Acta Veterinaria Scandinavica Supplement, 74*, 1-22.

Fohner, A. E., Sparreboom, A., Altman, R. B., & Klein, T. E. (2017). PharmGKB summary: Macrolide antibiotic pathway, pharmacokinetics/pharmacodynamics. *Pharmacogenetics and Genomics, 27*(4), 164-167.

Foley, J. (2015). Mini-review: Strategies for variation and evolution of bacterial antigens. *Computational and Structural Biotechnology Journal, 13*, 407-416. https://doi.org/10.1016/j.csbj.2015.07.002

Fourour, S., Fablet, C., Tocqueville, V., Dorenlor, V., Eono, F., Eveno, E., Kempf, I., & Marois-Créhan, C. (2018). A new multiplex real-time TaqMan® PCR for quantification of *Mycoplasma hyopneumoniae*, *M. hyorhinis* and *M. flocculare*: Exploratory epidemiological investigations to research mycoplasmal association in enzootic pneumonia-like lesions in slaughtered pigs. *Journal of Applied Microbiology, 125*, 345-355.

Fourour, S., Marois-Crehan, C., Martelet, L., Fablet, C., Kempf, I., Gottschalk, M., & Segura, M. (2019a). Intra-species and inter-species differences in cytokine production by porcine antigen-presenting cells stimulated by *Mycoplasma hyopneumoniae*, *M. hyorhinis*, and *M. flocculare*. *Pathogens, 8*(1), 34.

Fourour, S., Tocqueville, V., Paboeuf, F., Lediguerher, G., Morin, N., Kempf, I., & Marois-Crehan, C. (2019b). Pathogenicity study of *Mycoplasma hyorhinis* and *M. flocculare* in specific-pathogen-free pigs pre-infected with *M. hyopneumoniae*. *Veterinary Microbiology, 232*, 50-57.

Fraile, L., Alegre, A., Lopez-Jimenez, R., Nofrarias, M., & Segales, J. (2010). Risk factors associated with pleuritis and cranio-ventral pulmonary consolidation in slaughter-aged pigs. *Veterinary Journal, 184*(3), 326-333.

Frank, H. (1989). [For the therapy and prophylaxis of enzootic pneumonia of the pig with Baytril®] (in German). *Wiener Tierarztliche Monatschrift, 76*, 312.

French, G. L. (2006). Bactericidal agents in the treatment of MRSA infections - The potential role of daptomycin. *Journal of Antimicrobial Chemotherapy, 58*(6), 1107-1117.

Frey, J. (1995). Virulence in *Actinobacillus pleuropneumoniae* and RTX toxins. *Trends in Microbiology, 3*(7), 257-261.

Frey, J., Haldimann, A., & Nicolet, J. (1992). Chromosomal heterogeneity of various *Mycoplasma hyopneumoniae* field strains. *International Journal of Systematic Bacteriology, 42*, 275-280. https://doi.org/10.1099/00207713-42-2-275

Friis, N. F. (1971). Mycoplasmas cultivated from the respiratory tract of Danish pigs. *Acta Veterinaria Scandinavica, 12*, 69-79.

Friis, N. F. (1971). *Mycoplasma hyorhinis* as a causative agent in pneumonia of pigs. *Acta Veterinaria Scandinavica, 12*, 116-119.

Friis, N. F. (1973). The pathogenicity of *Mycoplasma flocculare*. *Acta Veterinaria Scandinavica, 14*, 344-346.

Friis, N. F. (1974). *Mycoplasma suipneumoniae* and *Mycoplasma flocculare* in comparative pathogenicity studies. *Acta Veterinaria Scandinavica, 15*, 507-518.

Friis, N. F. (1975). Some recommendations concerning primary isolation of *Mycoplasma suipneumoniae* and *Mycoplasma flocculare*, a survey. *Nordisk Veterinaermedicin, 27*, 337-339.

Friis, N. F. (1976). A serologic variant of *Mycoplasma hyorhinis* recovered from the conjunctiva of swine. *Acta Veterinaria Scandinavica, 17*, 343-353.

Friis, N. F., & Feenstra, A. A. (1994). *Mycoplasma hyorhinis* in the etiology of serositis among piglets. *Acta Veterinaria Scandinavica, 35*, 93-98.

Friis, N. F., & Szancer, J. (1994). Sensitivity of certain porcine and bovine mycoplasmas to antimicrobial agents in a liquid medium test compared to a disc assay. *Acta Veterinary Scandinavia, 35*(4), 389-394.

Friis, N. F., Ahrens, P., Hagedorn-Olsen, T., Nielsen, E. O., & Kokotovic, B. (2003). *Mycoplasma hyopharyngis* isolation from swine. *Acta Veterinaria Scandinavica, 44*, 103-104.

Fu, T., Ulmer, J. B., Caufield, M. J., Deck, R. R., Friedman, A., Wang, S., Liu, X., Donnelly, J. J., & Liu, M. A. (1997). Priming of cytotoxic T lymphocytes by DNA vaccines. Requirement for professional antigen presenting cells and evidences for antigen transfer from myocytes. *Molecular Medicine, 3*, 362-371.

Gabridge, M. G., Johnson, C. K., & Cameron, A. M. (1974). Cytotoxicity of *Mycoplasma pneumoniae* membranes. *Infection and Immunity, 10*, 1127-1134.

Galli, V., Simionatto, S., Marchioro, S., Fisch, A., Gomes, C., Conceição, F., &. Dellagostin, O. (2012). Immunisation of mice with *Mycoplasma hyopneumoniae* antigens P37, P42, P46 and P95 delivered as recombinant subunit or DNA vaccines. *Vaccine, 31*, 135-140.

Galli, V., Simionatto, S., Marchioro, S. B., Klabunde, G. H. F., Conceição, F. R., & Dellagostin, O. A. (2013). Recombinant secreted antigens from *Mycoplasma hyopneumoniae* delivered as a cocktail vaccine enhance the immune response of mice. *Clinical Vaccine Immunology CVI, 20*, 1370-1376. https://doi.org/10.1128/CVI.00140-13

Galvao Ferrarini, M., Mucha, S. G., Parrot, D., Meiffrein, G., Ruggiero Bachega, J. F., Comte, G., Zaha, A., & Sagot, M. F. (2018). Hydrogen peroxide production and myo-inositol metabolism as important traits for virulence of *Mycoplasma hyopneumoniae*. *Molecular Microbiology, 108*, 683-696.

Gamage, P. P.T., Khalili, F., Khurshidul Azad, M. D., & Mansy, H. A. (2018). Modeling inspiratory flow in a porcine lung airway. *Journal of Biomechanical Engineering, 140*(6).

Ganter, M., Dudziak, D., & Delbeck, F. (1995). [Treatment of swine with chronic pneumonia with chlortetracycline-medicated feed] (in German). *Deutsche Tierärztliche Wochenschrift, 102*, 44-49.

Garcia-Morante, B., Dors, A., León-Kempis, R., Pérez de Rozas, A., Segalés, J., & Sibila, M. (2018). Assessment of the in vitro growing dynamics and kinetics of the non-pathogenic J and pathogenic 11 and 232 *Mycoplasma hyopneumoniae* strains. *Veterinary Research, 49*(1), 45. doi:10.1186/s13567-018-0541-y

Garcia-Morante, B., Segalés, J., Fraile, L., Llarden, G., Coll, T., & Sibila, M. (2017b). Potential use of local and systemic humoral immune response parameters to forecast *Mycoplasma hyopneumoniae* associated lung lesions. *PLoS One, 12*(4), e0175034. doi:10.1371/journal.pone.0175034

Garcia-Morante, B., Segalés, J., Fraile, L., Pérez de Rozas, A., Maiti, H., Coll, T., & Sibila, M. (2016a). Assessment of *Mycoplasma hyopneumoniae*-induced Pneumonia using different lung lesion scoring systems: A comparative review. *Journal of Comparative Pathology, 154*, 125-1344.

Garcia-Morante, B., Segalés, J., Lopez-Soria, S., de Rozas, A. P., Maiti, H., Coll, T., & Sibila, M. (2016b). Induction of mycoplasmal pneumonia in experimentally infected pigs by means of different inoculation routes. *Veterinary Research, 47*(1), 54. doi:10.1186/s13567-016-0340-2

Garcia-Morante, B., Segalés, J., Serrano, E., & Sibila, M. (2017a). Determinants for swine mycoplasmal pneumonia reproduction under experimental conditions: A systematic review and recursive partitioning analysis. *PLoS One, 12*(7), e0181194. doi:10.1371/journal.pone.0181194

Garza-Moreno, L., Pieters, M., López-Soria, S., Carmona, M., Krejci R., Segalés, J., & Sibila, M. (2019a). Comparison of vaccination protocols against *Mycoplasma hyopneumoniae* during the gilt acclimation period. *Veterinary Microbiology, 229*, 7-13.

Garza-Moreno, L., Segalés, J., Aragón, V., Correa-Fiz, F., Pieters, M., Carmona, M., Krejci, R., & Sibila, M. (2019b). Characterization of *Mycoplasma hyopneumoniae* strains in vaccinated and non-vaccinated pigs from Spanish slaughterhouses. *Veterinary Microbiology, 231*, 18-23. doi:10.1016/j.vetmic.2019.02.023

Garza-Moreno, L., Segalés, J., Pieters, M., Romagosa, A., & Sibila, M. (2017). Survey on *Mycoplasma hyopneumoniae* gilt acclimation practices in Europe. *Porcine Health Management, 3*, 21.

Garza-Moreno, L., Segalés, J., Pieters, M., Romagosa, A., & Sibila, M. (2018). Acclimation strategies in gilts to control *Mycoplasma hyopneumoniae* infection. *Veterinary Microbiology, 219*, 23-29. doi:10.1016/j.vetmic.2018.04.005

Gaurivaud, P., Baranowski, E., Pau-Roblot, C., Sagne, E., Citti, C., & Tardy, F. (2016). *Mycoplasma agalactiae* Secretion of beta-(1-->6)-Glucan, a rare polysaccharide in prokaryotes, is governed by high-frequency phase variation. *Applied and Environmental Microbiology, 82*, 3370-3383.

Gaurivaud, P., Laigret, F., & Bove, J. M. (1996). Insusceptibility of members of the class *Mollicutes* to rifampin: Studies of the *Spiroplasma citri* RNA polymerase beta-subunit gene. *Antimicrobial Agents and Chemotherapy, 40*(4), 858-862.

Gautier-Bouchardon, A.V. (2018). Antimicrobial resistance in *Mycoplasma* spp. *Microbiology Spectrum, 6*(4). doi:10.1128/microbiolspec.ARBA-0030-2018

Gautier-Bouchardon, A.V., Reinhardt, A. K., Kobisch, M., & Kempf, I. (2002). *In vitro* development of resistance to enrofloxacin, erythromycin, tylosin, tiamulin and oxytetracycline in *Mycoplasma gallisepticum*, *Mycoplasma iowae* and *Mycoplasma synoviae*. *Veterinary Microbiology, 88*(1), 47-58.

Geary, S. J., & Walczak, E. M. (1983). Cytopathic effect of whole cells and purified membranes of *Mycoplasma hyopneumoniae*. *Infection and Immunity, 41*, 132-136.

Geers, R., Dellaert, B., Goedseels,V., Hoogerbrugge, A.,Vranken, E., Maes, F., & Berckmans, D. (1989). An assessment of optimal air temperature in pig houses by the quantification of behavioural and health related problems. *Animal Production, 48*, 571-578.

Geudeke, T., Junker, K., & Greijdanus, S. (2016). *Mycoplasma hyosynoviae* infections in finisher pigs with arthritis. Proceedings 24th International Pig Veterinary Society, Dublin, Ireland, 247.

Giguère, S. (2013). Macrolides, azalides and ketolides. In S. Giguère, J. F. Prescott, & P. M. Dowling (Eds.), *Antimicrobial therapy in veterinary medicine* (5th ed.; pp. 211-232). Ames, Iowa: Whiley Blackwell.

Giménez-Lirola, L. G., Meiroz-De-Souza-Almeida, H., Magtoto, R. L., McDaniel, A. J., Merodio, M. M., Matias Ferreyra, F. S., Poonsuk, K., Gatto, I. R. H., Baum, D. H., Ross, R. F., Arruda, P. H. E., Schwartz, K. J., Zimmerman, J. J., Derscheid, R. J., & Arruda, B. L. (2019). Early detection and differential serodiagnosis of *Mycoplasma hyorhinis* and *Mycoplasma hyosynoviae* infections under experimental conditions. *PLoS One, 14*, e0223459.

Girard, V., & Mourez, M. (2006). Adhesion mediated by autotransporters of Gram-negative bacteria: structural and functional features. *Research in Microbiology, 157*, 407-416. https://doi.org/10.1016/j.resmic.2006.02.001

Glasser, K. (1939). Zum Schweinegrippe problem. *Deutsche Tierärztliche Wochenschrift, 47*, 209.

Godinho, K. S. (2008). Susceptibility testing of tulathromycin: interpretative breakpoints and susceptibility of field isolates. *Veterinary Microbiology, 129*(3-4), 426-432.

Gois, M., Cerný, M., Rozkosný, V., & Sovadina M. (1969). Studies on the epizootiological significance of some species of mycoplasma isolated from nasal swabs and lungs of pigs. *Zentralblatt für Veterinärmedizin Series B, 16*, 253-265.

Gois, M., Pospisil, M., Cerny, M., & Mrva, V. (1971). Production of pneumonia after intranasal inoculation of gnotobiotic piglets with three strains of *Mycoplasma hyorhinis*. *Journal of Comparative Pathology, 81*, 401-411.

Gois, M., Valícek, L., & Sovadina, M. (1968). *Mycoplasma hyorhinis*, a causative agent of pig pneumonia. *Zentralblatt für Veterinärmedizin B, 15*(2), 230-240.

Goldstein, I., Wallet, F., Robert, J., Becquemin, M. H., Marquette, C. H., & Rouby, J. J. (2002). Lung tissue concentrations of nebulized amikacin during mechanical ventilation in piglets with healthy lungs. *American Journal of Respiratory and Critical Care Medicine, 165*(2), 171-175.

Gomes-Neto, J. C., Bower, L., Erickson, B. Z., Wang, C., Raymond, M., & Strait, E. L. (2015). Quantitative real-time polymerase chain reaction for detecting *Mycoplasma hyosynoviae* and *Mycoplasma hyorhinis* in pen-based oral, tonsillar, and nasal fluids. *Journal of Veterinary Science, 16*, 195-201.

Gomes-Neto, J. C., Gauger, P. C., Strait, E. L., Boyes, N., Madson, M., & Schwartz, K. J. (2012). *Mycoplasma*-associated arthritis: Critical points for diagnosis. *Journal of Swine Health and Production, 20*, 82-86.

Gomes-Neto, J. C., Raymond, M., Bower, L., Ramirez, A., Madson, D. M., Strait, E. L., Rosey, E. L., & Rapp-Gabrielson, V. J. (2016). Two clinical isolates of *Mycoplasma hyosynoviae* pattern of lameness and pathogen detection in experimentally challenged pigs. *Journal of Veterinary Science, 17*, 489-496.

Gomes-Neto, J. C., Strait, E. L., Raymond, M., Ramirez, A., & Minion, F. C. (2014). Antibody responses of swine following infection with *Mycoplasma hyopneumoniae, M. hyorhinis, M. hyosynoviae* and *M. flocculare*. *Veterinary Microbiology, 174*, 163-171. doi:10.1016/j.vetmic.2014.08.008

Gondaira, S., Higuchi, H., Nishi, K., Iwano, H., & Nagahata, H. (2017). Mycoplasma bovis escapes bovine neutrophil extracellular traps. *Veterinary Microbiology, 199*, 68-73. doi:10.1016/j.vetmic.2016.12.022

Goodison, S., Urquidi, V., Kumar, D., Reyes, L., & Rosser, C. J. (2013). Complete genome sequence of *Mycoplasma hyorhinis* strain SK76. *Genome Announcement, 1*, e00101-12.

Goodwin, R. F. (1979). Activity of tiamulin against *Mycoplasma suipneumoniae* and enzootic pneumonia of pigs. *Veterinary Record, 104*, 194-195.

Goodwin, R. F. (1985). Apparent reinfection of enzootic-pneumonia-free pig herds: Search for possible causes. *Veterinary Record, 116*(26), 690-694.

Goodwin, R. F., Hodgson, R. G., Whittlestone, P., & Woodhams, R. L. (1969). Some experiments relating to artificial immunity in enzootic pneumonia of pigs. *Journal of Hygiene (London), 67*(3), 465-476.

Goodwin, R. F. W., Pomeroy, A. P., & Whittlestone, P. (1965). Production of enzootic pneumonia in pigs with a mycoplasma. *Veterinary Record, 77*, 1247-1249.

Goodwin, R. F. W., Pomeroy, A. P., & Whittlestone, P. (1967). Characterization of *Mycoplasma suipneumonia*: A mycoplasma causing enzootic pneumonia of pigs. *Journal of Hygiene (London), 65*, 85-96.

Goodwin, R. F., & Whittlestone, P. (1963). Production of enzootic pneumonia in pigs with an agent grown in tissue culture from the natural disease. *British Journal of Experimental Pathology, 44*, 291-299.

Goodwin, R. F., & Whittlestone, P. (1964). Production of enzootic pneumonia in pigs with a microorganism grown in media free from living cells. *Veterinary Record, 76*, 611-613.

Goodwin, R. F., & Whittlestone, P. (1966). Enzootic pneumonia of pigs: Growth and behaviour of the causal mycoplasma in liquid media. *British Journal of Experimental Pathology*, *47*(5), 518-524.

Gordon, W. (1963). Environmental studies in pig housing. V. The effects of housing on the degree and incidence of pneumonia in bacon pigs. *British Veterinary Journal, 119*, 307-314.

Goret, J., Beven, L., Faustin, B., Contin-Bordes, C., Le Roy, C., Claverol, S., Renaudin, H., Bebear, C., & Pereyre, S. (2017). Interaction of *Mycoplasma hominis* PG21 with human dendritic cells: Interleukin-23-inducing mycoplasmal lipoproteins and inflammasome activation of the cell. *Journal of Bacteriology, 199*(15). doi:10.1128/jb.00213-17

Gottschalk, M. (2012a). Actinobacillosis. In J. J. Zimmerman, L. A. Karriker, A. Ramirez, K. J. Schwartz, & G. Stevenson (Eds.), *Diseases of swine* (10th ed.; pp. 653-669). Oxford: Wiley Blackwell.

Gottschalk, M. (2012b). Streptococcosis. In J. J. Zimmerman, L. A. Karriker, A. Ramirez, K. J. Schwartz, & G. Stevenson (Eds.), *Diseases of swine* (10th ed.; pp. 841-855). Oxford: Wiley Blackwell.

Gresham, A., Rogers, J., Tribe, H., & Phipps, L. P. (1994). *Eperythrozoon suis* in weaned pigs. *Veterinary Record, 134*, 71-72.

Groebel, K., Hoelzle, K., Wittenbrink, M. M., Ziegler U., & Hoelzle, L. E. (2009). *Mycoplasma suis* invades porcine erythrocytes. *Infection and Immunity*, 77, 576-584.

Grosse Beilage, E., & Schreiber, A. (2005). [Vaccination of sows against *Mycoplasma hyopneumoniae* with Hyoresp]. *Deutsche Tierärztliche Wochenschrift, 112*(7), 256-261.

Grover, R. K., Zhu, X., Nieusma, T., Jones, T., Boreo, I., MacLeod, A., Mark, A., Niessen, S., Kom, H., Kong, L., Assad-Garcia, N., Kwon, K., Chesi, M., et al. (2014). A structurally distinct human mycoplasma protein that generically blocks antigen-antibody union. *Science, 343*, 656-661.

Guimaraes, A. M., Biondo, A. W., Lara, A. C., & Messick, J. B. (2007). Exploratory study of *Mycoplasma suis* (*Eperythrozoon suis*) on four commercial pig farms in southern Brazil. *Veterinary Record, 160*, 50-53.

Guimaraes, A. M., Santos, A. P., SanMiguel, P., Walter, T., Timenetsky, J., & Messick, J. B. (2011a). Complete genome sequence of *Mycoplasma suis* and insights into its biology and adaption to an erythrocyte niche. *PLoS One*, 6, e19574.

Guimaraes, A. M., Vieira, R. F., Poletto, R., Vemulapalli, R., Santos, A. P., de Moraes, W., Cubas, Z. S., Santos, L. C., Marchant-Forde, J. N., Timenetsky, J., Biondo, A. W., & Messick, J. B. (2011b). A quantitative TaqMan PCR assay for the detection of *Mycoplasma suis*. *Journal of Applied Microbiology, 111*, 417-425.

Guimaraes, A. M., Santos, A. P., Timenetsky, J., Bower, L. P., Strait, E., & Messick, J. B. (2014). Identification of *Mycoplasma suis* antigens and development of a multiplex microbead immunoassay. *Journal of Veterinary Diagnostic Investigation, 26*, 203-212.

Gulliksen, S. M., Lium, B., Framstad, T., Jorgensen, A., Skomsoy, A., Gjestvang, M., Kjelvik, O., & Grontvedt, C. A. (2019). The successful eradication of *Mycoplasma hyopneumoniae* from Norwegian pig herds – 10 years later. European Symposium of Porcine Health Management, Utrecht, The Netherlands.

Gulrajani, T. S., & Beveridge, W. I. (1951). Studies on respiratory diseases of pigs. IV. Transmission of infectious pneumonia and its differentiation from swine influenza. *Journal of Comparative Pathology, 61*(2), 118-139.

Henthorn, C. R., Minion, C. F., & Sahin, O. (2018). Utilization of macrophage extracellular trap nucleotides by *Mycoplasma hyopneumoniae*. *Microbiology,* 164, 1394-1404.

Hernandez-Garcia, J., Robben, N., Magnée, D., Eley, T., Dennis, I., Kayes, S. M., Thomson, J. R., & Tucker, A. W. (2017). The use of oral fluids to monitor key pathogens in porcine respiratory disease complex. *Porcine Health Management, 3*, 7.

Hickman-Davis, J. M. (2002). Role of innate immunity in respiratory mycoplasma infection. *Frontiers in Bioscience, 7*, d1347-1355.

Hodge, L. M., & Simecka, J. W. (2002). Role of upper and lower respiratory tract immunity in resistance to *Mycoplasma* respiratory disease. *Journal of Infectious Diseases, 186*(2), 290-294. doi:10.1086/341280

Hodgins, D., Shewen, P., & Dewey, C. (2004). Influence of age and maternal antibodies on antibody responses of neonatal piglets vaccinated against *Mycoplasma hyopneumoniae*. *Journal of Swine Health and Production, 12*(1), 10-16.

Hoelzle, K., Doser, S., Ritzmann, M., Heinritzi, K., Palzer, A., Elicker, S., Kramer, M., Felder, K. M., & Hoelzle, L. E. (2009). Vaccination with the *Mycoplasma suis* recombinant adhesion protein MSG1 elicits a strong immune response but fails to induce protection in pigs. *Vaccine, 27*, 5376-5382.

Hoelzle, K., Engels, M., Kramer, M. M., Wittenbrink, M. M., Dieckmann, S. M., & Hoelzle, L. E. (2010). Occurrence of *Mycoplasma suis* in wild boars (*Sus scrofa* L.). *Veterinary Microbiology, 143*, 405-409.

Hoelzle, L. E., Hoelzle, K., Harder, A., Ritzmann, M., Aupperle, H., Schoon, H. A., Heinritzi, K., & Wittenbrink, M. M. (2007a). First identification and functional characterization of an immunogenic protein in unculturable haemotrophic mycoplasmas (*Mycoplasma suis* HspA1). *FEMS Immunology and Medical Microbiology, 49*, 215-223.

Hoelzle, L. E., Hoelzle, K., Helbling, M., Aupperle, H., Schoon, H. A., Ritzmann, M., Heinritzi, K., Felder, K. M., & Wittenbrink, M. M. (2007b). MSG1, a surface-localised protein of *Mycoplasma suis* is involved in the adhesion to erythrocytes. *Microbes and Infection, 9*, 466-474.

Hoelzle, L. E., Helbling, M., Hoelzle, K., Ritzmann, M., Heinritzi, K., & Wittenbrink, M. M. (2007c). First LightCycler real-time PCR assay for the quantitative detection of *Mycoplasma suis* in clinical samples. *Journal of Microbiological Methods, 70*, 346-354.

Hoelzle, K., Grimm, J., Ritzmann, M., Heinritzi, K., Torgerson, P., Hamburger, A., Wittenbrink, M. M. & Hoelzle, L. E. (2007d). Use of recombinant antigens to detect antibodies against Mycoplasma suis, with correlation of serological results to hematological findings. *Clinical and Vaccine Immunology, 14*, 1616-1622.

Hoelzle, L. E. (2008). Haemotrophic mycoplasmas: Recent advances in *Mycoplasma suis*. *Veterinary Microbiology, 130*, 215-226.

Hoelzle, L. E., Hoelzle, K., Ritzmann, M., Heinritzi, K., & Wittenbrink, M. M. (2006). *Mycoplasma suis* antigens recognized during humoral immune response in experimentally infected pigs. *Clinical and Vaccine Immunology, 13*, 116-122.

Hoelzle, L. E., Zeder, M., Felder, K. M., & Hoelzle, K. (2014). Pathobiology of *Mycoplasma suis*. *Veterinary Journal, 202*, 20-25.

Holmgren, N., Lundeheim, N., & Wallgren, P. (1999). Infections with *Mycoplasma hyopneumoniae* and *Actinobacillus pleuropneumoniae* in fattening pigs; Influence of piglet production system and influence on production parameters. *Journal of Veterinary Medicine Series B, 46*, 535-544.

Holst S., Yeske P., & Pieters M. (2015). Elimination of *Mycoplasma hyopneumoniae* from breed-to-wean farms: A review of current protocols with emphasis on herd closure and medication. *Journal of Swine Health and Production, 23*(6), 321-330.

Holtkamp, D., Rotto, R., & Garcia, R. (2007). The economic cost of major health challenges in large US swine production systems. Proceedings AASV, Orlando, Florida, 85-90.

Hooper, D. C. (2001). Emerging mechanisms of fluoroquinolone resistance. *Emerging and Infectious Disease, 7*(2), 337-341.

Howard, C. J. (1980). Variation in the susceptibility of bovine mycoplasmas to killing by the alternative complement pathway in bovine serum. *Immunology, 41*(3), 561-568.

Howard, C. J. (1984). Comparison of bovine IgG1, IgG2 and IgM for ability to promote killing of Mycoplasma bovis by bovine alveolar macrophages and neutrophils. *Veterinary Immunology and Immunopathology, 6*(3-4), 321-326.

Howard, C. J., Thomas, L. H., & Parsons, K. R. (1987). Immune response of cattle to respiratory mycoplasmas. *Veterinary Immunology and Immunopathology, 17*(1-4), 401-412.

Hsu, F. S., Yeh, T. P., & Lee, C. T. (1983). Tiamulin feed medication for the maintenance of weight gains in the presence of mycoplasmal pneumonia in swine. *Journal of Animal Science, 57*(6), 1474-1478.

Hsu, T., Artiushin, S., & Minion, F. C. (1997). Cloning and functional analysis of the P97 swine cilium adhesin gene of *Mycoplasma hyopneumoniae*. *Journal of Bacteriology, 179*, 1317-1323.

Hsu, T., & Minion, F. C. (1998). Identification of the cilium binding epitope of the *Mycoplasma hyopneumoniae* P97 adhesin. *Infection and Immunity, 66*, 4762-4766.

Hu, Z., Yin, J., Shen, K., Kang, W., & Chen, Q. (2009). Outbreaks of hemotrophic mycoplasma infections in China. *Emerging Infectious Diseases, 15*, 1139-1140.

Huang, D. S., Guan, P., Wu, W., Shen, T. F., Liu, H. L., Cao, S., & Zhou, H. (2012). Infection rate of *Eperythrozoon* spp. in Chinese population: A systematic review and meta-analysis since the first Chinese case reported in 1991. *BMC Infectious Disease, 12*, 171.

Huang, Y., Ladinig, A., Ashley, C., Haines, D., & Harding, J. (2014). Innate and adaptive immune responses of snatch-farrowed porcine-colostrum-deprived pigs to *Mycoplasma hyopneumoniae* vaccination. *BMC Veterinary Research, 10*, 29. doi:10.1186/s12917-014-0219-2

Huhn, R. G. (1971a). The action of certain antibiotics and ether on swine enzootic pneumonia. *Canadian Journal of Comparative Medicine, 35*(1), 1-4.

Huhn, R. G. (1971b). Swine enzootic pneumonia: Age susceptibility and treatment schemata. *Canadian Journal of Comparative Medicine, 35*(1), 77-81.

Hurnik, D., Dohoo, I. R., & Bate, L. A. (1994). Types of farm management as risk factors for swine respiratory disease. *Preventive Veterinary Medicine, 20*(1-2), 147-157.

Hyun, Y., Ellis, M., & Johnson, R. W. (1998). Effects of feeder type, space allowance, and mixing on the growth performance and feed intake pattern of growing pigs. *Journal of Animal Science, 76*(11), 2771-2778.

Hyun, Y., Ellis, M., Curtis, S. E., & Johnson, R. W. (2005). Environmental temperature, space allowance, and regrouping: Additive effects of multiple concurrent stressors in growing pigs. *Journal Swine Health and Production, 13*(3), 131-138.

Inamoto, T., Takahashi, H., Yamamoto, K., Nakai, Y., & Ogimoto, K. (1994). Antibiotic susceptibility of *Mycoplasma hyopneumoniae* isolated from swine. *Journal of Veterinary Medical Science, 56*(2), 393-394.

Inui, T., Taira, T., Matsushita, T., & Endo, T. (1998). Pharmacokinetic properties and oral bioavailabilities of difloxacin in pig and chicken. *Xenobiotica, 28*(9), 887-893.

Inzana, T. J., Ma, J., Workman, T., Gogolewski, R. P., & Anderson, P. (1988). Virulence properties and protective efficacy of the capsular polymer of *Haemophilus* (*Actinobacillus*) *pleuropneumoniae* serotype 5. *Infection and Immunity, 56*(8), 1880-1889.

Jang, J. C., Jin, X. H., Hong, J. S., & Kim, Y. Y. (2017). Effects of different space allowances on growth performance, blood profile and pork quality in a grow-to-finish production system. *Asian-Australasian Journal of Animal Science, 30*(12), 1796-1802.

Jansson, E., Backman, A., Hakkarainen, K., Miettinen, A., & Seniusová, B. (1983). Mycoplasmas and arthritis. *Rheumatology, 42*, 315-319.

Jarocki, V. M., Santos, J., Tacchi, J. L., Raymond, B. B., Deutscher, A. T., Jenkins, C., Padula, M. P., & Djordjevic, S. P. (2015). MHJ_0461 is a multifunctional leucine aminopeptidase on the surface of *Mycoplasma hyopneumoniae*. *Open Biology, 5*, 140175.

Jarocki, V. M., Steele, J. R., Widjaja, M., Tacchi, J. L., Padula, M. P., & Djordjevic, S. P. (2019). Formylated N-terminal methionine is absent from the *Mycoplasma hyopneumoniae* proteome: Implications for translation initiation. *International Journal of Medical Microbiology, 309*, 288-298.

Jarosz, L. S., Gradzki, Z., & Kalinowski, M. (2014). *Trueperella pyogenes* infections in swine: Clinical course and pathology. *Polish Journal of Veterinary Science, 17*(2), 395-404.

Jarvill-Taylor, K. J., VanDyk, C., & Minion, F. C. (1999). Cloning of *mnuA*, a membrane nuclease gene of *Mycoplasma pulmonis*, and analysis of its expression in *Escherichia coli*. *Journal of Bacteriology, 181*, 1853-1860.

Jayappa, H., Davis, R, Rapp-Gabrielsen, V. T. W., & Thacker, E. (2001). Evaluation of the efficacy of *Mycoplasma hyopneumoniae* bacterin following immunization of young pigs in the presence of varying levels of maternal antibodies. American Association of Swine Veterinarians - Annual Meeting, Nashville, Tennessee, United States, 237-241.

Jenkins, C., Wilton, J. L., Minion, F. C., Falconer, L., Walker, M. J., & Djordjevic, S. P. (2006). Two domains within the *Mycoplasma hyopneumoniae* cilium adhesin bind heparin. *Infection and Immunity, 74*, 481-487.

Jensen, C. S., Ersbøll, A. K., & Nielsen, J. P. (2002). A meta-analysis comparing the effect of vaccines against *Mycoplasma hyopneumoniae* on daily weight gain in pigs. *Preventive Veterinary Medicine, 54*, 265-278.

Jenvey, C. J., Reichel, M. P., & Cockcroft, P. D. (2015). *Erysipelothrix rhusiopathiae* and *Mycoplasma hyopneumoniae*: The sensitivities of enzyme-linked immunosorbent assays for detecting vaccinated sows of unknown disease status using serum and colostrum, and the correlation of the results for the sow serum, colostrum, and piglet serum. *Journal of Veterinary Diagnostic and Investigation, 27*, 211-216.

Jiang, F., He, J., Navarro-Alvarez, N., Xu, J., Li, X., Li, P., & Wu, W. (2016). Elongation factor Tu and heat shock protein 70 are membrane-associated proteins from *Mycoplasma ovipneumoniae* capable of inducing strong immune response in mice. *PLoS One, 11*, e0161170. https://doi.org/10.1371/journal.pone.0161170

Johansson, K. E., Heldtander, M. U., & Pettersson, B. (1998). Characterization of mycoplasmas by PCR and sequence analysis with universal 16S rDNA primers. *Methods in Molecular Biology, 104*, 145-165.

Johnson, C. M., & Grossman, A. D. (2015). Integrative and Conjugative Elements (ICEs): What they do and how they work. *Annual Review of Genetics, 49*, 577-601.

Jolie, R., Backstrom, L., Olson, L., & Chase, C. (1999). Respiratory and systemic health parameters in pigs raised in a conventional farm or in isolation. *Journal of Swine Health and Production*, 7(6), 269-275.

Jones, G. F., Rapp-Gabrielson, V., Wilke, R., Thacker, E. L., Thacker, B. J., Gergen, L., Sweeney, D., & Wasmoen, T. (2005). Intradermal vaccination for *Mycoplasma hyopneumoniae*. *Journal of Swine Health and Production, 13*, 19-27.

Jones, H. P., & Simecka, J. W. (2003). T lymphocyte responses are critical determinants in the pathogenesis and resistance to mycoplasma respiratory disease. *Frontiers in Bioscience, 8*, d930-945.

Jones, H. P., Tabor, L., Sun, X., Woolard, M. D., & Simecka, J. W. (2002). Depletion of CD8+ T cells exacerbates CD4+ Th cell-associated inflammatory lesions during murine mycoplasma respiratory disease. *Journal of Immunology, 168*(7), 3493-3501.

Jones, S. (1993). *The language of the genes*. London: Harper Collins.

Jores, J., Ma, L., Ssajjakambwe, P., Schieck, E., Liljander, A., Chandran, S., Stoffel, M., Cippa, V., Arfi, Y., Assad-Garcia, N., Falquet, L., Sirand-Pugnet, P., et al. (2019). Removal of a subset of non-essential genes fully attenuates a highly virulent mycoplasma strain. *Frontiers in Microbiology, 10*, Article 664.

Jores, J., Schieck, E., Liljander, A., Sacchini, F., Posthaus, H., Lartigue, C., Blanchard, A., Labroussaa, F., & Vashee, S. (2018). In vivo role of capsular polysaccharide in *Mycoplasma mycoides*. *Journal of Infectious Diseases*. doi:10.1093/infdis/jiy713

Jorge, S., de Oliveira, N. R., Marchioro, S. B., Fisch, A., Gomes, C. K., Hartleben, C. P., Conceição, F. R., & Dellagostin, O. A. (2014). The *Mycoplasma hyopneumoniae* recombinant heat shock protein P42 induces an immune response in pigs under field conditions. *Comparative Immunology Microbiology and Infectious Diseases, 37*, 229-236. https://doi.org/10.1016/j.cimid.2014.07.001

Jorsal, S. E., & Thomsen, B. L. (1988). A Cox regression analysis of risk factors related to *Mycoplasma suipneumoniae* reinfection in Danish SPF-herds. *Acta Veterinaria Scandinavica*, Suppl., *84*, 436-438.

Jouglar, J., Borne, P., Sionneau, G., Bardon, T., & Eclache, D. (1993). [Evaluation of the interest of tiamulin-oxytetracycline combination in the prevention of bronchopneumonia in swine] (in French). *Revue de Médécine Vétérinaire, 144*, 981-988.

Jungling, A., Erhard, M. H., Heinritzi, K., & Lösch, U. (1994). Bedeutung und Verlauf eines Kälteagglutinins bei der *Eperythrozoon suis*-Infektion des Schweines. *Berliner und Münchener Tierärztliche Wochenschrift, 107*, 271-275.

Karriker, L. A., Coetzee, J., Friendship, R. M., & Prescott, J. F. (2012). Drug pharmacology, therapy and prophylaxis. In J. J. Zimmerman, L. A. Karriker, A. Ramirez, K. J. Schwartz, & G. W. Stevenson (Eds.), *Diseases of swine* (10th ed.; pp. 106-118). Ames, Iowa: Wiley Blackwell.

Karuppannan, A. K., & Opriessnig, T. (2017). Porcine circovirus type 2 (PCV2) vaccines in the context of current molecular epidemiology. *Viruses, 9*(5), 99.

Kasza, L., Hodges, R. T., Betts, A. O., & Trexler, P. C. (1969). Pneumonia in gnotobiotic pigs produced by simultaneous inoculation of a swine adenovirus and *Mycoplasma hyopneumoniae*. *Veterinary Record, 84*(11), 262-267.

Kavanagh, N. (1994). The effect of pulse medication with a combination of tiamulin and oxytetracycline on the performance of fattening pigs in a herd infected with enzootic pneumonia. *Irish Veterinary Journal, 47*, 58-61.

Kearse, M., Moir, R., Wilson, A., Stones-Havas, S., Cheung, M., Sturrock, S., Buxton, S., Cooper, A., Markowitz, S., Duran, C., Thierer, T., Ashton, B., Meintjes, P., & Drummond, A. (2012). Geneious Basic: An integrated and extendable desktop software platform for the organization and analysis of sequence data. *Bioinformatics, 28*, 16471649. https://doi.org/10.1093/*bioinformatics*/bts199

Khan, A., Rilfors, L., Wieslander, A., & Lindblom, G. (1981). The effect of cholesterol on the phase structure of glucolipids from *Acholeplasma laidlawii* membranes. *European Journal of Biochemistry, 116*, 215-220.

Khare, S., Dorfleutner, A., Bryan, N. B., Yun, C., Radian, A. D., de Almeida, L., Rojanasakul, Y., & Stehlik, C. (2012). An NLRP7-containing inflammasome mediates recognition of microbial lipopeptides in human macrophages. *Immunity, 36*(3), 464-476. doi:10.1016/j.immuni.2012.02.001

Kim, B., Lee, K., Han, K., Kim, D., Ha, Y., Kim, C.H., Oh, Y., Kang, I., Lee, J., & Chae, C. (2010). Development of in situ hybridization for the detection of *Mycoplasma hyorhinis* in formalin-fixed paraffin-embedded tissues from naturally infected pigs with polyserositis. *The Journal of Veterinary Medical Science, 72*, 1225-1227.

Kim, J., Chung, H. K., & Chae, C. (2003). Association of porcine circovirus 2 with porcine respiratory disease complex. *Veterinary Journal, 166*(3), 251-256.

King, K. W., Faulds, D. H., Rosey, E. L., & Yancey, R. J., Jr. (1997). Characterization of the gene encoding Mhp1 from *Mycoplasma hyopneumoniae* and examination of Mhp1's vaccine potential. *Vaccine, 15*, 25-35.

Kinsley, A. T. (1932). Protozoan-like body in the blood of swine. *Veterinary Medicine, 27*, 196.

Kinsley, A. T., & Ray, J. D. (1934). Anaplasmosis-like disease in swine. *Journal of the American Veterinary Medical Association, 84*, 391-392.

Kishima, M., & Ross, R. F. (1985). Suppressive effect of nonviable *Mycoplasma hyopneumoniae* on phytohemagglutinin-induced transformation of swine lymphocytes. *American Journal of Veterinary Research, 46*(11), 2366-2368.

Kishima, M., Ross, R. F., & Kuniyasu, C. (1985). Cell-mediated and humoral immune response to *Mycoplasma hyopneumoniae* in pigs enhanced by dextran sulfate. *American Journal Veterinary Research, 46*(2), 456-462.

Kixmöller, M., Ritzmann, M., Eddicks, M., Saalmüller, A., Elbers, K., & Fachinger, V. (2008). Reduction of PMWS-associated clinical signs and co-infections by vaccination against PCV2. *Vaccine, 26*, 3443-3451.

Klein, U., de Jong, A., Moyaert, H., El Garch, F., Leon, R., Richard-Mazet, A., Rose, M., Maes, D., Pridmore, A., Thomson, J. R., & Ayling, R. D. (2017). Antimicrobial susceptibility monitoring of *Mycoplasma hyopneumoniae* and *Mycoplasma bovis* isolated in Europe. *Veterinary Microbiology, 204*, 188-193.

Kloster, A. M., Descarga, C. O., Davies, P., Piscitelli, H. G., Diaz, L. R., & Zielinski, G.C. (1987). *Eperythrozoonosis porcina*: Observaciones sobre la infeccion natural y experimental. *Veterinaria Argentina, 3*, 27-40.

Kobayashi, H., Morozumi, T., Munthali, G., Mitani, K., Ito, N., & Yamamoto, K. (1996). Macrolide susceptibility of *Mycoplasma hyorhinis* isolated from pigs. *Antimicrobial Agents and Chemotherapy, 40*, 1030-1032.

Kobayashi, H., Nakajima, H., Shimizu, Y., Eguchi, M., Hata, E., & Yamamoto, K. (2005). Macrolides and lincomycin susceptibility of *Mycoplasma hyorhinis* and variable mutation of domain II and V in 23S ribosomal RNA. *Journal of Veterinary Medical Science, 67*, 795-800.

Köbe, K. (1933). Die aetiologie der ferkelgrippe (enzootische pneumonic des Ferkels). *Zentralblatt für Bakteriologie, Mikrobiologie und Hygiene Orig., 129*, 161.

Kobisch, M., & Friis, N. F. (1996). Swine mycoplasmoses. *Scientific and Technical Review of the Office International des Epizooties / Revue Scientifique et Technique de l'Office International des Epizooties, 15*(4), 1569-1605.

Kobisch, M., & Sibelle, C. (1982). [Evaluation of the effectiveness of tiamulin in piglets experimentally infected with *Mycoplasma hyopneumoniae*] (in French). *Receuil de Médécine Vétérinaire, 158*, 375-381.

Kobisch, M., & Tillon, J. P. (1985). Respiratory diseases of swine: Some epidemiologic aspects. *Comparative Immunology, Microbiology and Infectious Diseases, 8*(1), 65-71.

Kobisch, M., Blanchard, B., & Le Potier, M. F. (1993). Mycoplasma hyopneumoniae infection in pigs: Duration of the disease and resistance to reinfection. *Veterinary Research, 24*(1), 67-77.

Kokotovic, B., Friis, N. F., Jensen, J. S., & Ahrens, P. (1999). Amplified-fragment length polymorphism fingerprinting of *Mycoplasma* species. *Journal of Clinical Microbiology, 37*, 8.

Konovalova, A., Søgaard-Andersen, L., & Kroos, L. (2014). Regulated proteolysis in bacterial development. *FEMS Microbiology Reviews, 38*, 493-522. https://doi.org/10.1111/1574-6976.12050

Kranker, S., Nielsen, J., Bille-Hansen, V., & Botner, A. (1998). Experimental inoculation of swine at various stages of gestation with a Danish isolate of porcine reproductive and respiratory syndrome virus (PRRSV). *Veterinary Microbiology, 61*(1-2), 21-31.

Krasteva, I., Liljander, A., Fischer, A., Smith, D. G., Inglis, N. F., Scacchia, M., Pini, A., Jores, J., & Sacchini, F. (2014). Characterization of the *in vitro* core surface proteome of *Mycoplasma mycoides* subsp. *mycoides*, the causative agent of contagious bovine pleuropneumonia. *Veterinary Microbiology, 168*, 116-123.

Kuhnert, P., & Overesch, G. (2014). Molecular epidemiology of *Mycoplasma hyopneumoniae* from outbreaks of enzootic pneumonia in domestic pig and the role of wild boar. *Veterinary Microbiology, 174*(1-2), 261-266. https://doi.org/10.1016/j.vetmic.2014.08.022

Kuhnert, P., Overesch, G., & Belloy, L. (2011). Genotyping of *Mycoplasma hyopneumoniae* in wild boar lung samples. *Veterinary Microbiology, 152*(1-2), 191-195.

Kunesh, J. (1981). A comparison of two antibodies in treating mycoplasma pneumonia in swine. *Veterinary Medicine. Small Animal Clinics, 76*, 871-872.

Kurata, S., Osaki, T., Yonezawa, H., Arae, K., Taguchi, H., & Kamiya, S. (2014). Role of IL-17A and IL-10 in the antigen induced inflammation model by *Mycoplasma pneumoniae*. *BMC Microbiology, 14*, 156. doi:10.1186/1471-2180-14-156

Kureljušić, B., Weissenbacher-Lang, C., Nedorost, N., Stixenberger, D., & Weissenböck, H. (2016). Association between *Pneumocystis* spp. and co-infections with *Bordetella bronchiseptica, Mycoplasma hyopneumoniae* and *Pasteurella multocida* in Austrian pigs with pneumonia. *Veterinary Journal, 207*, 177-179.

Kurth, K. T., Hsu, T., Snook, E. R., Thacker, E. L., Thacker, B. J., & Minion, F. C. (2002). Use of a *Mycoplasma hyopneumoniae* nested polymerase chain reaction test to determine the optimal sampling sites in swine. *Journal of Veterinary Diagnostic Investigation, 14*, 463-469.

Kuwano, K., Tanaka, N., Shimizu, T., & Kida, Y. (2006). Antimicrobial activity of inducible human beta defensin-2 against *Mycoplasma pneumoniae*. *Current Microbiology, 52*(6), 435-438. doi:10.1007/s00284-005-0215-7

Kwon, D., Choi, C., & Chae, C. (2002). Chronologic localization of *Mycoplasma hyopneumoniae* in experimentally infected pigs. *Veterinary Pathology, 39*, 584-587. doi:10.1354/vp.39-5-584

L'Ecuyer, C., & Switzer, W. P. (1963). Virus pneumonia of pigs; Attempts at propagation of the causative agent in cell cultures and chicken embryos. *Canadian Journal of Comparative Medicine & Veterinary Science, 27*, 91-99.

Lai, J. F., Zindl, C. L., Duffy, L. B., Atkinson, T. P., Jung, Y. W., van Rooijen, N., Waites, K. B., Krause, D. C., & Chaplin, D. D. (2010). Critical role of macrophages and their activation via MyD88-NFkappaB signaling in lung innate immunity to *Mycoplasma pneumoniae*. *PLoS One, 5*(12), e14417. doi:10.1371/journal.pone.0014417

Lallemand, E. A., Lacroix, M. Z., Toutain, P. L., Boullier, S., Ferran, A. A., & Bousquet-Melou, A. (2016). *In vitro* degradation of antimicrobials during use of broth microdilution method can increase the measured minimal inhibitory and minimal bactericidal concentrations. *Frontiers in Microbiology, 7*, 2051. doi:10.3389/fmicb.2016.02051

Lang, F., Lang, K. S., Lang, P. A., Huber, S. M., & Wieder, T. (2006). Mechanisms and significance of eryptosis. *Antioxidants and Redox Signalling, 8*, 1183-1192.

Lang, K. S., Lang, P. A., Bauer, C., Duranton, C., Wieder, T., Huber, S. M., & Lang, F. (2005). Mechanisms of suicidal erythrocyte death. *Cellular Physiology and Biochemistry, 15*, 195-202.

Lange, P. F., & Overall, C. M. (2013). Protein TAILS: When termini tell tales of proteolysis and function. *Current Opinion in Chemical Biology, 17*, 73-82. https://doi.org/10.1016/j.cbpa.2012.11.025

Langhoff, R., Stueckler, T., Ladinig, A., Barz, A., Spergser, J., Palzer, A., & Ritzmann, M. (2012). Untersuchung der Wirksamkeit von Tulathromycin (Draxxin®) gegen *Mycoplasma hyorhinis* in einem Feldversuch. *Tierärztliche Umschau, 67*, 3-9.

Lauritsen, K. T., Hagedorn-Olsen, T., Friis, N. F., Lind, P., & Jungersen, G. (2008). Absence of strictly age-related resistance to *Mycoplasma hyosynoviae* infection in 6-week-old pigs. *Veterinary Microbiology, 130*, 385-390.

Lauritsen, K. T., Hagedorn-Olsen, T., Jungersen, G., Riber, U., Stryhn, H., Friis, N. F., Lind, P., & Kristensen, B. (2017). Transfer of maternal immunity to piglets is involved in early protection against *Mycoplasma hyosynoviae*. *Veterinary Immunology and Immunopathology, 183*, 22-30.

Lauritsen, K. T., Nielsen, E. O., Christensen, D., & Jungersen, G. (2013). Novel *Mycoplasma hyosynoviae* vaccination of one herd failed to prevent lameness in finishing pigs. Proceedings 5th European Symposium of Porcine Health Management, Edinburgh, United Kingdom, 195.

Le Carrou, J., Laurentie, M., Kobisch, M., & Gautier-Bouchardon, A. V. (2006). Persistence of *Mycoplasma hyopneumoniae* in experimentally infected pigs after marbofloxacin treatment and detection of mutations in the parC gene. *Antimicrobial Agents and Chemotherapy, 50*(6), 1959-1966.

Le Floc'h, N., Deblanc, C., Cariolet, R., Gautier-Bouchardon, A.V., Merlot, E., & Simon, G. (2014). Effect of feed restriction on performance and postprandial nutrient metabolism in pigs co-infected with *Mycoplasma hyopneumoniae* and swine influenza virus. *PLoS One, 9*(8), e104605.

Le Grand, A., & Kobisch, M. (1996). [Comparison of the use of a vaccine and sequential antibiotic treatment in a herd infected with *Mycoplasma hyopneumoniae*] (in French). *Veterinary Research, 27*(3), 241-253.

Leal Zimmer, F. M. dos A., Paludo, G. P., Moura, H., Barr, J. R., & Ferreira, H. B. (2019). Differential secretome profiling of a swine tracheal cell line infected with mycoplasmas of the swine respiratory tract. *Journal of Proteomics, 192*, 147-159. https://doi.org/10.1016/j.jprot.2018.08.018

Leclercq, R. (2002). Mechanisms of resistance to macrolides and lincosamides: Nature of the resistance elements and their clinical implications. *Clinical Infectious Disease, 34*(4), 482-492.

Lee, J. A., Hwang, M. A., Han, J. H., Cho, E. H., Lee, J. B., Park, S. Y., Song, C. S., Choi, I. S., & Lee, S. W. (2018). Reduction of mycoplasmal lesions and clinical signs by vaccination against *Mycoplasma hyorhinis*. *Veterinary Immunology and Immunopathology, 196*, 14-17.

Lee, J. A., Oh, Y. R., Hwang, M. A., Lee, J. B., Park, S. Y., Song, C. S., Choi, I. S., & Lee, S. W. (2016). *Mycoplasma hyorhinis* is a potential pathogen of porcine respiratory disease complex that aggravates pneumonia caused by porcine reproductive and respiratory syndrome virus. *Veterinary Immunology and Immunopathology, 177*, 48-51.

Lee, S. H., Lee, S., Chae, C., & Ryu, D.Y. (2014). A recombinant chimera comprising the R1 and R2 repeat regions of *M. hyopneumoniae* P97 and the N-terminal region of *A. pleuropneumoniae* ApxIII elicits immune responses. *BMC Veterinary Research, 10*, 43.

Lehner, S., Meemken, D., Nathues, H., & grosse Beilage, E. (2008). Efficacy of vaccination against *Mycoplasma hyopneumoniae* in pigs from vaccinated and non-vaccinated sows. *Tierärztliche Praxis, 36*(G), 399-406.

Leon, E. A., Madec, F., Taylor, N. M., & Kobisch, M. (2001). Seroepidemiology of *Mycoplasma hyopneumoniae* in pigs from farrow-to-finish farms. *Veterinary Microbiology, 78*(4), 331-341.

Letunic, I., & Bork, P. (2019). Interactive Tree Of Life (iTOL) v4: Recent updates and new developments. *Nucleic Acids Research, 47*, W256-W259. https://doi.org/10.1093/nar/gkz239

Levison, M. E., & Levison, J. H. (2009). Pharmacokinetics and pharmacodynamics of antibacterial agents. *Infectious Disease Clinics of North America, 23*(4), 791.

Li, P., Li, Y., Shao, G., Yu, Q., & Yang, Q. (2015). Comparison of immune responses to intranasal and intrapulmonary vaccinations with the attenuated *Mycoplasma hyopneumoniae* 168 strain in pigs. *Journal of Veterinary Medical Science, 77*, 519-525.

Li, W., Raoult, D., & Fournier, P.-E. (2009a). Bacterial strain typing in the genomic era. *FEMS Microbiology Reviews, 33*, 892-916. https://doi.org/10.1111/j.1574-6976.2009.00182.x

Li, Y.-Z., Ho, Y.-P., Chen, S.-T., Chiou, T.-W., Li, Z.-S., & Shiuan, D. (2009b). Proteomic comparative analysis of pathogenic strain 232 and avirulent strain J of *Mycoplasma hyopneumoniae*. *Biochemistry Moscow, 74*, 215-220. https://doi.org/10.1134/S0006297909020138

Lien, E., Sellati, T. J., Yoshimura, A., Flo, T. H., Rawadi, G., Finberg, R. W., Carroll, J. D., Espevik, T., Ingalls, R. R., Radolf, J. D., & Golenbock, D. T. (1999). Toll-like receptor 2 functions as a pattern recognition receptor for diverse bacterial products. *Journal of Biological Chemistry, 274*(47), 33419-33425.

Lin, J., Weng, C., Liao, C., Yeh, K., & Pan, M. (2003). Protective effects of oral microencapsulated *Mycoplasma hyopneumoniae* vaccine prepared by co-spray drying method. *Journal of Veterinary Medical Science, 65*, 69-74.

Lin, J. H., Chen, S. P., Yeh, K. S., & Weng, C. N. (2006). *Mycoplasma hyorhinis* in Taiwan: Diagnosis and isolation of swine pneumonia pathogen. *Veterinary Microbiology, 115*, 111-116.

Linhares, D. (2017). Economics of *Mycoplasma hyopneumoniae* elimination. Swine health and research session, Allen D Leman Conference. St. Paul, MN.

Liu, J., Fung, K.-F., Chen, Z., Zeng, Z., & Zhang, J. (2003). Pharmacokinetics of florfenicol in healthy pigs and in pigs experimentally infected with *Actinobacillus pleuropneumoniae*. *Antimicrobial Agents and Chemotherapy, 47*(2), 820-823.

Liu, J., Zhou, D., Cheng, Z., Wang, Z., Wang, L., Wang, S., Yang, D., & Chai, T. (2012). Development and evaluation of enzyme-linked immunosorbent assay based on recombinant inorganic pyrophosphatase gene antigen for the detection of *Mycoplasma suis* antibodies. *Research in Veterinary Science, 93*, 48-50.

Liu, M., Du, G., Liu, B., Hu, Y., Liu, J., Jia, Y., Minion, F. C., Shao, G., & Zhao, R. (2017). Cholesterol exacerbates *Mycoplasma hyopneumoniae*-induced apoptosis via stimulating proliferation and adhesion to porcine alveolar macrophages. *Veterinary Microbiology, 211*, 112-118.

Liu, W., Fang, L., Li, S., Li, Q., Zhou, Z., Feng, Z., Luo, R., Shao, G., Wang, L., Chen, H., & Xiao S. (2010). Complete genome sequence of *Mycoplasma hyorhinis* strain HUB-1. *Journal of Bacteriology, 192*, 5844-5845.

Liu, W., Feng, Z., Fang, L., Zhou, Z., Li, Q., Li, S., Luo, R., Wang, L., Chen, H., Shao, G., & Xiao, S. (2011). Complete genome sequence of *Mycoplasma hyopneumoniae* strain 168. *Journal of Bacteriology, 193*, 1016-1017. https://doi.org/10.1128/JB.01305-10

Liu, W., Xiao, S., Li, M., Guo, S., Li, S., Luo, R., Feng, Z., Li, B., Zhou, Z., Shao, G., Chen, H., & Fang, L. (2013). Comparative genomic analyses of *Mycoplasma hyopneumoniae* pathogenic 168 strain and its high-passaged attenuated strain. *BMC Genomics, 14*, 80. https://doi.org/10.1186/1471-2164-14-80

Liu, W., Zhou, D., Yuan, F., Liu, Z., Duan, Z., Yang, K., Guo, R., Li, M., Li, S., Fang, L., Xiao, S., & Tian, Y. (2019). Surface proteins mhp390 (P68) contributes to cilium adherence and mediates inflammation and apoptosis in *Mycoplasma hyopneumoniae*. *Microbial Pathogenesis, 126*, 92-100.

Lium, B. (1992). An attempt to eradicate *Mycoplasma hyopneumoniae* from selected Norwegian farrowing to finishing herds. Proceedings 12th IPVS, 17-20 August 1992, The Hague, The Netherlands, 300.

Lloyd, L. C., Cottew, G. S., & Anderson, D. A. (1987). Early serological responses to Mycoplasma hyopneumoniae infection. *Israel Journal of Medical Sciences, 23*(6), 647-649.

Lorenzo, H., Quesada, O., Assuncao, P., Castro, A., & Rodriguez, F. (2006). Cytokine expression in porcine lungs experimentally infected with *Mycoplasma hyopneumoniae*. *Veterinary Immunology and Immunopathology, 109*(3-4), 199-207. doi:10.1016/j.vetimm.2005.07.021

Love, W., Dobbs, N., Tabor, L., & Simecka, J. W. (2010). Toll-like receptor 2 (TLR2) plays a major role in innate resistance in the lung against murine Mycoplasma. *PLoS One, 5*(5), e10739. doi:10.1371/journal.pone.0010739

Luehrs, A., Siegenthaler, S., Grutzner, N., Grosse Beilage, E., Kuhnert, P., & Nathues, H. (2017). Occurrence of *Mycoplasma hyorhinis* infections in fattening pigs and association with clinical signs and pathological lesions of Enzootic Pneumonia. *Veterinary Microbiology, 203*, 1-5.

Lukert, P., & Mulkey, G. (1982). Treatment of mycoplasmosis in young swine. *Modern Veterinary Practice, 63*, 107-110.

Lung, O., Ohene-Adjei, S., Buchanan, C., Joseph, T., King, R., Erickson, A., Detmer, S., & Ambagala A. (2017). Multiplex PCR and microarray for detection of swine respiratory pathogens. *Transboundary and Emerging Diseases, 64*, 834-848.

Luo, W., Yu, H., Cao, Z., Schoeb, T. R., Marron, M., & Dybvig, K. (2008). Association of *Mycoplasma* arthritidis mitogen with lethal toxicity but not with arthritis in mice. *Infection and Immunity, 76*, 4989-4998.

Lurette, A., Belloc, C., Touzeau, S., Hoch, T., Seegers, H., & Fourichon, C. (2008). Modelling batch farrowing management within a farrow-to-finish pig herd: Influence of management on contact structure and pig delivery to the slaughterhouse. *Animal, 2*(1), 105-116.

Macedo, N., Arruda, B., Canning, P., Karriker, L., Mora, J., & Clavijo, M. (2019). *Mycoplasma hyosynoviae* diagnostics: What experimental data con tell us about field strains. Proceedings American Association of Swine Veterinarians Annual Meeting, Orlando, USA, 58-60.

Maddison, J. E., Watson, A. D. J., & Elliott, J. (2008). Antibacterial drugs. In J. E. Maddison, S. W. Page, & D. B. Church (Eds.), *Small animal clinical pharmacology* (2nd ed.; pp. 148-185). Nottingham, UK: Saunders Ltd.

Madec, F., & Kobisch, M. (1984). *[Etat sanitaire du porcelet et evolution des lesions au niveau de l'arbre respiratoire au cours des differentes phases d'elevage]*. Paper presented at the Journées de la Recherche Porcine en France.

Madsen, M. L., Oneal, M. J., Gardner, S. W., Strait, E. L., Nettleton, D., Thacker, E. L., & Minion, F. C. (2007). Array-based genomic comparative hybridization analysis of field strains of *Mycoplasma hyopneumoniae*. *Journal of Bacteriology, 189*, 7977-7982. https://doi.org/10.1128/JB.01068-07

Maes, D., Deluyker, H., Verdonck, M., Castryck, F., Miry, C., Lein, A., Vrijens, B., & de Kruif, A. (1998). The effect of vaccination against *Mycoplasma hyopneumoniae* in pig herds with a continuous production system. *Zoonoses and Public Health, 45*, 495-505.

Maes, D., Deluyker, H., Verdonck, M., Castryck, F., Miry, C., Vrijens, B., & de Kruif, A. (1999). Risk indicators for the seroprevalences of *Mycoplasma hyopneumoniae*, porcine infuenza viruses and Aujeszky's disease virus in slaughter pigs from fattening pig herds. *Journal of Veterinary Medicine B, 46*(5), 341-352.

Maes, D., Deluyker, H., Verdonck, M., Castryck, F., Miry, C., Vrijens, B., & de Kruif, A. (2000). Risk indicators for the seroprevalences of *Mycoplasma hyopneumoniae*, porcine influenza viruses and Aujeszky's disease virus in slaughter pigs from farrow-to-finish pig herds in Belgium. *Veterinary Research, 31*(3), 313-327.

Maes, D., Deluyker, H., Verdonck, M., Castryck, F., Miry, C., Vrijens, B., Ducatelle, R., & de Kruif, A. (2001). Non-infectious herd factors associated with macroscopic and microscopic lung lesions in slaughter pigs from farrow-to-finish pig herds. *Veterinary Record, 148*, 41-46.

Maes, D., Deluyker, H., Verdonck, M., Castryck, F., Miry, C., Vrijens, B., Verbeke, W., Viaene, J., & de Kruif, A. (1999). Effect of vaccination against *Mycoplasma hyopneumoniae* in pig herds with an all-in/all-out production system. *Vaccine, 17*, 1024-1034.

Maes, D., Segales, J., Meyns, T., Sibila, M., Pieters, M., & Haesebrouck, F. (2008). Control of *Mycoplasma hyopneumoniae* infections in pigs. *Veterinary Microbiology, 126*(4), 297-309.

Maes, D., Sibila, M., Kuhnert, P., Segales, J., Haesebrouck, F., & Pieters, M. (2018). Update on *Mycoplasma hyopneumoniae* infections in pigs: Knowledge gaps for improved disease control. *Transboundary and Emerging Diseases, 65(Suppl 1)*, 110-124. doi:10.1111/tbed.12677

Maes, D., Verbeke, W., Vicca, J., Verdonck, M., & de Kruif, A. (2003). Benefit to cost of vaccination against *Mycoplasma hyopneumoniae* in pig herds under Belgian market conditions from 1996 to 2000. *Livestock Production Science, 83*, 85-93.

Maes, D., Verdonck, M., Deluyker, H., & de Kruif, A. (1996). Enzootic pneumonia in pigs. *Veterinary Quarterly, 18*(3), 104-109.

Maniloff, J. (1992). Phylogeny of mycoplasmas. In J. Maniloff, R. McElhaney, L. Finch, & J. Baseman (Ed.), *Mycoplasmas: Molecular biology and pathogenesis*. Washington, D.C.: American Society for Microbiology.

Marchioro, S., Del Pozo Sacristán, R., Michiels, A., Haesebrouck, F., Conceição, F., Dellagostin, O., & Maes, D. (2014a). Immune responses of a chimeric protein vaccine containing *Mycoplasma hyopneumoniae* antigens and LTB against experimental *M. hyopneumoniae* infection in pigs. *Vaccine, 32*, 4689-4694.

Marchioro, S., Maes, D., Flahou, B., Pasmans, F., Del Pozo Sacristán, R., Vranckx, K., Melkebeek, V., Cox, E., Wuyts, N., & Haesebrouck, F. (2013). Local and systemic immune responses in pigs intramuscularly injected with an inactivated *Mycoplasma hyopneumoniae* vaccine. *Vaccine, 31*(9), 1305-1311.

Marchioro, S. B., Fisch, A., Gomes, C. K., Jorge, S., Galli, V., Haesebrouck, F., Maes, D., Dellagostin, O., & Conceição, F. R. (2014b). Local and systemic immune responses induced by a recombinant chimeric protein containing *Mycoplasma hyopneumoniae* antigens fused to the B subunit of *Escherichia coli* heat-labile enterotoxin LTB. *Veterinary Microbiology, 173*, 166-171. https://doi.org/10.1016/j.vetmic.2014.07.009

Maré, C., & Switzer, W. (1965). New species: *Mycoplasma hyopneumoniae*. A causative agent of virus pig pneumonia. *Veterinary Medicine, 60*, 841-846.

Maré, C. J., & Switzer, W. P. (1966a). Virus pneumonia of pigs: Drug and ether sensitivity of a causative agent. *American Journal of Veterinary Research, 27*(121), 1671-1675.

Maré, C. J., & Switzer, W. P. (1966b). Virus pneumonia of pigs: Filtration and visualization of a causative agent. *American Journal of Veterinary Research, 27*(121), 1677-1686.

Maré, C. J., & Switzer, W. P. (1966c). Virus pneumonia of pigs: Propagation and characterization of a causative agent. *American Journal of Veterinary Research, 27*(121), 1687-1693.

Marois, C., Le Carrou, J., Kobisch, M., & Gautier-Bouchardon, A.V. (2007). Isolation of *Mycoplasma hyopneumoniae* from different sampling sites in experimentally infected and contact SPF piglets. *Veterinary Microbiology, 120*, 96-104.

Marois, C., Dory, D., Fablet, C., Madec, F., & Kobisch, M. (2010). Development of a quantitative Real-Time TaqMan PCR assay for determination of the minimal dose of *Mycoplasma hyopneumoniae* strain 116 required to induce pneumonia in SPF pigs. *Journal of Applied Microbiology, 108*(5), 1523-1533.

Marois, C., Gottschalk, M., Morvan, H., Fablet, C., Madec, F., & Kobisch, M. (2009). Experimental infection of SPF pigs with *Actinobacillus pleuropneumoniae* serotype 9 alone or in association with *Mycoplasma hyopneumoniae*. *Veterinary Microbiology, 135*(3-4), 283-291.

Marois, C., Le Carrou, J., Kobisch, M., & Gautier-Bouchardon, A.V. (2007). Isolation of *Mycoplasma hyopneumoniae* from different sampling sites in experimentally infected and contact SPF piglets. *Veterinary Microbiology, 120*, 96-104. https://doi.org/10.1016/j.vetmic.2006.10.015

Marois, P., Di Franco, E., Boulay, G., & Assaf, R. (1989a). Enzootic pneumonia in feeder pigs: Association with transmissible gastroenteritis virus infection. *Canadian Veterinary Journal, 30*(4), 328-330.

Marois, P., Di Franco, E., Boulay, G., Flipot, P., Assaf, R., & Lamoureux, G. (1989b). The influence of previous infections on the development of experimental swine pleuropneumonia in specific pathogen free pigs. *Canadian Journal of Veterinary Research, 53*, 394-399.

Martelli, P., Saleri, R., Cavalli, V., De Angelis, E., Ferrari, L., Benetti, M., Ferrarini, G., Merialdi, G., & Borghetti, P. (2014). Systemic and local immune response in pigs intradermally and intramuscularly injected with inactivated *Mycoplasma hyopneumoniae* vaccines. *Veterinary Microbiology, 168*, 357-364.

Martelli, P., Terreni, M., Guazzetti, S., & Cavirani, S. (2006). Antibody response to *Mycoplasma hyopneumoniae* infection in vaccinated pigs with or without maternal antibodies induced by sow vaccination. *Journal of Veterinary Medicine Series B Infectious Diseases and Veterinary Public Health, 53*, 229-233. doi:10.1111/j.1439-0450.2006.00952.x

Martineau, G. P. (1980). [Enzootic bronchopneumonia of pigs: improvement of the pulmonary index after treatment with tiamulin] (in French). *Annales de Médécine Vétérinaire, 124*, 281.

Martínez-Miró, S., Tecles, T., Ramón, M., Escribano, D., Hernández, F., Madrid, J., Orengo, J., Martínez-Subiela, S., Manteca, X., & Cerón, J. J. (2017). Causes, consequences and biomarkers of stress in swine: An update. *BMC Veterinary Research, 12*, 171.

Martinson, B., Minion, F. C., & Jordan, D. (2018a). Development and optimization of a cell-associated challenge model for *Mycoplasma hyorhinis* in 7-week-old caesarean-derived colostrum-deprived pigs. *Canadian Journal of Veterinary Research, 82*, 12-23.

Martinson, B., Minion, F. C., Kroll, J., & Hermann, J. (2017). Age susceptibility of caesarean derived colostrum deprived pigs to *Mycoplasma hyorhinis* challenge. *Veterinary Microbiology, 210*, 147-152.

Martinson, B., Zoghby, W., Barrett, K., Bryson, L., Christmas, R., Minion, F.C., & Kroll, J. (2018b). Efficacy of an inactivated *Mycoplasma hyorhinis* vaccine in pigs. *Vaccine, 36*, 408-412.

Mateusen, B., Maes, D., Hoflack, G., Verdonck, M., & de Kruif, A. (2001). A comparative study of the preventive use of tilmicosin phosphate (Pulmotil premix) and *Mycoplasma hyopneumoniae* vaccination in a pig herd with chronic respiratory disease. *Journal of Veterinary Medicine B Infectious Diseases and Veterinary Public Health, 48*(10), 733-741.

Mateusen, B., Maes, D., Van Goubergen, M., Verdonck, M., & de Kruif, A. (2002). Effectiveness of treatment with lincomycin hydrochloride and/or vaccination against *Mycoplasma hyopneumoniae* for controlling chronic respiratory disease in a herd of pigs. *Veterinary Record, 151*(5), 135-140.

Matthijs, A. M. F., Auray, G., Boyen, F., Michiels, A., Garcia-Nicolas, O., Barut, G. T., Barut, G., Barnier-Quer, C., Jakob, V., Collin, N., Devriendt, N., Summerfield, A., Haesebrouck, F., & Maes, D. (2019b). Efficacy of three innovative bacterin vaccines against experimental infection with *Mycoplasma hyopneumoniae*. *Veterinary Research*, Nov 8, *50*(1), 91. doi:10.1186/s13567-019-0709-0

Matthijs, A., Auray, G., Jakob, V., García-Nicolás, O., Braun, R., Keller, I., Bruggman, R., Devriendt, B., Boyen, F., Guzman, C., Michiels, A., Haesebrouck, F., Collin, N., Barnier-Quer, C., Maes, D., & Summerfield, A. (2019a). System immunology characterization of novel vaccine formulations for *Mycoplasma hyopneumoniae* bacterins. *Frontiers in Immunology, 10*, 1087. doi:10.3389/fimmu.2019.01087. eCollection 2019

Mayor, D., Jores, J., Korczak, B. M., & Kuhnert, P. (2008). Multilocus sequence typing (MLST) of *Mycoplasma hyopneumoniae*: A diverse pathogen with limited clonality. *Veterinary Microbiology, 127*, 63-72. https://doi.org/10.1016/j.vetmic.2007.08.010

Mayor, D., Zeeh, F., Frey, J., & Kuhnert, P. (2007). Diversity of *Mycoplasma hyopneumoniae* in pig farms revealed by direct molecular typing of clinical material. *Veterinary Research, 38*, 391-398. https://doi.org/10.1051/vetres:2007006

McAuliffe, L., Ellis, R. J., Lawes, J. R., Ayling, R. D., & Nicholas, R. A. (2005). 16S rDNA PCR and denaturing gradient gel electrophoresis; a single generic test for detecting and differentiating *Mycoplasma* species. *Journal of Medical Microbiology, 54*, 731-739.

McCormack, W. (1993). Susceptibility of mycoplasmas to antimicrobial agents: Clinical implications. *Clinical Infectious Diseases, 17*(Suppl 1), S200-S201.

McGowin, C. L., Popov, V. L., & Pyles, R. B. (2009). Intracellular *Mycoplasma genitalium* infection of human vaginal and cervical epithelial cells elicits distinct patterns of inflammatory cytokine secretion and provides a possible survival niche against macrophage-mediated killing. *BMC Microbiology, 9*, 139.

McGowin, C. L., & Totten, P. A. (2017). The unique microbiology and molecular pathogenesis of *Mycoplasma genitalium*. *Journal of Infectious Diseases, 216*, S382-S388. https://doi.org/10.1093/infdis/jix172

McKelvie, J., Morgan, J. H., Nanjiani, I. A., Sherington, J., Rowan, T. G., & Sunderland, S. J. (2005). Evaluation of tulathromycin for the treatment of pneumonia following experimental infection of swine with *Mycoplasma hyopneumoniae*. *Veterinary Therapy, 6*(2), 197-202.

McLaughlin, B. G., McLaughlin, P. S., & Evans, C.N. (1991). An Eperythrozoon-like parasite of llamas: Attempted transmission to swine, sheep, and cats. *Journal of Veterinary Diagnostic Investigation, 3*, 352-353.

Meng, K., Sun, W., Zhao, P., Zhang, L., Cai, D., Cheng, Z., Guo, H., Liu, J., Yang, D., Wang, S., & Chai, T. (2014). Development of colloidal gold-based immunochromatographic assay for rapid detection of *Mycoplasma suis* in porcine plasma. *Biosensors and Bioelectronics, 55*, 396-399.

Messick, J. B. (2004). Hemotrophic mycoplasmas (hemoplasmas): A review and new insights into pathogenic potential. *Veterinary Clinical Pathology, 33*, 2-13.

Messick, J. B., Cooper, S. K., & Huntley, M. (1999). Development and evaluation of a polymerase chain reaction assay using the 16S rRNA gene for detection of *Eperythrozoon suis* infection. *Journal of Veterinary Diagnostic Investigation, 11*, 229-236.

Messier, S., & Ross, R. F. (1991). Interactions of *Mycoplasma hyopneumoniae* membranes with porcine lymphocytes. *American Journal Veterinary Research, 52*(9), 1497-1502.

Messier, S., Ross, R. F., & Paul, P. S. (1990). Humoral and cellular immune responses of pigs inoculated with *Mycoplasma hyopneumoniae*. *American Journal of Veterinary Research, 51*(1), 52-58.

Meyling, A., & Friis, N. F. (1972). Serological identification of a new porcine *Mycopasma* species, *M. flocculare*. *Acta Veterinaria Scandinavica, 13*, 287-289.

Meyns, T., Dewulf, J., de Kruif, A., Calus, D., Haesebrouck, F., & Maes, D. (2006). Comparison of transmission of *Mycoplasma hyopneumoniae* in vaccinated and non-vaccinated populations. *Vaccine, 24*, 7081-7086.

Meyns, T., Maes, D., Calus, D., Ribbens, S., Dewulf, J., Chiers, K., de Kruif, A., Cox, E., Decostere, A., & Haesebrouck, F. (2007). Interactions of highly and low virulent *Mycoplasma hyopneumoniae* isolates with the respiratory tract of pigs. *Veterinary Microbiology, 120*(1-2), 87-95. https://doi.org/10.1016/j.vetmic.2006.10.010

Meyns, T., Maes, D., Dewulf, J., Vicca, J., Haesebrouck, F., & de Kruif, A. (2004). Quantification of the spread of *Mycoplasma hyopneumoniae* in nursery pigs using transmission experiments. *Preventive Veterinary Medicine, 66*(1-4), 265-275. https://doi.org/10.1016/j.prevetmed.2004.10.001

Meyns, T., Vansteelant, J., Rolly, E., Dewulf, J., Haesebrouck, F., & Maes, D. (2011). A cross-sectional study of risk factors associated with pulmonary lesions in pigs at slaughter. *Veterinary Journal, 187*, 388-392.

Michiels, A., Arsenakis, I., Matthijs, A., Boyen, F., Haesaert, G., Audenaert, K., Eeckhout, M., Croubels, S., Haesebrouck, F., & Maes, D. (2018). Clinical impact of deoxynivalenol, 3-acetyl-deoxynivalenol and 15-acetyl-deoxynivalenol on the severity of an experimental *Mycoplasma hyopneumoniae* infection in pigs. *BMC Veterinary Research, 14*(1), 190.

Michiels, A., Piepers, S., Ulens, T., Van Ransbeeck, N., Del Pozo Sacristán, R., Sierens, A., Haesebrouck, F., Demeyer, P., & Maes, D. (2015). Impact of particulate matter and ammonia on average daily weight gain, mortality and lung lesions in pigs. *Preventive Veterinary Medicine, 121*, 99-107.

Michiels, A., Vranckx, K., Piepers, S., Del Pozo Sacristán, R., Arsenakis, I., Boyen, F., Haesebrouck, F., & Maes, D. (2017). Impact of diversity of *Mycoplasma hyopneumoniae* strains on lung lesions in slaughter pigs. *Veterinary Research, 48*(1), 1-14. https://doi.org/10.1186/s13567-016-0408-z

Milovanović, A., Petrović, T., Apić, J., Lupulović, D., Barna, T., & Lazić, G. (2017). Boar quality semen testing and presence of *Mycoplasma* organism. *Journal of Veterinary Science and Research, 2*(3).

Minion, F. C., Adams, C., & Hsu, T. (2000). R1 region of P97 mediates adherence of *Mycoplasma hyopneumoniae* to swine cilia. *Infection and Immunity, 68*, 3056-3060.

Minion, F. C., & Goguen, J. D. (1986). Identification and preliminary characterization of external membrane-bound nuclease activities in *Mycoplasma pulmonis*. *Infection and Immunity, 51*, 352-354.

Minion, F. C., Jarvill-Taylor, K. J., Billings, D. E., & Tigges, E. (1993). Membrane-associated nuclease activities in mycoplasmas. *Journal of Bacteriology, 175*, 7842-7847.

Minion, F. C., Lefkowitz, E. J., Madsen, M. L., Cleary, B. J., Swartzell, S. M., & Mahairas, G. G. (2004). The genome sequence of *Mycoplasma hyopneumoniae* strain 232, the agent of swine mycoplasmosis. *Journal of Bacteriology, 186*, 7123-7133. https://doi.org/10.1128/JB.186.21.7123-7133.2004

Mitiku, F., Hartley, C. A., Sansom, F. M., Coombe, J. E., Mansell, P. D., Beggs, D. S., & Browning, G. F. (2018). The major membrane nuclease MnuA degrades neutrophil extracellular traps induced by *Mycoplasma bovis*. *Veterinary Microbiology, 218*, 13-19. doi:10.1016/j.vetmic.2018.03.002

Mize, M. T., Sun, X. L., & Simecka, J. W. (2018). Interleukin-17A exacerbates disease severity in BALB/c mice susceptible to lung infection with *Mycoplasma pulmonis*. *Infection and Immunity, 86*(9). doi:10.1128/iai.00292-18

Moges, R., Desmonts De Lamache, D., Sajedy, S., Renaux B. S., Hollenberg, M. D., Muench, G., Abbott, E. M., & Buret, A. G. (2018). Anti-inflammatory benefits of antibiotics: Tylvalosin induces apoptosis of porcine neutrophils and macrophages, promotes efferocytosis, and inhibits pro-Inflammatory CXCL-8, IL1α, and LTB4 production, while inducing the release of pro-resolving lipoxin A4 and resolvin D1. *Frontiers in Veterinary Science, 5*, 57.

Moore, C. (2005). Parity segregation. London Swine Conference – Production at the Leading Edge, 61-67.

Moreau, I. A., Miller, G. Y., & Bahnson, P. B. (2004). Effects of *Mycoplasma hyopneumoniae* vaccine on pigs naturally infected with *M. hyopneumoniae* and porcine reproductive and respiratory syndrome virus. *Vaccine, 22*(17-18), 2328-2333.

Morita, T., Fukuda, H., Awakura, T., Shimada, A., Umemura, T., Kazama, S., & Yagihashi, T. (1995). Demonstration of *Mycoplasma hyorhinis* as a possible primary pathogen for porcine otitis media. *Veterinary Pathology, 32*, 107-111.

Morita, T., Ohiwa, S., Shimada, A., Kazama, S., Yagihashi, T., & Umemura, T. (1999). Intranasally inoculated *Mycoplasma hyorhinis* causes eustachitis in pigs. *Veterinary Pathology, 36*, 174-178.

Morita, T., Sasaki, A., Kaji, N., Shimada, A., Kazama, S., Ygihashi, T., & Umemura, T. (1998). Induction of temporary otitis media in specific-pathogen-free pigs by intratympanic inoculation of *Mycoplasma hyorhinis*. *American Journal of Veterinary Research, 59*, 869-873.

Morris, C., Gardner, I., Hietala, S., Carpenter, T., Anderson, R., & Parker, K. (1994). Persistence of passively acquired antibodies to *Mycoplasma hyopneumoniae* in a swine herd. *Preventive Veterinary Medicine, 21*, 29-41.

Morris, C. R., Gardner, I. A., Hietala, S. K., Carpenter, T. E., Anderson, R. J., & Parker, K. M. (1995). Seroepidemiologic study of natural transmission of *Mycoplasma hyopneumoniae* in a swine herd. *Preventive Veterinary Medicine, 21*(4), 323-337.

Mouquet, H., & Nussenzweig, M. C. (2012). Polyreactive antibodies in adaptive immune responses to viruses. *Cellular and Molecular Life Sciences, 69*(9), 1435-1445. doi:10.1007/s00018-011-0872-6

Mrázek, J. (2006). Analysis of distribution indicates diverse functions of simple sequence repeats in mycoplasma genomes. *Molecular Biology Evolution, 23*, 1370-1385. https://doi.org/10.1093/molbev/msk023

Müller, E., & Neddenriep, G. (1979). Eperythrozoonose in einem Ferkelerzeugerbetrieb in Norddeutschland. *Praktischer Tierärzt, 8*, 662-665.

Muneta, Y., Minagawa, Y., Shimoji, Y., Nagata, R., Markham, P. F., Browning, G. F., & Mori, Y. (2006). IL-18 expression in pigs following infection with *Mycoplasma hyopneumoniae*. *Journal of Interferon and Cytokine Research, 26*(9), 637-644. doi:10.1089/jir.2006.26.637

Muneta, Y., Minagawa, Y., Shimoji, Y., Ogawa, Y., Hikono, H., & Mori, Y. (2008). Immune response of gnotobiotic piglets against *Mycoplasma hyopneumoniae*. *Journal of Veterinary Medical Sciences, 70*, 1065-1070.

Muneta, Y., Uenishi, H., Kikuma, R., Yoshihara, K., Shimoji, Y., Yamamoto, R., Hamashima, N., Yokomizo, Y., & Mori, Y. (2003). Porcine TLR2 and TLR6: Identification and their involvement in *Mycoplasma hyopneumoniae* infection. *Journal of Interferon and Cytokine Research, 23*, 583-590. doi:10.1089/107999003322485080

Murphy, T. (2011). The effects of individual and combinations of airborne pollutants on feed intake, immune function and physiology of the pig. PhD Thesis, University of Adelaide Faculty of Science, School of Animal and Veterinary Science, Discipline of Animal Science, 176-180.

Murphy, T., Cargill, C., & Carr, J. (2000). The effects of stocking density on air quality. Proceedings 16th International Pig Veterinary Society Congress, Melbourne, Australia, 326.

Murphy, T., Cargill, C., Rutley, D., & Stott, P. (2012). Pig-shed air polluted by α-haemolytic cocci and ammonia causes subclinical disease and production losses. *Veterinary Record,* 10.1136/vr.100413.

Namiki, K., Goodison, S., Porvasnik, S., Allan, R. W., Iczkowski, K. A., Urbanek, C., Reyes, L., Sakamoto, N., & Rosser, C. J. (2009). Persistent exposure to *Mycoplasma* induces malignant transformation of human prostate cells. *PLos One, 4*, e6872.

Nanjiani, I. A., McKelvie, J., Benchaoui, H. A., Godinho, K. S., Sherington, J., Sunderland, S. J., Weatherley, A. J., & Rowan, T. G. (2005). Evaluation of the therapeutic activity of tulathromycin against swine respiratory disease on farms in Europe. *Veterinary Therapy, 6*(2), 203-213.

Nathues, H., Doehring, S., Woeste, H., Fahrion, A. S., Doherr, M. G., & Grosse Beilage, E. (2013a). Individual risk factors for *Mycoplasma hyopneumoniae* infections in suckling pigs at the age of weaning. *Acta Veterinaria Scandinavica, 55*(1), 44.

Nathues, H., Fournie, G., Wieland, B., Pfeiffer, D. U., & Stärk, K. D. (2016). Modelling the within-herd transmission of *Mycoplasma hyopneumoniae* in closed pig herds. *Porcine Health Management, 2*(1), 10.

Nathues, H., Grosse Beilage, E., Kreienbrock, L., Rosengarten, R., & Spergser, J. (2011). RAPD and VNTR analyses demonstrate genotypic heterogeneity of *Mycoplasma hyopneumoniae* isolates from pigs housed in a region with high pig density. *Veterinary Microbiology, 152*(3-4), 338-345. https://doi.org/10.1016/j.vetmic.2011.05.029

Nathues, H., Spergser, J., Rosengarten, R., Kreienbrock, L., & Grosse Beilage, E. (2012). Value of the clinical examination in diagnosing enzootic pneumonia in fattening pigs. *Veterinary Journal, 193*, 443-447.

Nathues, H., Strutzberg-Minder, K., Kreienbrock, L., & Grosse, E. B. (2006). Possibilities and limits of serological diagnosis in pig stocks in the case of *Mycoplasma hyopneumoniae* infection. *DTW. Deutsche Tierärztliche Wochenschrift, 113*(12), 448-452.

Nathues, H., Woeste, H., Doehring, S., Fahrion, A. S., Doherr, M. G., & Beilage, E. (2013b). Herd specific risk factors for *Mycoplasma hyopneumoniae* infections in suckling pigs at the age of weaning. *Acta Veterinaria Scandinavica, 55*, 30.

Negrete-Abascal, E., Tenorio, V. R., Guerrero, A. L., Garcia, R. M., Reyes, M. E., & de la Garza, M. (1998). Purification and characterization of a protease from *Actinobacillus pleuropneumoniae* serotype 1, an antigen common to all the serotypes. *Canadian Journal of Veterinary Research, 62*(3), 183-190.

Negrete-Abascal, E., Tenorio, V. R., Serrano, J. J., Garcia, C., & de la Garza, M. (1994). Secreted proteases from *Actinobacillus pleuropneumoniae* serotype 1 degrade porcine gelatin, hemoglobin and immunoglobulin A. *Canadian Journal of Veterinary Research, 58*(2), 83-86.

Neil, D. H., McKay, K. A., L'Ecuyer, C., & Corner, A. H. (1969). Glässer's disease of swine produced by intratracheal inoculation of *Haemophilus suis*. *Canadian Journal of Comparative Medicine, 33*, 187-193.

Neimark, H. Johansson, K. E., Rikihisa, Y., & Tully, J., G. (2001). Proposal to transfer some members of the genera *Haemobartonella* and *Eperythrozoon* to the genus *Mycoplasma* with descriptions of '*Candidatus* Mycoplasma haemofelis', '*Candidatus* Mycoplasma haemomuris', '*Candidatus* Mycoplasma haemosuis' and 'Candidatus Mycoplasma wenyonii'. *International Journal of Systematic and Evolutionary Microbiology, 51*, 891-899.

Neimark, H., Johannson, K. E., Rikihisa, Y., & Tully, J. G. (2002). Revision of haemotrophic *Mycoplasma* species names. *International Journal of Systematic and Evolutionary Microbiology, 52*, 683.

Ni, B., Bai, F. F., Wei, Y., Liu, M. J., Feng, Z. X., Xiong, Q. Y., Hua, L. Z., & Shao, G. Q. (2015). Apoptosis induced by lipid-associated membrane proteins from *Mycoplasma hyopneumoniae* in a porcine lung epithelial cell line with the involvement of caspase 3 and the MAPK pathway. *Genetics and Molecular Research Journal, 14*, 11429-11443.

Nielsen, E. O., Lauritsen, K. T., Friis, N. F., Enge, C., Hagedorn-Olsen, T., & Jungersen, G. (2005). Use of a novel serum ELISA method and the tonsil-carrier state for evaluation of *Mycoplasma hyosynoviae* distribution in pig herds with or without clinical arthritis. *Veterinary Microbiology, 111*, 41-50.

Nielsen, E. O., Nielsen, N. C., & Friis, N. F. (2001). *Mycoplasma hyosynoviae* arthritis in grower-finisher pigs. *Journal of Veterinary Medicine Series A, 48*, 475-486.

Nielsen, P., & Gyrd-Hansen, N. (1996). Bioavailability of oxytetracycline, tetracycline and chlortetracycline after oral administration to fed and fasted pigs. *Journal of Veterinary Pharmacology and Therapeutics, 19*(4), 305-311.

Nielsen, P., & Gyrd-Hansen, N. (1997). Bioavailability of enrofloxacin after oral administration to fed and fasted pigs. *Pharmacology and Toxicology, 80*(5), 246-250.

Nonaka, N., Thacker, B. J., Schillhorn van Veen, T. W., & Bull, R. W. (1996). In vitro maintenance of *Eperythrozoon suis*. *Veterinary Parasitology, 61*, 181-199.

Novotny, G. W., Jakobsen, L., Andersen, N. M., Poehlsgaard, J., & Douthwaite, S. (2004). Ketolide antimicrobial activity persists after disruption of interactions with domain II of 23S rRNA. *Antimicrobial Agents and Chemotherapy, 48*(10), 3677-3683.

Noyes, E. P., Feeney, D. A., & Pijoan, C. (1990). Comparison of the effect of pneumonia detected during lifetime with pneumonia detected at slaughter on growth in swine. *Journal of the American Veterinary Medical Association, 197*(8), 1025-1029.

Nutsch, R. G., Hart, F. J., Rooney, K. A., Weigel, D. J., Kilgore, W. R., & Skogerboe, T. L. (2005). Efficacy of tulathromycin injectable solution for the treatment of naturally occurring Swine respiratory disease. *Veterinary Therapy, 6*(2), 214-224.

Oberst, R. D., Gwaltney, S. M., Hays, M. P., Morgan, S., & Stair, E. L. (1993). Experimental infections and natural outbreaks of eperythrozoonosis in pigs identified by PCR-DNA hybridizations. *Journal of Veterinary Diagnostic Investigation, 5*, 351-358.

Oberst, R. D., Hall S. M., & Schoneweis D. A. (1990). Detection of *Eperythrozoon suis* DNA from swine blood by whole organism DNA hybridizations. *Veterinary Microbiology, 24*, 127-134.

Obi, C. C., Vayla, S., de Gannes, V., Berres, M. E., Walker, J., Pavelec, D., Hyman, J., & Hickey, W. J. (2018). The Integrative Conjugative Element clc (ICEclc) of Pseudomonas aeruginosa JB2. *Frontiers in Microbiology, 9*. https://doi.org/10.3389/fmicb.2018.01532

Odeh, A. N., & Simecka, J. W. (2016). Regulatory CD4+CD25+ T cells dampen inflammatory disease in murine *Mycoplasma* pneumonia and promote IL-17 and IFN-gamma responses. *PLoS One, 11*(5), e0155648. doi:10.1371/journal.pone.0155648

Oehlerking, J., Kube, M., Felder, K. M., Matter, D., Wittenbrink, M. M., Schwarzenbach, S., Kramer, M. M., Hoelzle, K., & Hoelzle, L. E. (2011). Complete genome sequence of the hemotrophic *Mycoplasma suis* strain KI3806. *Journal of Bacteriology, 193*, 2369-2370.

Ogawa, Y., Oishi, E., Muneta, Y., Sano, A., Hikono, H., Shibahara, T., Yagi, Y., & Shimoji, Y. (2009). Oral vaccination against mycoplasmal pneumonia of swine using a live *Erysipelothrix rhusiopathiae* vaccine strain as a vector. *Vaccine, 27*, 4543-4550.

Okamba, F., Arella, M., Music, N, Jia, J., Gottschalk, M., &. Gagnon, C. (2010). Potential use of a recombinant replication-defective adenovirus vector carrying the C-terminal portion of the P97 adhesin protein as a vaccine against *Mycoplasma hyopneumoniae* in swine. *Vaccine, 28*, 4802-4809.

Okamba, F., Moreau, E., Cheikh Saad, B., Gagnon, C., Massie, B., &. Arella, M. (2007). Immune responses induced by replication-defective adenovirus expressing the C-terminal portion of the *Mycoplasma hyopneumoniae* P97 adhesin. *Clinical Vaccine Immunology, 14*, 767-774.

Okusawa, T., Fujita, M., Nakamura, J., Into, T., Yasuda, M., Yoshimura, A., et al., Hara, Y., Hasebe, A., Golenbock, D., Morita, M., Kuroki, Y., Ogawa, T., & Shibata, K. (2004). Relationship between structures and biological activities of mycoplasmal diacylated lipopeptides and their recognition by toll-like receptors 2 and 6. *Infection and Immunity, 72*(3), 1657-1665.

Olaitan, A. O., Morand, S., & Rolain, J. M. (2014). Mechanisms of polymyxin resistance: Acquired and intrinsic resistance in bacteria. *Frontiers in Microbiology, 5*, 643. doi:10.3389/fmicb.2014.00643

Oliveira Filho, J. X., Mores, M. A. Z., Rebellato, R., Kich, J. D., Cantao, M. E., Klein, C. S., Guedes, R. M. C., Coldebella, A., Barcellos, D., & Mores, N. (2018). Pathogenic variability among *Pasteurella multocida* type A isolates from Brazilian pig farms. *BMC Veterinary Research, 14*(1), 244.

Opriessnig, T., Gimenez-Lirola, L. G., & Halbur, P. G. (2011a). Polymicrobial respiratory disease in pigs. *Animal Health Research Review, 12*(2), 133-148.

Opriessnig, T., Kuster, C., & Halbur, P. G. (2006). Demonstration of porcine circovirus type 2 in the testes and accessory sex glands of a boar. *Journal of Swine Health and Production, 14*(1), 42-45.

Opriessnig, T., Madson, D. M., Schalk, S., Brockmeier, S., Shen, H. G., Beach, N. M., Meng, X. J., Baker, R. B., Zanella, E. L., & Halbur, P. G. (2011b). Porcine circovirus type 2 (PCV2) vaccination is effective in reducing disease and PCV2 shedding in semen of boars concurrently infected with PCV2 and *Mycoplasma hyopneumoniae*. *Theriogenology, 76*(2), 351-360.

Opriessnig, T., Thacker, E. L., Yu, S., Fenaux, M., Meng, X.-J., & Halbur, P. G. (2004). Experimental reproduction of postweaning multisystemic wasting syndrome in pigs by dual infection with *Mycoplasma hyopneumoniae* and porcine circovirus type 2. *Veterinary Pathology, 41*, 624-640.

Oswald, I. P. (2011). Risk of mycotoxin to pig health: Immunosuppressive effects. *Pig Journal, 66*, 50-54.

Otake, S., Dee, S., Corzo, C., Oliveira, S., & Deen, J. (2010). Long-distance airborne transport of infectious PRRSV and *Mycoplasma hyopneumoniae* from a swine population infected with multiple viral variants. *Veterinary Microbiology, 145*(3-4), 198-208.

Overesch, G., & Kuhnert, P. (2017). Persistence of *Mycoplasma hyopneumoniae* sequence types in spite of a control program for enzootic pneumonia in pigs. *Preventive Veterinary Medicine, 145*, 67-72. https://doi.org/10.1016/j.prevetmed.2017.06.007

Paes, J. A., Lorenzatto, K. R., de Moraes, S. N., Moura, H., Barr, J. R., & Ferreira, H. B. (2017a). Secretomes of *Mycoplasma hyopneumoniae* and *Mycoplasma flocculare* reveal differences associated to pathogenesis. *Journal of Proteomics, 154*, 69-77.

Paes, J. A., Machado, L. D. P. N., dos Anjos Leal, F. M., de Moraes, S. N., Moura, H., Barr, J. R., & Ferreira, H. B. (2018). Comparative proteomics of two *Mycoplasma hyopneumoniae* strains and *Mycoplasma flocculare* identified potential porcine enzootic pneumonia determinants. *Virulence*. https://doi.org/10.1080/21505594.2018.1499379

Paes, J. A., Virginio, V. G., Cancela, M., Leal, F. M. A., Borges, T. J., Jaeger, N., Bonorino, C., Schrank, I. S., & Ferreira, H. B. (2017b). Pro-apoptotic effect of a *Mycoplasma hyopneumoniae* putative type I signal peptidase on PK(15) swine cells. *Veterinary Microbiology, 201*, 170-176.

Pallarés, F. J., Halbur, P. G., Opriessnig, T., Sorden, S. D., Villar, D., Janke, B. H., Yaeger, M. J., Larson, D. J., Schwartz, K. J., Yoon, K. J., & Hoffman, L. J. (2002). Porcine circovirus type 2 (PCV-2) coinfections in US field cases of postweaning multisystemic wasting syndrome (PMWS). *Journal of Veterinary Diagnostic and Investigation, 14*(6), 515-519.

Pallarés, F. J., Lasa, C., Roozen, M., & Ramis, G. (2015). Use of tylvalosin in the control of porcine enzootic pneumonia. *Veterinary Record Open, 2*(1), e000079. doi:10.1136/vetreco-2014-000079

Palzer, A., Austin-Busse, R. L., Ladinig, A., Balka, G., Zoels, S., & Ritzmann, M. (2015a). Histopathological lesions in conventional pigs experimentally infected with *Haemophilus parasuis* serovar 5. *Tierärztliche Praxis Ausgabe Grosstiere Nutztiere, 43*, 91-96.

Palzer, A., Haedke, K., Heinritzi, K., Zoels, S., Ladinig, A., & Ritzmann, M. (2015b). Associations among *Haemophilus parasuis*, *Mycoplasma hyorhinis*, and porcine reproductive and respiratory syndrome virus infections in pigs with polyserositis. *Canadian Veterinary Journal, 56*, 285-287.

Palzer, A., Kolb, K., Strutzberg-Minder, K., Zoels, S., Eddicks, M., Heinritzi, K., & Ritzmann, M. (2015c). Serologische Verlaufsuntersuchungen auf *Haemophilus parasuis* und *Mycoplasma hyorhinis* in drei schweinehaltenden Betrieben. *Schweizer Archiv für Tierheilkunde, 157*, 97-103.

Palzer, A., Ritzmann, M., Hafner-Marx, A., Wolf, G., & Heinritzi, K. (2006a). Nachweis von *Haemophilus parasuis* und *Mycoplasma hyorhinis* bei Schweinen sowie Assoziation dieser Erreger mit klinischen und pathologisch-anatomischen Befunden. *Deutsche Tierärztliche Wochenschrift, 113*, 227-230.

Palzer, A., Ritzmann, M., & Heinritzi K. (2007a). Frühe Vakzinierung gegen die Glässersche Krankheit mit Porcilis® Glässer in einem Feldversuch. *Schweizer Archiv für Tierheilkunde, 149*, 389-394.

Palzer, A., Ritzmann, M., Heinritzi, K., & Hafner-Marx, A. (2006b). Improving diagnosis of Glaesser's disease using PCR on serosal swabs. Proceedings 19th International Pig Veterinary Society Congress, Copenhagen, Denmark, Volume 1, 202.

Palzer, A., Ritzmann, M., Majzoub, M., Wolf, G., Hermanns, W., & Heinritzi, K. (2007b). Häufigkeit des Vorkommens pneumonie-assoziierter Erreger und deren Korrelation mit klinischen und pathologisch-anatomischen Befunden bei Schweinen. *Berliner und Münchener Tierärztliche Wochenschrift, 120*, 483-489.

Palzer, A., Ritzmann, M., Wolf, G., & Heinritzi, K. (2008). Association between pathogens in healthy pigs and pigs with pneumonia. *Veterinary Record, 162*, 267-271.

Pantoja, L. G., Pettit, K., Dos Santos, L. F., Tubbs, R., & Pieters, M. (2016). *Mycoplasma hyopneumoniae* genetic variability within a swine operation. *Journal of Veterinary Diagnostic Investigation, 28*, 175-179. https://doi.org/10.1177/1040638716630767

Papich, M. G. (2017). Chloramphenicol and derivatives, macrolides, lincosamides, and miscellaneous antimicrobials. In J. E. Rivière & M. G. Papich (Eds.), *Veterinary pharmacology and therapeutics* (10th ed.; pp. 905-952). Hoboken, USA: John Wiley & Sons.

Park, C., Jeong, J., Choi, K., & Chae, C. (2016a). Efficacy of a new bivalent vaccine of porcine circovirus type 2 and *Mycoplasma hyopneumoniae* (FosteraTM PCV MH) under experimental conditions. *Vaccine, 34*, 270-272.

Park, C., Jeong, J., Kang, I., Choi, K., Park, S. J., & Chae, C. (2016b). Increased fucosyl glycoconjugate by *Mycoplasma hyopneumoniae* enhances adherences of *Pasteurella multocida* type A in the ciliated epithelial cells of the respiratory tract. *BMC Veterinary Research, 12*, 25.

Park, H. J., Kang, K. M., Dybvig, K., Lee, B. L., Jung, Y. W., & Lee, I. H. (2013). Interaction of cationic antimicrobial peptides with *Mycoplasma pulmonis*. *FEBS Letters, 587*(20), 3321-3326. doi:10.1016/j.febslet.2013.08.016

Park, S. J., Seo, H. W., Park, C., & Chae, C. (2014). Interaction between single-dose *Mycoplasma hyopneumoniae* and porcine reproductive and respiratory syndrome virus vaccines on dually infected pigs. *Research in Veterinary Science, 96*(3), 516-522.

Paukner, S., & Riedl, R. (2017). Pleuromutilins: Potent drugs for resistant bugs - Mode of action and resistance. *Cold Spring Harbor Perspectives in Medicine, 7*(1), a027110.

Pedersen, S., & Sällvik, K. (2002). 4th report of Working Group on Climatization of Animal Houses Heat and moisture production at animal and house levels. Research Centre Bygholm, Danish Institute of Agricultural Sciences. ISBN 87-88976-60-2, p. 37-40.

Pedersen, S., Nonnenmann, M., Rautiainen, R., Demmers, T. G. M., Banhazi, T., & Lyngbye M. (2000). Dust in pig buildings. *Journal of Agricultural Safety and Health, 6*(4), 261-274.

Pendarvis, K., Padula, M. P., Tacchi, J. L., Petersen, A. C., Djordjevic, S. P., Burgess, S. C., & Minion, F. C. (2014). Proteogenomic mapping of Mycoplasma hyopneumoniae virulent strain 232. *BMC Genomics, 15*, 576. https://doi.org/10.1186/1471-2164-15-576

Pereira, C. E. R., Vannucci, F. A., Gabardo, M. P., dos Santos, L. F., Mores, N., & Guedes, R. M. C. (2017). *Mycoplasma hyorhinis* infection in early cases of mycoplasmal pneumonia in swine and evaluation of diagnostic assays. *Pesquisa Veterinária Brasileira, 37*, 1057-1063.

Peterhans, E., Bertoni, G., Koppel, P., Wyler, R., & Keller, R. (1984). Antibody-free target cells stimulate chemiluminescence in polymorphonuclear leukocytes: An artifact due to mycoplasma contamination. *European Journal of Immunology, 14*(2), 201-203. doi:10.1002/eji.1830140217

Peters, I.R., Helps, C. R., McAuliffe, L., Neimark, H., Lappin, M. R., Gruffydd-Jones, T. J. Day, M. J., Hoelzle, L. E., Willi, B., Meli, M., Hofmann-Lehmann, R., & Tasker, S. (2008). RNase P RNA Gene (*rnpB*) Phylogeny of hemoplasmas and other *Mycoplasma* species. *Journal of Clinical Microbiology, 46*, 1873-1877.

Petersen, A. C., Clampitt, J. M., & Minion, F. C. (2019). Analysis of swine antigen-specific antibody responses to *Mycoplasma hyopneumoniae* infection determined by protein microarray. *Veterinary Microbiology*, *230*, 195-201. https://doi.org/10.1016/j.vetmic.2019.02.010

Petersen, A. C., Oneal, D. C., Seibel, J. R., Poel, K., Daum, C. L., Djordjevic, S. P., & Minion, F. C. (2016). Cross reactivity among the swine mycoplasmas as identified by protein microarray. *Veterinary Microbiology, 192*, 204-212.

Pettersson, B., Tully, J. G., Bolske, G., & Johansson, K-E. (2000). Updated phylogenetic description of the *Mycoplasma hominis* cluster (Weisburg et al. 1989) based on 16S rDNA sequences. *International Journal of Systematic and Evolutionary Microbiology, 50*, 291-301.

Pettersson, B., Tully, J.G., Bolske, G., & Johansson, K-E. (2001). Re-evaluation of the classical *Mycoplasma lipophilum* cluster (Weisburg et al. 1989) and description of two new clusters in the hominis group based on 16S rDNA sequences. *International Journal of Systematic and Evolutionary Microbiology, 51*, 633-643.

Piboonpocanun, S., Chiba, H., Mitsuzawa, H., Martin, W., Murphy, R. C., Harbeck, R. J., & Voelker, D. R. (2005). Surfactant protein A binds *Mycoplasma pneumoniae* with high affinity and attenuates its growth by recognition of disaturated phosphatidylglycerols. *Journal of Biological Chemistry, 280*(1), 9-17. doi:10.1074/jbc.M411570200

Pierron, A., Alassane-Kpembi, I., & Oswald, I. P. (2016). Impact of mycotoxin on immune response and consequences for pig health. *Animal Nutrition, 2*, 63-68.

Pieters, M. (2014). Risk factors for *M. hyopneumoniae* piglet colonization during the lactation period. Proceedings 23rd International Pig Veterinary Society, Cancun, Mexico, 93.

Pieters, M. (2018). Achieving day 0 for *Mycoplasma hyopneumoniae*: What is it? Why should it be done? Retrieved in May 2018, from www.pig333.com.

Pieters, M., Cline, G. S., Payne, B. J., Prado, C., Ertl, J. R., & Rendahl, A. K. (2014). Intra-farm risk factors for *Mycoplasma hyopneumoniae* colonization at weaning age. *Veterinary Microbiology, 172*, 575-580.

Pieters, M., Daniels, J., & Rovira, A. (2017). Comparison of sample types and diagnostic methods for in vivo detection of *Mycoplasma hyopneumoniae* during early stages of infection. *Veterinary Microbiology*, *203*, 103-109.

Pieters, M., & Fano, E. (2016). *Mycoplasma hyopneumoniae* management in gilts. *Veterinary Record, 178*(5), 122-123.

Pieters, M., Fano, E., Pijoan, C., & Dee, S. (2010). An experimental model to evaluate *Mycoplasma hyopneumoniae* transmission from asymptomatic carriers to unvaccinated and vaccinated sentinel pigs. *Canadian Journal of Veterinary Research, 74*(2), 157-160.

Pieters, M., Pijoan, C., Fano, E., & Dee, S. (2009). An assessment of the duration of *Mycoplasma hyopneumoniae* infection in an experimentally infected population of pigs. *Veterinary Microbiology, 134*(3-4), 261-266.

Pieters, M. G., & Maes, D. (2019). Mycoplasmosis. In J. J. Zimmermann, L. A. Karriker, A. Ramirez, K. J. Schwartz, G. W. Stevenson, & J. Zhang, *Diseases of swine* (11th ed.; pp. 863-883). Hoboken, NJ, USA: John Wiley & Sons, Inc. https://doi.org/10.1002/9781119350927.ch56

Pijoan, C. (2001). Disease eradication: Should we go there? Proceedings AD Leman Conf. Saint Paul, Minnesota, 7-9.

Pijoan, C., Lastra, A., Ramirez, C., & Leman, A. D. (1984). Isolation of toxigenic strains of *Pasteurella multocida* from lungs of pneumonic swine. *Journal of American Veterinary and Medical Association, 185*(5), 522-523.

Pillman, D., Surendran Nair, M., Schwartz, J., & Pieters, M. (2019). Detection of *Mycoplasma hyorhinis* and *Mycoplasma hyosynoviae* in oral fluids and correlation with pig lameness scores. *Veterinary Microbiology, 239*, 108448.

Pilo, P., Vilei, E. M., Peterhans, E., Bonvin-Klotz, L., Stoffel, M. H., Dobbelaere, D., & Frey, J. (2005). A metabolic enzyme as a primary virulence factor of *Mycoplasma mycoides* subsp. *mycoides* small colony. *Journal of Bacteriology, 187*, 6824-6831.

Pinto, P. M., Carvalho, M. O. de, Alves-Junior, L., Brocchi, M., & Schrank, I. S. (2007a). Molecular analysis of an integrative conjugative element, ICEH, present in the chromosome of different strains of *Mycoplasma hyopneumoniae*. *Genetics and Molecular Biology, 30*, 256-263. https://doi.org/10.1590/S1415-47572007000200014

Pinto, P. M., Chemale, G., de Castro, L. A., Costa, A. P. M., Kich, J. D., Vainstein, M. H., Zaha, A., & Ferreira, H. B. (2007b). Proteomic survey of the pathogenic *Mycoplasma hyopneumoniae* strain 7448 and identification of novel post-translationally modified and antigenic proteins. *Veterinary Microbiology, 121*, 83-93. https://doi.org/10.1016/j.vetmic.2006.11.018

Pinto, P. M., Klein, C. S., Zaha, A., & Ferreira, H. B. (2009). Comparative proteomic analysis of pathogenic and non-pathogenic strains from the swine pathogen *Mycoplasma hyopneumoniae*. *Proteome Science, 7*, 45. https://doi.org/10.1186/1477-5956-7-45

Plank, G., & Heinritzi, K. (1990). [Disseminated intravascular coagulation in eperythrozoonosis of swine]. *Berliner und Münchener Tierärztliche Wochenschrift, 103*, 13-18.

Poehlsgaard, J., Andersen, N. M., Warrass, R., & Douthwaite, S. (2012). Visualizing the 16-membered ring macrolides tildipirosin and tilmicosin bound to their ribosomal site. *ACS Chemical Biology*, 7(8), 1351-1355.

Pointon, A., Cargill, C., & Slade, J. (1995). *The good health manual for pigs.* Canberra, Australia: Pig Research and Development Corporation.

Poland, J., Edington, N., Gois, M., & Betts, A. O. (1971). The production of pneumonia with or without pleurisy in gnotobiotic piglets with pure cultures of strain TR 32 of *Mycoplasma hyorhinis*. *Journal of Hygiene, 69*, 145-154.

Posa, R., Magyar, T., Stoev, S. D., Glavits, R., Donko, T., Repa, I., & Kovacs, M. (2013). Use of computed tomography and histopathologic review for lung lesions produced by the interaction between *Mycoplasma hyopneumoniae* and fumonisin mycotoxins in pigs. *Veterinary Pathology, 50*(6), 971-979.

Posa, R., Stoev, S., Kovacs, M., Donko, T., Repa, I., & Magyar, T. (2016). A comparative pathological finding in pigs exposed to fumonisin B1 and/or *Mycoplasma hyopneumoniae. Toxicology and Industrial Health, 32*(6), 998-1012.

Pospischil, A., & Hoffmann, R. (1982). *Eperythrozoon suis* in naturally infected pigs: A light and electron microscopic study. *Veterinary Pathology, 19*, 651-657.

Poveda, J. B., & Nicholas, R. (1998). Serological identification of mycoplasmas by growth and metabolic inhibition tests. *Methods in Molecular Biology, 104*, 105-111.

Preston, K. S., & Greve, J. H. (1965). Eperythrozoonosis in 4-week-old pigs. *Iowa State University Veterinarian, 27*, 119.

Price, M. N., Dehal, P. S., & Arkin, A. P. (2010). FastTree 2 – Approximately Maximum-Likelihood Trees for Large Alignments. *Plos One, 5*, e9490. https://doi.org/10.1371/journal.pone.0009490

Priebe, S., & Schwarz, S. (2003). *In vitro* activities of florfenicol against bovine and porcine respiratory tract pathogens. *Antimicrobial Agents and Chemotherapy, 47*(8), 2703-2705.

Prullage, J. B., Williams, R. E., & Gaafar, S. M. (1993). On the transmissibility of *Eperythrozoon suis* by *Stomoxys calcitrans* and *Aedes aegypti. Veterinary Parasitology, 50*, 125-135.

Pulgarón, Y. B., Marques, L. M., Campos, A. C. de A., Rodriguez, J. P., Márquez, O. P., Ortiz, A. D., Rivero, E. L., & Timenetsky, J. (2015). Detection, quantification and genetic variability of *Mycoplasma hyopneumoniae* from apparently healthy and pneumonic swine. *Brazilian Journal of Veterinary Research and Animal Science, 52*, 310. https://doi.org/10.11606/issn.1678-4456.v52i4p310-318

Qiu, G., Rui, Y., Zhang, J., Zhang, L., Huang, S., Wu, Q., Li, K., Han, Z., Liu, S., & Li, J. (2018). Macrolide-resistance selection in tibetan pigs with a high load of *Mycoplasma hyopneumoniae. Microbial Drug Resistance, 24*(7), 1043-1049.

Quesnel, H. (2011). Colostrum: Roles in piglet performance and production by the sow. VI Suinsui-Simposio internacional de Suinocultura, Porto Alegre (Brasil), 1-12.

Quin, A. H. (1938). A herd condition of swine characterized by icterus and anemia. *Journal of the American Veterinary Medical Association, 93*, 327-328.

Ramirez, A. Differential diagnosis. In J. J. Zimmermann, L. A. Karriker, A. Ramirez, K. J. Schwartz, G. W. Stevenson, & J. Zhang, *Diseases of swine* (11th ed.; pp. 59-74). Ames, Iowa: State University Press.

Rammohan, L., Xue, L., Wang, C., Chittick, W., Ganesan, S., & Ramamoorthy, S. (2012). Increased prevalence of torque teno viruses in porcine respiratory disease complex affected pigs. *Veterinary Microbiology, 157*(1-2), 61-68.

Rapp-Gabrielson, V. J., Oliveira, S., & Pijoan, C. (2006). *Haemophilus parasuis.* In B. Straw, J. Zimmerman, S. D'Allaire, & D. J. Taylor (Eds.), *Diseases of swine* (9th ed.; pp. 681-690). Ames, Iowa: Blackwell Publishing.

Rautiainen, E., Oravainen, J., Virolainen, J.V., & Tuovinen, V. (2001). Regional eradication of *Mycoplasma hyopneumoniae* from pig herds and documentation of freedom of the disease. *Acta Veterinaria Scandinavica, 42*(3), 355-364.

Rautiainen, E., Virtala, A. M., Wallgren, P., & Saloniemi, H. (2000). Varying effects of infections with *Mycoplasma hyopneumoniae* on the weight gain recorded in three different multisource fattening pig herds. *Journal of Veterinary Medicine B, Infectious Diseases and Veterinary Public Health, 47*(6), 461-469.

Rawal, G., Arruda, P., Rademancher, C., & Linhares, D.C. (2018). General overview of the detection of *Mycoplasma hyopneumoniae* DNA by quantitative polymerase chain reaction in diagnostic cases submitted to the Iowa state University Veterinary Diagnostic Laboratory from 2004 to 2016. *Journal of Swine Health Production, 26*, 309-315.

Raymond, B. B., & Djordjevic, S. (2015). Exploitation of plasmin(ogen) by bacterial pathogens of veterinary significance. *Veterinary Microbiology, 178*, 1-13.

Raymond, B. B. A., Jenkins, C., Seymour, L. M., Tacchi, J. L., Widjaja, M., Jarocki, V. M., Deutscher, A. T., Turnbull, L., Whitchurch, C. B., Padula, M. P., & Djordjevic, S. P. (2015). Proteolytic processing of the cilium adhesin MHJ_0194 (P123J) in *Mycoplasma hyopneumoniae* generates a functionally diverse array of cleavage fragments that bind multiple host molecules. *Cellular Microbiology, 17*, 425-444. https://doi.org/10.1111/cmi.12377

Raymond, B. B. A., Jenkins, C., Turnbull, L., Whitchurch, C.,B., & Djordjevic, S.,P. (2018a). Extracellular DNA release from the genome-reduced pathogen *Mycoplasma hyopneumoniae* is essential for biofilm formation on abiotic surfaces. *Scientific Reports, 8*, 10373.

Raymond, B. B. A., Madhkoor, R., Schleicher, I., Uphoff, C. C., Turnbull, L., Whitchurch, C. B., Rohde, M., Padula, M. P., & Djordjevic, S. P. (2018b). Extracellular actin is a receptor for *Mycoplasma hyopneumoniae*. *Frontiers in Cellular and Infection Microbiology, 8*, 54.

Raymond, B. B. A., Tacchi, J. L., Jarocki, V. M., Minion, F. C., Padula, M. P., & Djordjevic, S. P. (2013). P159 from *Mycoplasma hyopneumoniae* binds porcine cilia and heparin and is cleaved in a manner akin to ectodomain shedding. *Journal of Proteome Research, 12*, 5891-5903. https://doi.org/10.1021/pr400903s

Raymond, B. B. A., Turnbull, L., Jenkins, C., Madhkoor, R., Schleicher, I., Uphoff, C. C., Whitchurch, C. B., Rohde, M., & Djordjevic, S. P. (2018c). *Mycoplasma hyopneumoniae* resides intracellularly within porcine epithelial cells. *Scientific Reports, 8*, 17697. https://doi.org/10.1038/s41598-018-36054-3

Razin, S. (1992). Mycoplasma taxonomy and ecology. In J. Maniloff, R. N. McElhaney, L. R. Finch, & J. B. Baseman (Eds.), *Mycoplasmas, molecular biology and pathogenesis*. Washington, D.C.: American Society for Microbiology.

Razin, S., Knyszynski, A., & Lifshitz, Y. (1964). Nucleases of *Mycoplasma*. *Journal of General Microbiology, 36*, 323-332.

Rebaque, F., Camacho, P., Parada, J., Lucchesi, P., Ambrogi, A., & Tamiozzo, P. (2018). Persistence of the same genetic type of *Mycoplasma hyopneumoniae* in a closed herd for at least two years. *Revista Argentina de Microbiología, 50*, 147-150. https://doi.org/10.1016/j.ram.2017.05.002

Redgrave, L. S., Sutton, S. B., Webber, M. A., & Piddock, L. J. (2014). Fluoroquinolone resistance: Mechanisms, impact on bacteria, and role in evolutionary success. *Trends in Microbiology, 22*(8), 438-445.

Redondo, E., Masot, A. J., Fernandez, A., & Gazquez, A. (2009). Histopathological and immunohistochemical findings in the lungs of pigs infected experimentally with *Mycoplasma hyopneumoniae*. *Journal of Comparative Pathology, 140*(4), 260-270.

Regia Silva, S. K., Mauric Frossard, R. A., Roberto Nunes, G. P., Eliza Facioni, G. S., Lopes, P. S., Veroneze, R., & Gasparino, E. (2011). Toll-like receptor 6 differential expression in two pig genetic groups vaccinated against *Mycoplasma hyopneumoniae*. *BMC Proceedings, 5*(Suppl 4), S9.

Register, K. B., Brockmeier, S. L., de Jong, M. F., & Pijoan, C. (2012). Pasteurellosis. In J. J. Zimmerman, L. A. Karriker, A. Ramirez, K. J. Schwartz, & G. Stevenson (Eds.), *Diseases of swine* (10th ed.; pp. 798-810). Oxford: Wiley Blackwell.

Regula, G., Lichtensteiger, C. A., Mateus-Pinilla, N. E., Scherba, G., Miller, G. Y., & Weigel, R. M. (2000). Comparison of serologic testing and slaughter evaluation for assessing the effects of subclinical infection on growth in pigs. *Journal of American Veterinary Medical Association, 217*(6), 888-895.

Reinhardt, A. K., Bebear, C. M., Kobisch, M., Kempf, I., & Gautier-Bouchardon, A. V. (2002a). Characterization of mutations in DNA gyrase and topoisomerase IV involved in quinolone resistance of *Mycoplasma gallisepticum* mutants obtained *in vitro*. *Antimicrobial Agents and Chemotherapy, 46*(2), 590-593.

Reinhardt, A. K., Kempf, I., Kobisch, M., & Gautier-Bouchardon, A. V. (2002b). Fluoroquinolone resistance in *Mycoplasma gallisepticum*: DNA gyrase as primary target of enrofloxacin and impact of mutations in topoisomerases on resistance level. *Journal of Antimicrobial Chemotherapy, 50*(4), 589-592.

Resende, T. P., Pieters, M., & Vannucci, F. A. (2019). Swine conjunctivitis outbreaks associated with *Mycoplasma hyorhinis*. *Journal of Veterinary Diagnostic Investigation, 31*, 766-769.

Ritzmann, M., Grimm, J., Heinritzi, K., Hoelzle, K., & Hoelzle, L. E. (2009). Prevalence of *Mycoplasma suis* in slaughter pigs, with correlation of PCR results to hematological findings. *Veterinary Microbiology, 133*, 84-91.

Ro, L. H., & Ross, R. F. (1983). Comparison of *Mycoplasma hyopneumoniae* strains by serologic methods. *American Journal of Veterinary Research, 44*, 2087-2094.

Robb, A. D. (1943). Ictero-anemia in growing swine. *Veterinary Medicine, 38*, 271-274.

Robbins, R., Betlach, A., Mondragon-Evans, M., & Pieters, M. (2019). Development of a herd-specific lung homogenate for exposure to *Mycoplasma hyopneumoniae* under field conditions. *Journal of Swine Health and Production, 27*(4), 221-227.

Roberts, D. H., & Little, T. W. (1970). An auto-immune response associated with porcine enzootic pneumonia. *Veterinary Records, 86*(11), 328.

Roberts, E. D., Switzer, W. P., & Ramsey, F. K. (1963). Pathology of the visceral organs of swine inoculated with *Mycoplasma hyorhinis*. *American Journal of Veterinary Research, 24*, 9-18.

Robinson, M. W., Buchtmann, K. A., Jenkins C. Tacchi, J., Raymond, B., To, J., Chowdhury, P., Woolley, L., Labbate, M., Turnbull, L., Whitchurch, C., Padula, M., & Djordjevic, S. (2013). MHJ_0125 is an M42 glutamyl aminopeptidase that moonlights as a multifunctional adhesin on the surface of *Mycoplasma hyopneumoniae*. *Open Biology, 3*, 130017.

Rocha, E. P. C., & Blanchard, A. (2002). Genomic repeats, genome plasticity and the dynamics of Mycoplasma evolution. *Nucleic Acids Research, 30*, 2031-2042.

Rodriguez, F., Ramirez, G. A., Sarradell, J., Andrada, M., & Lorenzo, H. (2004). Immunohistochemical labelling of cytokines in lung lesions of pigs naturally infected with Mycoplasma hyopneumoniae. *Journal of Comparative Pathology, 130*(4), 306-312. doi:10.1016/j.jcpa.2003.12.008

Roos, L., Fano, E., Homwong, N., Payne, B., & Pieters, M. (2016). A model to investigate the optimal seeder-to-naïve ratio for successful natural *Mycoplasma hyopneumoniae* gilt exposure prior to entering the breeding herd. *Veterinary Microbiology, 184*, 51-58.

Roos, L. R., Surendran Nair, M., Rendahl, A. K., & Pieters, M. (2019). *Mycoplasma hyorhinis* and *Mycoplasma hyosynoviae* dual detection patterns in dams and piglets. *PLoS One, 14*, e0209975.

Roques, E., Girard, A., Gagnon, C., & Archambault, D. (2013). Antibody responses induced in mice immunized with recombinant adenovectors expressing chimeric proteins of various porcine pathogens. *Vaccine, 31*, 2698-2704.

Rosales, R. S., Churchward, C. P., Antunes, N. T., Suarez, A., Ramirez, A. S., Assunçao, P., Poveda, C., Nicholas, R. A. J., & Poveda, J. B. (2008). Molecular typing and immunological characterization of field strains of *Mycoplasma hyorhinis*. Proceedings 17th International Congress of the International Organization for Mycoplasmology, Tianjin, China, 59.

Rosengarten, R., & Wise, K. S. (1991). The Vlp system of *Mycoplasma hyorhinis*: combinatorial expression of distinct size variant lipoproteins generating high-frequency surface antigenic variation. *Journal of Bacteriology, 173*, 4782-4793.

Rosengarten, R., Citti, C., Glew, M., Lischewski, A., Droeße, M., Much, P., Winner, F., Brank, M., & Spergser, J. (2000). Host pathogen interactions in mycoplasma pathogenesis: Virulence and survival strategies of minimalist prokaryotes. *International Journal of Medical Microbiology, 290*, 15-25.

Ross, R. (1999). Mycoplasmosis. In B. E. Straw, S. D'Allaire, W. L. Mengeling, & D. J. Taylor (Eds.), *Diseases of swine* (8th ed.). Iowa: State University Press.

Ross, R. F., & Cox, D. F. (1988). Evaluation of tiamulin for treatment of mycoplasmal pneumonia in swine. *Journal of the American Veterinary Medicine Association, 193*(4), 441-446.

Ross, R. F., & Duncan, J. R. (1970). *Mycoplasma hyosynoviae* arthritis of swine. *Journal of the American Veterinary Association, 157*, 1515-1518.

Ross, R. F., & Karmon, J. A. (1970). Heterogeneity among strains of *Mycoplasma granularum* and identification of *Mycoplasma hyosynoviae*, sp. n. *Journal of Bacteriology, 103*, 707-713.

Ross, R. F., & Young, T. F. (1993). The nature and detection of mycoplasmal immunogens. *Veterinary Microbiology, 37*, 369-380.

Rossow, K. D. (1998). Porcine reproductive and respiratory syndrome. *Veterinary Pathology, 35*(1), 1-20.

Rottem, S. (2003). Interaction of mycoplasmas with host cells. *Physiological Reviews, 83*, 417-432.

Roy Chowdhury, P., Scott, M., Worden, P., Huntington, P., Hudson, B., Karagiannis, T., Charles, I. G., & Djordjevic, S. P. (2016). Genomic islands 1 and 2 play key roles in the evolution of extensively drug-resistant ST235 isolates of *Pseudomonas aeruginosa*. *Open Biology, 6*. https://doi.org/10.1098/rsob.150175

Ruiz, A., Galina, L., & Pijoan, C. (2002). *Mycoplasma hyopneumoniae* colonization by pigs sired by different boars. *The Canadian Journal of Veterinary Research, 66*, 79-85.

Ruiz, A., Utrera, V., & Pijoan, C. (2003). Effect of *Mycoplasma hyopneumoniae* sow vaccination on piglet 819 colonization at weaning. *Journal of Swine Health and Production, 11*, 131-135.

Salyers, A. A., & Whitt, D. D. (2002). *Mycoplasma pneumoniae* and *Chamydia pneumoniae*. In A. A. Salyers, & D. D. Whitt (Eds.), *Bacterial pathogenesis. A molecular approach* (2nd ed.). Washington: ASM Press.

Sarradell, J., Andrada, M., Ramirez, A. S., Fernandez, A., Gomez-Villamandos, J. C., Jover, A., Lorenzo, H., Herraez, P., & Rodriguez, F. (2003). A morphologic and immunohistochemical study of the bronchus-associated lymphoid tissue of pigs naturally infected with *Mycoplasma hyopneumoniae*. *Veterinary Pathology, 40*, 395-404.

Sato, T., Okamura, T., Kojima-Shibata, C., Kadowaki, H., Suzuki, E., Uenishi, H., & Suzuki, K. (2016). Correlated response of peripheral blood cytokines with selection for reduced mycoplasma pneumonia of swine lesions in Landrace pigs. *Animal Science Journal, 87*(4), 477-483. doi:10.1111/asj.12462

Sautter, C. A., Auray, G., Python, S., Liniger, M., & Summerfield, A. (2018). Phenotypic and functional modulations of porcine macrophages by interferons and interleukin-4. *Developmental and Comparative Immunology, 84*, 181-192. doi:10.1016/j.dci.2018.01.018

Scarman, A. L., Chin, J. C., Eamens, G. J., Delaney, S. F., & Djordjevic, S. P. (1997). Identification of novel species-specific antigens of *Mycoplasma hyopneumoniae* by preparative SDS-PAGE ELISA profiling. *Microbiology, 143*(Pt 2), 663-673. https://doi.org/10.1099/00221287-143-2-663

Schachtschneider, K. M., Yeoman, C. J., Isaacson, R. E., White, B. A., Schook, L. B., & Pieters, M. (2013). Modulation of systemic immune responses through commensal gastrointestinal microbiota. *PLoS One, 8*, e53969.

Scheepens, C. J. M. (1996). Climatic stress in swine: hazards for health. *Pig Journal, 37*, 130-136.

Scheidt, A. B., Cline, T. R., Clark, L. K., Mayrose, V. B., Van Alstine, W. G., Diekman, M. A., & Singleton, W. L. (1995). The effect of all-in-all-out growing-finishing on the health of pigs. *Journal of Swine Health and Production, 3*(5), 202-205.

Schieck, E., Lartigue, C., Frey J., Vozza, N., Hegermann, J., Miller, R., Valguarnera, E., Muriuki, C., Meens, J., Nene, V., Naessens, J., Weber, J., et al. (2016). Galactofuranose in *Mycoplasma mycoides* is important for membrane integrity and conceals adhesins but does not contribute to serum resistance. *Molecular Microbiology, 99*, 55-70.

Schleper, M., Garza-Moreno, L., Tousignant, S., & Pieters, M. (2019). *Methods for achieving successful and safe gilt inoculation with* Mycoplasma hyopneumoniae. Orlando, Florida: American Association of Swine Veterinarians.

Schmitt, C. (2014). Management of *Mycoplasma hyosynoviae*. Proceedings American Association of Swine Veterinarians Annual Meeting, Dallas, USA, 545-546.

Schneider P. (2006). Experiences with *Mycoplasma hyopneumoniae* and transmissible gastroenteritis eradication from sow herds. Proceedings AD Leman Conf. Saint Paul, Minnesota, 2006, 82-86.

Schreiner, S. A., Hoelzle, K., Hofmann-Lehmann, R., Hamburger, A., Wittenbrink, M. M., Kramer, M. M., Sokoli, A., Felder, K. M., Groebel, K., & Hoelzle, L.E. (2012a). Nanotransformation of the haemotrophic *Mycoplasma suis* during in vitro cultivation attempts using modified cell free Mycoplasma media. *Veterinary Microbiology, 160*, 227-232.

Schreiner, S. A., Sokoli, A., Felder, K. M., Wittenbrink, M. M., Schwarzenbach, S., Guhl, B., Hoelzle, K., & Hoelzle, L. E. (2012b). The surface-localised α-enolase of *Mycoplasma suis* is an adhesion protein. *Veterinary Microbiology, 156*, 88-95.

Schuller, W., Heinritzi, K., al-Nuktha, S., Kolbl, S., & Schuh, M. (1990). [Serologic progression studies using CF and ELISA for the detection of antibodies against *Eperythrozoon suis* infection of swine]. *Berliner und Münchener Tierärztliche Wochenschrift, 103*, 9-12.

Schuller, W., Neumeister, E., & Vogl, D. (1977). [For the remediation of pigs infected with enzootic pneumonia] (in German). *Wiener Tierärztliche Monatschrift, 64*, 156-160.

Schulmann, A., Estola, T., & Garry-Anderson, A. S. (1970). On the occurrence of *Mycoplasma hyorhinis* in the respiratory organs of pigs with special reference to enzootic pneumonia. *Zentralblatt für Veterinärmedizin Series B, 17*, 549-553.

Schultz, K. K., Strait, E. L., Erickson, B. Z., & Levy, N. (2012). Optimization of an antibiotic sensitivity assay for *Mycoplasma hyosynoviae* and susceptibility profiles of field isolates from 1997 to 2011. *Veterinary Microbiology, 158*(1-2), 104-108.

Schwartz, J., Bruner, L., Evelsizer, B., Konz, B., Rovira, A., & Pieters, M. (2014). Dynamics of *Mycoplasma hyosynoviae* detection and clinical presentation. Proceedings American Association of Swine Veterinarians Annual Meeting, Dallas, USA, 115-116.

Schwartz, M. (2015). Cost of *M. hyopneumoniae* in growing pigs. Allen D. Leman Swine Conference, St. Paul, MN.

Schwartz, M., Schwartz, J., Leuwerke, B., & Pieters, M. (2018). A production system approach for ultimate *Mycoplasma hyopneumoniae* control. Proceedings International Pig Veterinary Society Congress, Chongqing, China.

Schwarz, S., Shen, J., Kadlec, K., Wang, Y., Brenner Michael, G., Fessler, A. T., & Vester, B. (2016). Lincosamides, streptogramins, phenicols, and pleuromutilins: Mode of action and mechanisms of resistance. *Cold Spring Harbor Perspectives in Medicine, 6*(11), a027037. doi:10.1101/cshperspect.a027037

Schwarz, S., Silley, P., Simjee, S., van Duijkeren, E., Johnson, A.P., & Gaastra, W. (2010). Assessing the antimicrobial susceptibility of bacteria obtained from animals. *Veterinary Microbiology, 141*, 1-4.

Schweighardt, H., Fellner, A., Pechan, P., & Lauermann, E. (1986). Eperythrozoonose beim Schwein - ein Fallbericht. *Wiener Tierärztliche Monatschrift, 73*, 250-253.

Scollo, A., Gottardo, F., Contiero, B., Mazzoni, C., Leneveu, P., Edwards, S. A. (2017). Benchmarking of pluck lesions at slaughter as a health monitoring tool for pigs slaughtered at 170 kg (heavy pigs). *Preventive Veterinary Medicine, 144*, 20-28.

Scott, H. M., Acuff, G., Bergeron, G., Bourassa, M. W., Gill, J., Graham, D. W., Kahn, L.H., Morley, P.S., Salois, M.J., Simjee, S., Singer, R. S., Smith, T. C., Storrs, C., & Wittum, T. E. (2019). Critically important antibiotics: criteria and approaches for measuring and reducing their use in food animal agriculture. *Annals of the New York Academy of Science, 1441*(1), 8-16.

Segalés, J., Domingo, M., Chianini, F., Majo, N., Dominguez, J., Darwich, L., & Mateu, E. (2004). Immunosuppression in postweaning multisystemic wasting syndrome affected pigs. *Veterinary Microbiology, 98*(2), 151-158.

Segalés, J., Valero, O., Espinal, A., López-Soria, S., Nofradías, M., Calsamiglia, M., & Sibila, M. (2012). Exploratory study on the influence of climatological parameters on *Mycoplasma hyopneumoniae* infection dynamics. International Journal of Biometeorology, *56*(6), 1167-1171.

Segovia, J. A., Chang, T. H., Winter, V. T., Coalson, J. J., Cagle, M. P., Pandranki, L., Bose, S., Baseman, J. B., & Kannan, T. R. (2018). NLRP3 Is a critical regulator of inflammation and innate immune cell response during *Mycoplasma pneumoniae* infection. *Infection and Immunity, 86*(1). doi:10.1128/iai.00548-17

Selbitz, H. J. (2006). [Bakterielle Krankheiten der Tiere.] In A. Mayr (Ed.), *Medizinische Mikrobiologie, Infektions- und Seuchenlehre*. 9. Auflage, Enke Verlag, Stuttgart, 543.

Seo, H. W., Park, S. J., Park, C., & Chae, C. (2014). Interaction of porcine circovirus type 2 and *Mycoplasma hyopneumoniae* vaccines on dually infected pigs. *Vaccine, 32*(21), 2480-2486.

Seo, M. G., Kwon, O. D., & Kwak. D. (2019). Prevalence and phylogenetic analysis of hemoplasma species in domestic pigs in Korea. *Parasites & Vectors, 12*, 378.

Seymour, L. M., Deutscher, A. T., Jenkins, C., Kuit, T. A., Falconer, L., Minion, F. C., Crossett, B., Padula, M., Dixon, N. E., Djordjevic, S. P., & Walker, M. J. (2010). A processed multidomain *Mycoplasma hyopneumoniae* adhesin binds fibronectin, plasminogen, and swine respiratory cilia. *Journal of Biological Chemistry, 285*, 33971-33978. https://doi.org/10.1074/jbc.M110.104463

Seymour, L. M., Falconer, L., Deutscher, A. T., Minion, F. C., Padula, M. P., Dixon, N. E., Djordjevic, S. P., & Walker, M .J. (2011). Mhp107 is a member of the multifunctional adhesin family of *Mycoplasma hyopneumoniae*. *Journal of Biological Chemistry, 286*, 10097-10104. https://doi.org/10.1074/jbc.M110.208140

Seymour, L. M., Jenkins, C., Deutscher, A. T., Raymond, B. B. A., Padula, M. P., Tacchi, J. L., Bogema, D. R., Eamens, G. J., Woolley, L. K., Dixon, N. E., Walker, M. J., & Djordjevic, S. P. (2012). Mhp182 (P102) binds fibronectin and contributes to the recruitment of plasmin(ogen) to the *Mycoplasma hyopneumoniae* cell surface. *Cellular Microbiology, 14*, 81-94. https://doi.org/10.1111/j.1462-5822.2011.01702.x

Shen, Y., Hu, W., Wei, Y., Feng, Z., & Yang, Q. (2017). Effects of *Mycoplasma hyopneumoniae* on porcine nasal cavity dendritic cells. *Veterinary Microbiology, 198*, 1-8.

Shibata, I., Okada, M., Urono, K., Samegai, Y., Ono, M., Sakano, T., & Sato, S. (1998). Experimental dual infection of cesarean-derived, colostrum-deprived pigs with *Mycoplasma hyopneumoniae* and pseudorabies virus. *Journal of Veterinary and Medical Sciences, 60*(3), 295-300.

Shibata, K-I., Sasaki, T., & Watanabe, T. (1995). AIDS-associated mycoplasmas possess phospholipases C in the membrane. *Infection and Immunity, 63*, 4174-4177.

Shimazu, T., Borjigin, L., Katayama, Y., Li, M., Satoh, T., Watanabe, K., Kitazawa, H., Roh, S., Aso, H., Kazuo, K., Suda, Y., Sakuma, A., Nakajo, M., & Suzuki, K. (2013). Immunological characterization of peripheral blood leukocytes using vaccine for mycoplasmal pneumonia of swine (MPS) in swine line selected for resistance to MPS. *Animal Science Journal, 84*(10), 683-692. doi:10.1111/asj.12058

Shimazu, T., Borjigin, L., Katayama, Y., Li, M., Satoh, T., Watanabe, K., Kitazawa, H., Roh, S., Aso, H., Kazuo, K., Suda, Y., Sakuma, A., Nakajo, M., & Suzuki, K. (2014). Genetic selection for resistance to mycoplasmal pneumonia of swine (MPS) in the Landrace line influences the expression of soluble factors in blood after MPS vaccine sensitization. *Animal Science Journal, 85*(4), 365-373. doi:10.1111/asj.12158

Shimizu, T. (2016). Inflammation-inducing factors of *Mycoplasma pneumoniae*. *Frontiers in Microbiology,* 7, 414. doi:10.3389/fmicb.2016.00414

Shimizu, T., Kida, Y., & Kuwano, K. (2005). A dipalmitoylated lipoprotein from *Mycoplasma pneumoniae* activates NF-kappa B through TLR1, TLR2, and TLR6. *Journal of Immunology, 175*(7), 4641-4646.

Shimoji, Y., Oishi, E., Muneta, Y., Nosaka, H., &. Mori, Y. (2003). Vaccine efficacy of the attenuated *Erysipelothrix rhusiopathiae* YS-19 expressing a recombinant protein of *Mycoplasma hyopneumoniae* P97 adhesin against mycoplasmal pneumonia of swine. *Vaccine, 21*, 532-537.

Shin, S. J., Kang, S. G., Nabin, R., Kang, M. L., & Yoo, H. S. (2005). Evaluation of the antimicrobial activity of florfenicol against bacteria isolated from bovine and porcine respiratory disease. *Veterinary Microbiology, 106*(1-2), 73-77.

Sibila, M., Bernal, R., Torrents, D., Riera, P., Llopart, D., Calsamiglia, M., & Segalés, J. (2008). Effect of sow vaccination against *Mycoplasma hyopneumoniae* on sow and piglet colonization and seroconversion, and pig lung lesions at slaughter. *Veterinary Microbiology, 127*(1-2), 165-170. doi:10.1016/j.vetmic.2007.07.027

Sibila, M., Calsamiglia, M., Nofrarías, M., López-Soria, S., Espinal, A., Segalés, J., Riera, P., Llopart, D., & Artigas, C. (2004b). Longitudinal study of *Mycoplasma hyopneumoniae* infection in naturally infected pigs. Proceedings 18th IPVS Congress, Hamburg, Germany, June 27 to July 1, 169.

Sibila, M., Calsamiglia, M., Segalés, J., & Rossell, C. (2004c). Association between *Mycoplasma hyopneumoniae* at different respiratory sites and presence of histopathological lung lesions. *Veterinary Record, 155*, 55-58.

Sibila, M., Calsamiglia, M., Vidal, D., Badiella, L., Aldaz, A., & Jensen, J. C. (2004a). Dynamics of *Mycoplasma hyopneumoniae* infection in 12 farms with different production systems. *Canadian Journal of Veterinary Research, 68*, 12-18.

Sibila, M., Fort, M., Nofrarías, M., Pérez de Rozas, A., Galindo-Cardiel, I., Mateu, E., & Segalés, J. (2012). Simultaneous porcine circovirus type 2 and *Mycoplasma hyopneumoniae* co-inoculation does not potentiate disease in conventional pigs. *Journal of Comparative Pathology, 147*(2-3), 285-295.

Sibila, M., Nofrarias, M., Lopez-Soria, S., Segalés, J., Valero, O., Espinal, A., & Calsamiglia, M. (2007). Chronological study of *Mycoplasma hyopneumoniae* infection, seroconversion and associated lung lesions in vaccinated and non-vaccinated pigs. *Veterinary Microbiology, 122*, 97-107.

Sibila, M., Pieters, M., Molitor, T., Maes, D., Haesebrouck, F., & Segalés, J. (2009). Current perspectives on the diagnosis and epidemiology of *Mycoplasma hyopneumoniae* infection. *Veterinary Journal, 181*, 221-231.

Sieve, A. N., Meeks, K. D., Bodhankar, S., Lee, S., Kolls, J. K., Simecka, J. W., & Berg, R. E. (2009). A novel IL-17-dependent mechanism of cross protection: respiratory infection with mycoplasma protects against a secondary listeria infection. *European Journal of Immunology, 39*(2), 426-438. doi:10.1002/eji.200838726

Silva, S. S., Yeske P., Morrison, R. B., & Linhares, D. L. C. (2019). Benefit-cost analysis to estimate the payback time and the economic value of two *Mycoplasma hyopneumoniae* elimination methods in breeding herds. *Preventive Veterinary Medicine, 168*, 95-102.

Simionatto, S., Marchioro, S. B., Maes, D., & Dellagostin, O. A. (2013). *Mycoplasma hyopneumoniae*: From disease to vaccine development. *Veterinary Microbiology, 165*(3-4), 234-242.

Simionatto, S., Marchioro, S., Galli, V., Brum, C., Klein, C., Rebelatto, R., Silva, E., Borsuk, S., Conceição, F., &. Dellagostin, O. (2012). Immunological characterization of *Mycoplasma hyopneumoniae* recombinant proteins. *Comparative Immunology and Microbiology, 35*, 209-216.

Siqueira, F. M., Gerber, A. L., Guedes, R. L. M., Almeida, L. G., Schrank, I. S., Vasconcelos, A. T. R., & Zaha, A. (2014). Unravelling the transcriptome profile of the swine respiratory tract mycoplasmas. *PLoS One, 9.*

Siqueira, F. M., Thompson, C. E., Virginio, V. G., Gonchoroski, T., Reolon, L., Almeida, L. G., da Fonsêca, M. M., de Souza, R., Prosdocimi, F., Schrank, I. S., Ferreira, H. B., de Vasconcelos, A. T., & Zaha, A. (2013). New insights on the biology of swine respiratory tract mycoplasma from a comparative genome analysis. *BMC Genomics, 14*, 175. https://doi.org/10.1186/1471-2164-14-175

Sisk, D. B., Cole, J. R., & Pursell, A. R. (1980). Serologic incidence of eperythrozoonosis in Georgia swine. Proceedings American Association ofVeterinary Laboratory Diagnosticans, 13, 91-99.

Sjaastad, Ø. V., Sand, O., & Hove, K. (2016). The respiratory system. *Physiology of Domestic Animals* (pp. 508-546). Scandinavian Veterinary Press.

Smith, A. R., & Rahn, T. (1975). An indirect hemagglutination test for the diagnosis of *Eperythrozoon suis* infection in swine. *American Journal of Veterinary Research, 36*, 1319-1321.

Smith, A. R., Cipriano, J. E., & Hall, S. M. (1990). In vitro and in vivo glucose consumption in swine eperythrozoonosis. *Zentralblatt für Veterinärmedizin Reihe B, 37*, 587-592.

Sokoli, A., Groebel, K., Hoelzle, K., Amselgruber, W. M., Mateos, J. M., Schneider, M. K. J., Ziegler, U., Felder, K. M., & Hoelzle, L. E. (2013). *Mycoplasma suis* infection results endothelial cell damage and activation: New insight into the cell tropism and pathogenicity of hemotrophic mycoplasma. *Veterinary Research, 44*, 6-28.

Song, Q., Zhang, W., Song, W., Liu, Z., Khan, M. K., He, L., Fang, R., Li, P., Zhou, Y., Hu, M., & Zhao, J. (2014). Seroprevalence and risk factors of *Mycoplasma suis* infection in pig farms in central China. *Preventive Veterinary Medicine, 117*, 215-221.

Sosa, C., Blois, A., Ibáñez, F., Tamiozzo, P. (2019). Genetic diversity of *Mycoplasma hyopneumoniae* in Mendoza province. *Revista Argentina de Microbiología*. https://doi.org/10.1016/j.ram.2018.07.004

Spencer, R. (1940). Anaplasmosis-like disease of swine. *Veterinary Medicine, 35*, 294-295.

Spergser, J., Hess, C., Loncaric, I., & Ramírez, A. S. (2019). Matrix-assisted laser desorption ionization-time of flight mass spectrometry is a superior diagnostic tool for the identification and differentiation of mycoplasmas isolated from animals. *Journal of Clinical Microbiology, 57*, pii: e00316-19.

Splitter, E. J. (1950). *Eperythrozoon suis* n. sp. and *Eperythrozoon parvum* n. sp., 2 new blood parasites of swine. *Science, 111*, 513-514.

Splitter, E. J., & Williamson, R. L. (1950). Eperythrozoonosis in swine. A preliminary report. *Journal of the American Veterinary Medical Association, 116,* 360-364.

Splitter, E. J. (1958). The complement-fixation test in diagnosis of eperythrozoonosis in swine. *Journal of the American Veterinary Medical Association, 132*, 47-49.

Sponheim, A. (2017). A diagnostic approach to confirm day zero. Allen D. Leman Swine Conference, St. Paul, MN.

Sponheim, A., Alvarez, J., Fano, E., Schmaling, E., Dee, S., Hanson, D., Wetzell, T., & Pieters, M. (2019). Comparison of the sensitivity of laryngeal swabs and deep tracheal catheters for detection of *Mycoplasma hyopneumoniae* in experimentally and naturally infected pigs early and late after infection. *Veterinary Microbiology*, Nov 5, 108500. doi:10.1016/j.vetmic.2019.108500

Stadler, J., Jannasch, C., Mack, S. L., Dietz, S., Zöls, S., Ritzmann, M., Hoelzle, K., & Hoelzle, L. E. (2014). Clinical and haematological characterisation of *Mycoplasma suis* infections in splenectomised and non-splenectomised pigs. *Veterinary Microbiology, 172*, 294-300.

Stadler, J., Willi, S., Ritzmann, M., Eddicks, M., Ade, J., Hoelzle, K., & Hoelzle, L. E. (2019). Detection of *Mycoplasma suis* in pre-suckling piglets indicates a vertical transmission. *BMC Veterinary Research, 15*, 252.

Stakenborg, T., Vicca, J., Butaye, P., Imberechts, H., Peeters, J., De Kruif, A., Haesebrouck, F., & Maes, D. (2006a). A multiplex PCR to identify porcine mycoplasmas present in broth cultures. *Veterinary Research Communication, 30*, 239-247.

Stakenborg, T., Vicca, J., Butaye, P., Maes, D., Minion, F. C., Peeters, J., De Kruif, A., & Haesebrouck, F. (2005a). Characterization of *in vivo* acquired resistance of *Mycoplasma hyopneumoniae* to macrolides and lincosamides. *Microbial Drug Resistance, 11*(3), 290-294.

Stakenborg, T., Vicca, J., Butaye, P., Maes, D., Peeters, J., Kruif, A. de, Haesebrouck, F. (2005b). The diversity of *Mycoplasma hyopneumoniae* within and between herds using pulsed-field gel electrophoresis. *Veterinary Microbiology, 109*, 29-36. https://doi.org/10.1016/j.vetmic.2005.05.005

Stakenborg, T., Vicca, J., Maes, D., Peeters, J., de Kruif, A., Haesebrouck, F., Butaye, P. (2006b). Comparison of molecular techniques for the typing of *Mycoplasma hyopneumoniae* isolates. *Journal of Microbiological Methods, 66*, 263-275. https://doi.org/10.1016/j.mimet.2005.12.002

Stärk, K. D., Miserez, R., Siegmann, S., Ochs, H., Infanger, P., & Schmidt, J. (2007). A successful national control programme for enzootic respiratory diseases in pigs in Switzerland. *Revue Scientifique et Technique, 26*(3), 595-606. Review.

Stärk, K. D., Nicolet, J., & Frey, J. (1998). Detection of *Mycoplasma hyopneumoniae* by air sampling with a nested PCR assay. *Applied and Environmental Microbiology, 64*(2), 543-548.

Stärk, K. D. C. (2000). Epidemiological investigation of the influence of environmental risk factors on respiratory diseases in swine - a literature review. *Veterinary Journal, 159*, 37-56.

Stärk, K. D. C., Keller, H., & Eggenberg, E. (1992). Risk factors for the reinfection of specific pathogen free pig breeding herds with enzootic pneumonia. *Veterinary Record, 131*, 512-535.

Stärk, K. D. C., Pfeiffer, D. U., & Morris, R. S. (1998). Risk factors for respiratory disease in New Zealand. *New Zealand Veterinary Journal, 46*, 3-10.

Steenhard, N. R., Jungersen, G., Kokotovic, B., Beshah, E., Dawson, H. D., Urban, J. F. J., Roepstorff, A., & Thamsborg, S. M. (2009). *Ascaris suum* infection negatively affects the response to a *Mycoplasma hyopneumoniae* vaccination and subsequent challenge infection in pigs. *Vaccine, 27*(37), 5161-5169.

Stemke, G. W. (1997). Gene amplification (PCR) to detect and differentiate mycoplasmas in porcine mycoplasmal pneumonia. *Letters in Applied Microbiology, 25*, 327-330.

Stemke, G. W., Laigret, F., Grau, O., & Bové, J. M. (1992). Phylogenetic relationships of three porcine mycoplasmas, *Mycoplasma hyopneumoniae*, *Mycoplasma flocculare* and *Mycoplasma hyorhinis*, and complete 16S rRNA sequence of *M. flocculare*. *International Journal of Systematic Bacteriology, 42*, 220-225.

Stemke, G. W., Phan, R., Young, T. F., & Ross, R. F. (1994). Differentiation of *Mycoplasma hyopneumoniae*, *M. flocculare*, and *M. hyorhinis* on the basis of amplification of a 16S rRNA gene sequence. *American Journal of Veterinary Research, 55*, 81-84.

Stipkovits, L., Laky, Z., Abonyi, T., Siugzdaite, J., & Szabo, I. (2003). Reduction of economic losses caused by mycoplasmal pneumonia of pigs by vaccination with Respisure and by Tiamutin treatment. *Acta Veterinaria Hungarica, 51*(3), 259-271.

Stipkovits, L., Miller, D., Glavits, R., Fodor, L., & Burch, D. (2001). Treatment of pigs experimentally infected with *Mycoplasma hyopneumoniae*, *Pasteurella multocida*, and *Actinobacillus pleuropneumoniae* with various antibiotics. *Canadian Journal of Veterinary Research, 65*, 213-222.

Stipkovits, L., Nicolet, J., Haldimann, A., & Frey, J. (1991). Use of antibodies against the P36 protein of *Mycoplasma hyopneumoniae* for the identification of *M. hyopneumoniae* strains. *Molecular and Cellular Probes, 5*, 451-457. https://doi.org/10.1016/S0890-8508(05)80017-9

Strait, E. L., Hawkins, P. A., & Wilson, W. D. (2012). Dysgalactia associated with *Mycoplasma suis* infection in a sow herd. *Journal of the American Veterinary Medical Association, 241*, 1666-1667.

Strait, E. L., Madsen, M. L., Minion, F. C., Christopher-Hennings, J., Dammen, M., Jones, K. R., & Thacker, E. L. (2008). Real-time PCR assays to address genetic diversity among strains of *Mycoplasma hyopneumoniae*. *Journal of Clinical Microbiology, 46*, 2491-2498.

Strasser, M., Abiven, P., Kobisch, M., & Nicolet, J. (1992). Immunological and pathological reactions in piglets experimentally infected with *Mycoplasma hyopneumoniae* and/or *Mycoplasma flocculare*. *Veterinary Immunology and Immunopathology, 31*(1-2), 141-153.

Stricker, T. M., Weissenbacher-Lang, C., Ritzmann, M., & Ladinig, A. (2013). A study to assess the compatibility of simultaneous use of porcine reproductive and respiratory syndrome virus (EU-type) vaccine and *Mycoplasma hyopneumoniae* vaccine under field conditions. *Berliner und Münchener Tierärztliche Wochenschrift, 126*(5-6), 189-196.

Suter, M., Kobisch, M., & Nicolet, J. (1985). Stimulation of immunoglobulin-containing cells and isotype-specific antibody response in experimental *Mycoplasma hyopneumoniae* infection in specific-pathogen-free pigs. *Infection and Immunity, 49*(3), 615-620.

Swiss Federal Food Safety and Veterinary Office website, https://www.infosm.blv.admin.ch/public/?lang=en. Accessed on December 2019.

Switzer, W. P. (1955). Studies on infectious atrophic rhinitis. IV. Characterization of a pleuropneumonia-like organism isolated from the nasal cavities of swine. *American Journal of Veterinary Research, 16*, 540-544.

Syrovets, T., Lunov, O., & Simmet, T. (2012). Plasmin as a proinflammatory cell activator. *Journal of Leukocyte Biology, 92*, 509-519.

Szczepanek, S. M., Boccaccio, M., Pflaum, K., Liao, X., & Geary, S. J. (2014). Hydrogen peroxide production from glycerol metabolism is dispensable for virulence of *Mycoplasma gallisepticum* in the tracheas of chickens. *Infection and Immunity, 82*, 4915-4920.

Tacchi, J. L., Raymond, B. B. A., Haynes, P. A., Berry, I. J., Widjaja, M., Bogema, D. R., Woolley, L. K., Jenkins, C., Minion, F. C., Padula, M. P., & Djordjevic, S. P. (2016). Post-translational processing targets functionally diverse proteins in *Mycoplasma hyopneumoniae*. *Open Biology, 6*, 150210. https://doi.org/10.1098/rsob.150210

Tacchi, J. L., Raymond, B. B. A., Jarocki, V. M., Berry, I. J., Padula, M. P., & Djordjevic, S. P. (2014). Cilium adhesin P216 (MHJ_0493) is a target of ectodomain shedding and aminopeptidase activity on the surface of *Mycoplasma hyopneumoniae*. *Journal of Proteome Research, 13*, 2920-2930. https://doi.org/10.1021/pr500087c

Tadee, P., Chanrittisen, K., Thainoi, N., Thongkamkoon, P., Patchanee, P. (n.d.). Genetic diversity and global relationships of *Mycoplasma hyopneumoniae* from slaughter age pigs in Chiang Mai and Lamphun Provinces, Thailand, 10.

Tajima, M., & Yagihashi, T. (1982). Interaction of *Mycoplasma hyopneumoniae* with the porcine respiratory epithelium as observed by electron microscopy. *Infection and Immunity, 37*, 1162-1169.

Tajima, F., & Nei, M. (1984). Estimation of evolutionary distance between nucleotide sequences. *Molecular Biology and Evolution, 1*, 269-285.

Tajima, M., Yagihashi, T., Nunoya, T., Takeuchi, A., & Ohashi, F. (1984). Mycoplasma hyopneumoniae infection in pigs immunosuppressed by thymectomy and treatment with antithymocyte serum. *American Journal of Veterinary Research, 45*(10), 1928-1932.

Takahashi, K., Kuniyasu, C., Yoshida, Y., & Momotani, E. (1978). Sensitivity *in vitro* to macrolide antibiotics and tetracyclines of *Mycoplasma hyopneumoniae* isolated from porcine pneumonic lungs. *National Institute of Animal Health Quarterly (Tokyo), 18*, 41-42.

Takeda, K., Takeuchi, O., & Akira, S. (2002). Recognition of lipopeptides by Toll-like receptors. *Journal of Endotoxin Research, 8*(6), 459-463. doi:10.1179/096805102125001073

Takeuchi, O., & Akira, S. (2007). Signaling pathways activated by microorganisms. *Current Opinions in Cell Biology, 19*(2), 185-191.

Takeuti, K. L., de Barcellos, D. E. S. N., & Pieters, M. (2017a). *Mycoplasma hyopneumoniae* detection in nylon-flocked and rayon-bud swabs. *Journal of Microbiology Methods, 141*, 118-120. doi:10.1016/j.mimet.2017.08.007

Takeuti, K. L., de Barcellos, D. E. S. N., de Andrade, C. P., de Almeida, L. L., & Pieters, M. (2017b). Infection dynamics and genetic variability of *Mycoplasma hyopneumoniae* in self-replacement gilts. *Veterinary Microbiology, 208*, 18-24.

Takeuti, K. L., de Barcellos, D. E. S. N., de Lara, A. C., Kunrath, C. F., & Pieters, M. (2017c). Detection of *Mycoplasma hyopneumoniae* in naturally infected gilts over time. *Veterinary Microbiology, 203*, 215-220.

Tamiozzo, P., Lucchesi, P. M. A., & Ambrogi, A. (2013). Monitoring for *Mycoplasma hyopneumoniae* before and after a partial depopulation program using a typing scheme based on the polyserine repeat motif of p 146. *Journal of Swine Health and Production, 21*, 4.

Tamiozzo, P., Zamora, R., Lucchesi, P. M. A., Estanguet, A., Parada, J., Carranza, A., Camacho, P., & Ambrogi, A. (2015). MLVA typing of Mycoplasma hyopneumoniae bacterins and field strains. *Veterinary Record Open, 2*. https://doi.org/10.1136/vetreco-2015-000117

Tang, J., Hu, M., Lee, S., & Roblin, R. (2000). A polymerase chain reaction based method for detecting *Mycoplasma/Acholeplasma* contaminants in cell culture. *Journal of Microbiological Methods, 39*, 121-126.

Tani, K., Shimizu, T., Kida, Y., & Kuwano, K. (2011). Mycoplasma pneumoniae infection induces a neutrophil-derived antimicrobial peptide, cathelin-related antimicrobial peptide. *Microbiology and Immunology, 55*(8), 582-588. doi:10.1111/j.1348-0421.2011.00353.x

Tanner, A. C., Erickson, B. Z., & Ross, R. F. (1993). Adaptation of the Sensititre broth microdilution technique to antimicrobial susceptibility testing of *Mycoplasma hyopneumoniae. Veterinary Microbiology, 36*(3-4), 301-306.

Taranu, I., Marin, D. E., Bouhet, S., Pascale, F., Bailly, J. D., Miller, J. D., Pinton, P., & Oswald, I. P. (2005). Mycotoxin fumonisin B1 alters the cytokine profile and decreases the vaccinal antibody titter in pigs. *Toxicological Sciences, 84*, 301-307.

Tarigan, S., Slocombe, R. F., Browning, G. F., & Kimpton, W. (1994). Functional and structural changes of porcine alveolar macrophages induced by sublytic doses of a heat-labile, hemolytic, cytotoxic substance produced by *Actinobacillus pleuropneumoniae. American Journal Veterinary Research, 55*(11), 1548-1557.

Tasker, S., Helps, C. R., Day, M. J., Harbour, D. A., Shaw, S. E., Harrus, S. , Baneth, G., Lobetti, R. G., Malik, R., Beaufils, J. P., Belford, C. R., & Gruffydd-Jones, T. J. (2003). Phylogenetic analysis of *Hemoplasma* species: an international study. *Journal of Clinical Microbiology, 41*, 3877-3880.

Tassis, P. D., Papatsiros, V. G., Nell, T., Maes, D., Alexopoulos, C., Kyriakis, S. C., & Tzika, E. D. (2012). Clinical evaluation of intradermal vaccination against porcine enzootic pneumonia (*Mycoplasma hyopneumoniae*). *Veterinary Record, 170*, 261.

Tavio, M. M., Poveda, C., Assuncao, P., Ramirez, A. S., & Poveda, J. B. (2014). *In vitro* activity of tylvalosin against Spanish field strains of *Mycoplasma hyopneumoniae. Veterinary Record, 175*(21), 539.

Taylor, D. J. (2012). Miscellaneous bacterial infections. In J. J. Zimmerman, L. A. Karriker, A. Ramirez, K. J. Schwartz, & G. Stevenson (Eds.), *Diseases of swine* (10th ed.; pp. 866-881). Oxford: Wiley Blackwell.

Taylor, G., Taylor-Robinson, D., & Fernald, G. W. (1974). Reduction in the severity of *Mycoplasma pneumoniae*-induced pneumonia in hamsters by immunosuppressive treatment with antithymocyte sera. *Journal of Medical Microbiology,* 7(3), 343-348. doi:10.1099/00222615-7-3-343

Taylor, M. A., Ferrell, R. V., Wise, K. S., & McIntosh, M. A. (1988). Reiterated DNA sequences defining genomic diversity within the species *Mycoplasma hyorhinis*. *Molecular Microbiology, 2*, 665-672.

Taylor, M. A., Wise, K. S., & McIntosh, M. A. (1985). Selective detection of *Mycoplasma hyorhinis* using cloned genomic DNA fragments. *Infection and Immunity, 4*, 827-830.

Taylor-Robinson, D., & Bebear, C. (1997). Antibiotic susceptibilities of mycoplasmas and treatment of mycoplasmal infections. *Journal of Antimicrobial Chemotherapy, 40*(5), 622-630.

Ter Laak, E. A., Pijpers, A., Noordergraaf, J. H., Schoevers, E. C., & Verheijden, J. H. (1991). Comparison of methods for *in vitro* testing of susceptibility of porcine *Mycoplasma* species to antimicrobial agents. *Antimicrobial Agents Chemotherapy, 35*(2), 228-233.

Thacker, E. L. (2001). Immunology of the porcine respiratory disease complex. *Veterinary Clinics North American Food Animal Practice, 17*(3), 551-565.

Thacker, E. L. (2004). Diagnosis of *Mycoplasma hyopneumoniae*. *Journal of Swine Health and Production, 12*, 252-254.

Thacker, E. L. (2006). Mycoplasmal diseases. In B. E. Straw, J. J. Zimmerman, S. D'Allaire, & D. J. Taylor (Eds.), *Diseases of swine* (9th ed.; pp. 701-717). Oxford, UK: Blackwell Publishing Ltd.

Thacker, E. L., Halbur, P. G., Ross, R. F., Thanawongnuwech, R., & Thacker, B. J. (1999). *Mycoplasma hyopneumoniae* potentiation of porcine reproductive and respiratory syndrome virus-induced pneumonia. *Journal of Clinical Microbiology, 37*(3), 620-627.

Thacker, E. L., & Minion, F. C. (2012). Mycoplasmosis. In J. J. Zimmerman, L. A. Karriker, A. Ramirez, K. J. Schwartz, & G. Stevenson (Eds.), *Diseases of swine* (10th ed.; pp. 779-797). Ames: Wiley-Blackwell.

Thacker, E. L., Thacker, B. J., Boettcher, T. F., & Jayappa, H. (1998). Comparison of antibody production, lymphocyte stimulation, and protection induced by four commercial *Mycoplasma hyopneumoniae* bacterins. *Journal of Swine Health and Production, 6*, 107-112.

Thacker, E. L., Thacker, B. J., & Janke, B. H. (2001). Interaction between *Mycoplasma hyopneumoniae* and swine influenza virus. *Journal of Clinical Microbiology, 39*(7), 2525-2530.

Thacker, E. L., Thacker, B. J., Kuhn, M., Hawkins, P. A., & Waters, W. R. (2000a). Evaluation of local and systemic immune responses induced by intramuscular injection of a *Mycoplasma hyopneumoniae* bacterin to pigs. *American Journal of Veterinary Research, 61*, 1384-1389.

Thacker, E. L., Thacker, B. J., Young, T. F., & Halbur, P. G. (2000b). Effect of vaccination on the potentiation of porcine reproductive and respiratory syndrome virus (PRRSV)-induced pneumonia by *Mycoplasma hyopneumoniae*. *Vaccine, 18*, 1244-1252.

Thacker, E. L., Thacker, B. J., & Wolff, T. (2004). Efficacy of a chlortetracycline feed additive in reducing pneumonia and clinical signs induced by experimental *Mycoplasma hyopneumoniae* challenge. *Journal of Swine Health and Production, 14*, 140-144.

Thanawongnuwech, R., Thacker, B., Halbur, P., & Thacker, E. L. (2004). Increased production of proinflammatory cytokines following infection with porcine reproductive and respiratory syndrome virus and *Mycoplasma hyopneumoniae*. *Clinical Diagnostic and Laboratory Immunology, 11*(5), 901-908.

Thanawongnuwech, R., Young, T. F., Thacker, B., & Thacker, E. L. (2001). Differential production of proinfammatory cytokines: in vitro PRRSV and *Mycoplasma hyopneumoniae* co-infection model. *Veterinary Immunology and Immunopathology, 79*, 115-127.

Thomas, E., Grandemange, E., Pommier, P., Wessel-Robert, S., & Davot, J. L. (2000). Field evaluation of efficacy and tolerance of a 2% marbofloxacin injectable solution for the treatment of respiratory disease in fattening pigs. *The Veterinary Quarterly, 22*(3), 131-135.

Thomas, R. J. (2013). Particle size and pathogenicity in the respiratory tract. *Virulence, 4*(8), 847-858.

Thongkamkoon, P., Narongsak, W., Kobayashi, H., Pathanasophon, P., Kishima, M., & Yamamoto, K. (2013). *In vitro* susceptibility of *Mycoplasma hyopneumoniae* field isolates and occurrence of fluoroquinolone, macrolides and lincomycin resistance. *Journal of the Veterinary Medical Science, 75*(8), 1067-1070.

Tielen, M. J. M., Trujen, W. T., van der Groes, C. A. M., Verstegen, M. A. W., de Bruin, J. J. M., & Conbey, R. A. P. H. (1978). Conditions of management and the construction of piggeries on pig-fattening farms as fac-tors in the incidence of diseases of the lung and liver in slaughtered pigs. *Tijdschrift voor Diergeneeskunde, 103*, 1155-1165.

Timmerman, T., Dewulf, J., Catry, B., Feyen, B., Opsomer, G., de Kruif, A., & Maes, D. (2006). Quantification and evaluation of antimicrobial drug use in group treatments for fattening pigs in Belgium. *Preventive Veterinary Medicine, 74*(4), 251-263.

Tjørnehøj, K., Eriksen, L., Aalbaek, B., & Nansen, P. (1992). Interaction between *Ascaris suum* and *Pasteurella multocida* in the lungs of mice. *Parasitology Research, 78*(6), 525-528.

Tocqueville,V., Ferré, S., Nguyen, N. H., Kempf, I., & Marois-Créhan, C. (2014). Multilocus sequence typing of *Mycoplasma hyorhinis* strains identified by a real-time TaqMan PCR assay. *Journal of Clinical Microbiology, 52*, 1664-1671.

Tocqueville, V., Kempf, I., Paboeuf, F., & Marois-Créhan, C. (2017). Quantification of *Pasteurella multocida* in experimentally infected pigs using a real-time PCR assay. *Research in Veterinary Science, 112*, 177-184.

Tølbøll Lauritsen, K., Hagedorn-Olsen,T., Jungersen, G., Riber, U., Stryhn, H., Friis, N. F., Lind, P., & Kristensen, B. (2017). Transfer of maternal immunity to piglets is involved in early protection against *Mycoplasma hyosynoviae* infection. *Veterinary Immunology & Immunopathology, 183*, 22-30.

Toohill, E. (2017). Achieving day zero in large swine operations. Allen D. Leman Swine Conference, St. Paul, MN.

Toth, K. F., Harrison, N., & Sears, B. B. (1994). Phylogenetic relationships among members of the class *Mollicutes* deduced from rps3 gene sequences. *International Journal of Systematic Bacteriology, 44*, 119-124.

Toutain, P.-L., Bousquet-Mélou, A., Damborg, P., Ferran, A. A., Mevius, D., Pelligand, L.,Veldman, K. T., & Lees, P. (2017). En route towards European clinical breakpoints for veterinary antimicrobial susceptibility testing: A position paper explaining the VetCAST approach. *Frontiers in Microbiology, 8*, 2344.

Trüeb, B., Catelli, E., Luehrs, A., Nathues, H., & Kuhnert, P. (2014). Genetic variability and limited clonality of *Mycoplasma hyorhinis* in pig herds. *Veterinary Microbiology, 191*, 9-14.

Trueeb, B. S., Gerber, S., Maes, D., Gharib, W. H., & Kuhnert, P. (2019). Tn-sequencing of *Mycoplasma hyopneumoniae* and *Mycoplasma hyorhinis* mutant libraries reveals non-essential genes of porcine mycoplasmas differing in pathogenicity. *Veterinary Research, 50*, 55.

Tuovinen,V.K., Gröhn,Y.T., & Straw, B.E. (1997). Farrowing unit housing and management factors associated with diseases and disease signs of importance for feeder pig quality. *Acta Agriculturae Scandinavica Section A, Animal Science, 47*, 117-125.

Uilenberg, G., Thiaucourt, F., & Jongejan, F. (2004). On molecular taxonomy: What is in a name? *Experimental and Applied Acarology, 32*, 301-312.

Underdahl, N. R., Kennedy, G. A., & Ramos, A. S. Jr. (1980). Duration of *Mycoplasma hyopneumoniae* infection in gnotobiotic pigs. *Canadian Veterinary Journal, 21*(9), 258-256.

Van Alstine, W. G., Stevenson, G. W., & Kanitz, C. L. (1996). Porcine reproductive and respiratory syndrome virus does not exacerbate *Mycoplasma hyopneumoniae* infection in young pigs. *Veterinary Microbiology, 49*(3-4), 297-303.

van der Peet-Schwering, C. M. C., Kuijken, N., Raymakers, R., Stockhofe, N., van Leengoed, L. A. M. G., Binnendijk, G. P., Cruijsen, T., & Augustijn, M. (2008). Effect of less animal contacts on pleuritis in growing and finishing pigs. Animal Science Group von Wageningen UR, Rapport 111, March 2008.

van Duijkeren, E., Greko, C., Pringle, M., Baptiste, K. E., Catry, B., Jukes, H., Moreno, M. A., Pomba, M. C., Pyörälä, S., Rantala, M., Ružauskas, M., Sanders, P., Teale, C., Threlfall, E. J., Torren-Edo, J., & Törneke, K. (2014). Pleuromutilins: Use in food-producing animals in the European Union, development of resistance and impact on human and animal health. *Journal of Antimicrobial Chemotherapy*, *69*(8), 2022-2031.

Van't Klooster, C. E., & Heitlager, B. P. (1994). Determination of minimum ventilation rate in pig houses with natural ventilation based on carbon dioxide balance. *Journal of Agricultural Engineering Research*, *57*(4), 279-287.

Vangroenweghe, F., Karriker, L., Main, R., Christianson, E., Marsteller, T., Hammen, K., Bates, J., Thomas, P., Ellingson, J., Harmon, K., Abate, S., & Crawford, K. (2015a). Assessment of litter prevalence of *Mycoplasma hyopneumoniae* in preweaned piglets utilizing an antemortem tracheobronchial mucus collection technique and a real-time polymerase chain reaction assay. *Journal of Veterinary Diagnostic Investigation, 25*, 606-610.

Vangroenweghe F., Willems E., Thas O., & Maes D. (2018). Confirmation of *Mycoplasma hyopneumoniae* in a breeding herd through tracheo-bronchial swab sampling and PCR. *Veterinary Record, 183*. doi 10.1136/vr.104712

Vangroenweghe, F. A., Labarque, G. G., Piepers, S., Strutzberg-Minder, K., & Maes D. (2015b). *Mycoplasma hyopneumoniae* infections in peri-weaned and post-weaned pigs in Belgium and The Netherlands: Prevalence and associations with climatic conditions. *Veterinary Journal, 205*(1), 93-97.

Vangroenweghe, F. A., Suls, L., Van Driessche, E., Maes, D., & De Graef, E. (2012). Health advantages of transition to batch management system in farrow-to-finish pig herds. *Veterinarni Medicina, 57*(2), 83-91.

Vasconcelos, A., Ferreira, H., Bizarro, C., Bonatto, S., Carvalho, M., Pinto, P., & Zaha, A. (2015). Swine and poultry pathogens: The complete genome sequences of two strains of *Mycoplasma hyopneumoniae* and a strain of *Mycoplasma synoviae*. *Journal of Bacteriology, 187*, 5568-5577.

Vasconcelos, A. T. R., Ferreira, H. B., Bizarro, C.V., Bonatto, S. L., Carvalho, M. O., Pinto, P. M., et al. (2005). Swine and poultry pathogens: The complete genome sequences of two strains of *Mycoplasma hyopneumoniae* and a strain of *Mycoplasma synoviae*. *Journal of Bacteriology, 187*, 5568-5577. https://doi.org/10.1128/JB.187.16.5568-5577.2005

Vergnaud, G., & Pourcel, C. (2009). Multiple locus variable number of tandem repeats analysis. In D. A. Caugant (Ed.), *Molecular epidemiology of microorganisms: Methods and protocols, Methods in Molecular Biology™* (pp. 141-158). Totowa, NJ: Humana Press. https://doi.org/10.1007/978-1-60327-999-4_12

Vermeulen, L., Van Beirendonck, S., Bulens, A., Van Thielen, J., & Driessen, B. (2017). The perception about batch management production systems among pig producers. *Canadian Journal of Animal Sciences*, *97*, 109-117.

Vicca, J. (2005). *Virulence and antimicrobial susceptibility of* Mycoplasma hyopneumoniae *isolates from pigs.* PhD, Ghent University, 219 p.

Vicca, J., Maes, D., Jonker, L., de Kruif, A., & Haesebrouck, F. (2005). Efficacy of in-feed medication with tylosin for the treatment and control of *Mycoplasma hyopneumoniae* infections. *Veterinary Record, 156*(19), 606-610.

Vicca, J., Maes, D., Stakenborg, T., Butaye, P., Minion, F., Peeters, J., de Kruif, A., Decostere, A., & Haesebrouck, F. (2007). Resistance mechanism against fluoroquinolones in *Mycoplasma hyopneumoniae* field isolates. *Microbial Drug Resistance, 13*(3), 166-170.

Vicca, J., Stakenborg, T., Maes, D., Butaye, P., Peeters, J., de Kruif, A., & Haesebrouck, F. (2003). Evaluation of virulence of *Mycoplasma hyopneumoniae* field isolates. *Veterinary Microbiology, 97*, 177-190. https://doi.org/10.1016/j.vetmic.2003.08.008

Vicca, J., Stakenborg, T., Maes, D., Butaye, P., Peeters, J., de Kruif, A., & Haesebrouck, F. (2004). *In vitro* susceptibilities of *Mycoplasma hyopneumoniae* field isolates. *Antimicrobial Agents Chemotherapy, 48*(11), 4470-4472.

Vilalta, C., Sanhueza, J. M., Murray, D., Johnson, L., & Pieters, M. (2019). Detection of *Mycoplasma hyopneumoniae* in piglet processing fluids. *Veterinary Record*. doi:10.1136/vr.105475

Villarreal, I., Maes, D., Meyns, T., Gebruers, F., Calus, D., Pasmans, F., & Haesebrouck, F. (2009). Infection with a low virulent *Mycoplasma hyopneumoniae* isolate does not protect piglets against subsequent infection with a highly virulent *M. hyopneumoniae* isolate. *Vaccine, 27*(12), 1875-1879. doi:10.1016/j.vaccine.2008.12.005

Villarreal, I., Maes, D., Vranckx, K., Calus, D., Pasmans, F. &Haesebrouck, F. (2011b). Effect of vaccination of pigs against experimental infection with high and low virulence *Mycoplasma hyopneumoniae* strains. *Vaccine, 29*(9), 1731-1735. https://doi.org/10.1016/j.vaccine.2011.01.002

Villarreal, I., Meyns, T., Haesebrouck, F., Dewulf, J., Vranckx, K., Calus, D., Pasmans, F., & Maes, D. (2011a). Effect of vaccination against *Mycoplasma hyopneumoniae* on the transmission of *M. hyopneumoniae* under field conditions. *Veterinary Journal, 188*, 48-52.

Villarreal, I.,Vranckx, K., Calus, D., Pasmans, F., Haesebrouck, F., & Maes, D. (2012). Effect of challenge of pigs previously immunised with inactivated vaccines containing homologous and heterologous *Mycoplasma hyopneumoniae* strains. *BMC Veterinary Research, 8*, 2.

Villarreal, I., Vranckx, K., Duchateau, L., Pasmans, F., Haesebrouck, F., Jensen, J., Nanjiani, I., & Maes, D. (2010). Early *Mycoplasma hyopneumoniae* infections in European suckling pigs in herds with respiratory problems: Detection rate and risk factors. *Veterinary Medicine Czech, 55*, 318-324.

Vink, C., Rudenko, G., & Seifert, H.S. (2012). Microbial antigenic variation mediated by homologous DNA recombination. *FEMS Microbiology Reviews, 36*, 917-948. https://doi.org/10.1111/j.1574-6976.2011.00321.x

Virginio, V., Bandeira, N., Leal, F., Lancelotti, M., Zaha, A., & Ferreira, H. (2017). Assessment of the adjuvant activity of mesoporous silica nanoparticles in recombinant *Mycoplasma hyopneumoniae* antigen vaccines. *Heliyon, 3*(1), e00225. doi:10.1016/j.heliyon.2016.e00225. eCollection 2017.

Virginio, V., Gonschoroski, T., Paes, J., Schuck, D., Zaha, A., &. Ferreira, H. (2014). Immune responses elicited by *Mycoplasma hyopneumoniae* recombinant antigens and DNA constructs with potential for use in vaccination against porcine enzootic pneumonia. *Vaccine, 44*, 5832-5838.

Volokhov, D. V., Simonvan, V., Davidson, M. K., & Chizhikov, V. E. (2012). RNA polymerase beta subunit (*rpoB*) gene and the 16S-23S rRNA intergenic transcribed spacer region (ITS) as complementary molecular markers in addition to the 16S rRNA gene for phylogenetic analysis and identification of the species of the family *Mycoplasmataceae*. *Molecular Phylogenetics and Evolution, 62*, 515-528.

Vranckx, K., Maes, D., Calus, D., Villarreal, I., Pasmans, F., & Haesebrouck, F. (2011). Multiple-locus variable-number tandem-repeat analysis is a suitable tool for differentiation of *Mycoplasma hyopneumoniae* strains without cultivation. *Journal of Clinical Microbiology, 49*(5), 2020-2023.

Vranckx, K., Maes, D., Sacristán, R. D. P., Pasmans, F., & Haesebrouck, F. (2012a). A longitudinal study of the diversity and dynamics of *Mycoplasma hyopneumoniae* infections in pig herds. *Veterinary Microbiology, 156*(3-4), 315-321. https://doi.org/10.1016/j.vetmic.2011.11.007

Vranckx, K., Maes, D., Villarreal, I., Chiers, K., Pasmans, F., & Haesebrouck, F. (2012b). Vaccination reduces macrophage infiltration in bronchus-associated lymphoid tissue in pigs infected with a highly virulent *Mycoplasma hyopneumoniae* strain. *BMC Veterinary Research, 8*, 24.

Vyt, P., De Bie, S., & Kanora, A. (2004). Correlation between white liver spots and pneumonia: A slaughterhouse survey. Proceedings 18th International Pig Veterinary Society Congress, Hamburg, Germany, Volume 2, 582.

Waites, K. B., Duffy, L. B., Bebear, C. M., Matlow, A., Talkington, D. F., Kenny, G. E., Totten, P. A., Bade, D. J., Zheng, X., Davidson, M. K., Shortridge, V. D., Watts, J. L., & Brown, S. D. (2012). Standardized methods and quality control limits for agar and broth microdilution susceptibility testing of *Mycoplasma pneumoniae*, *Mycoplasma hominis*, and *Ureaplasma urealyticum*. *Journal of Clinical Microbiology, 50*(11), 3542-3547.

Walker, J., Lee, R., Mathy, N., Doughty, S., & Conlon, J. (1996). Restricted B-cell responses to microbial challenge of the respiratory tract. *Veterinary Immunology and Immunopathology, 54*(1-4), 197-204.

Wallgren, P., Bölske, G., & Fossum, C. (1992). In vitro stimulation of antibody production to Mycoplasma hyopneumoniae by porcine peripheral blood mononuclear cells. *Veterinary Microbiology, 32*(3-4), 363-374.

Wallgren, P., Bölske, G., Gustafsson, S., Mattsson, S., & Fossum, C. (1998). Humoral immune responses to *Mycoplasma hyopneumoniae* in sows and offspring following an outbreak of mycoplasmosis. *Veterinary Microbiology, 60*(2-4), 193-205.

Wallgren, P., Nörregård, E., Molander, B., Persson, M., & Ehlorsson, C.J. (2016). Serological patterns of *Actinobacillus pleuropneumoniae, Mycoplasma hyopneumoniae, Pasteurella multocida* and *Streptococcus suis* in pig herds affected by pleuritis. *Acta Veterinaria Scandinavica, 58*(1), 71.

Wallgren, P., Vallgårda, J., Lindberg, M., Eliasson-Selling, L., &. Segall, T. (2000). The efficacy of different vaccination strategies against *Mycoplasma hyopneumoniae*. Proceedings 16th IPVS Congress, Melbourne, Australia, 461.

Wallgren, P., Wilén, I.-L., &. Fossum, C. (1994). Influence of experimentally induced endogenous production of cortisol on the immune capacity in swine. *Veterinary Immunology and Immunopathology, 42*, 301-316.

Wang, H., Feng, Z., Wu, Y., Wei, Y., Gan, Y., Hua, L., Li, B., Wang, X., Liu, M., Xiong, Q., & Shao, G. (2016a). The effects of *Mycoplasma hyopneumoniae* on porcine circovirus type 2 replication in vitro PK-15 cells. *Research in Veterinary Science, 105*, 56-61.

Wang, X., Chen, X., Tang, H., Zhu, J., Zhou, S., Xu, Z., Liu, F., & Su, C. (2016b). Increased frequency of Th17 cells in children with *Mycoplasma pneumoniae* pneumonia. *Journal of Clinical Laboratory Analysis, 30*(6), 1214-1219. doi:10.1002/jcla.22005

Washburn, L. R., Miller, E. J., & Weaver, K. E. (2000). Molecular characterization of *Mycoplasma arthritidis* membrane lipoprotein MAA1. *Infection and Immunity, 68*, 437-442.

Wattam, A. R., Davis, J. J., Assaf, R., Boisvert, S., Brettin, T., Bun, C., Conrad, N., Dietrich, E. M., Disz, T., Gabbard, J. L., Gerdes, S., Henry, C. S., Kenyon, R. W., Machi, D., Mao, C., Nordberg, E. K., Olsen, G. J., Murphy-Olson, D. E., Olson, R., Overbeek, R., Parrello, B., Pusch, G. D., Shukla, M., Vonstein, V., Warren, A., Xia, F., Yoo, H., & Stevens, R. L. (2017). Improvements to PATRIC, the all-bacterial Bioinformatics Database and Analysis Resource Center. *Nucleic Acids Research, 45*, D535-D542. https://doi.org/10.1093/nar/gkw1017

Webster, A. D., Furr, P. M., Hughes-Jones, N. C., Gorick, B. D., & Taylor-Robinson, D. (1988). Critical dependence on antibody for defence against mycoplasmas. *Clinical & Experimental Immunology, 71*(3), 383-387.

Weisburg, W. G., Tully, J. G., Rose, D. L., Petzel, J. P., Oyaizu, H., Yang, D., Mandelco, L. Sechrest, J., Lawrence, T. G., Van Etten, J., Maniloff, J., & Woese, C. R. (1989). A phylogenetic analysis of the mycoplasmas: basis for their classification. *Journal of Bacteriology, 171*, 6455-6467.

Weissenbacher-Lang, C., Kureljusic, B., Nedorost, N., Matula, B., Schiessl, W., Stixenberger, D., & Weissenbock, H. (2016). Retrospective analysis of bacterial and viral co-infections in *Pneumocystis* spp. positive lung samples of Austrian pigs with pneumonia. *PLoS One, 11*(7), e0158479.

Weldearegay, Y. B., Pich, A., Schieck, E., Gicheru, N., Wesonga, H. O., Thiaucourt, F., Kiirika, L. M., Valentin-Weigand, P., Jores, J., & Meens, J. (2016). Proteomic characterization of pleural effusion, a specific host niche of *Mycoplasma mycoides* subsp. *mycoides* from cattle with contagious bovine pleuropneumonia (CBPP). *Journal of Proteomics, 131*, 93-103.

Whittlestone, P. (1957). *Enzootic pneumonia of pigs and related conditions.* PhD thesis, Magdelane College, University of Cambridge.

Whittlestone, P. (1973). Enzootic pneumonia of pigs (EPP). *Advances in Veterinary Science and Comparative Medicine, 17*, 1-55.

Whittlestone, P. (1976). Effect of climatic conditions on enzootic pneumonia of pigs. *International Journal of Biometeorology, 20*(1), 42-48.

Wick, W. E. (1964). Influence of antibiotic stability on the results of *in vitro* testing procedures. *Journal of Bacteriology, 87*(5), 1162-1170.

Widjaja, M., Harvey, K. L., Hagemann, L., Berry, I., Jarocky, V., Raymond, B., Tacchi, J., Gründel, A., Steele, J., Padula, M., Charles, I., Dumke, R., & Djordjevic, S. (2017). Elongation factor Tu is a multifunctional and processed moonlighting protein. *Scientific Reports, 7*, 11227.

Williams, P. P. (1978). *In vitro* susceptibility of *Mycoplasma hyopneumoniae* and *Mycoplasma hyorhinis* to fifty-one antimicrobial agents. *Antimicrobial Agents and Chemotherapy, 14*(2), 210-213.

Wilson, S., Van Brussel, L., Saunders, G., Runnels, P., Taylor, L., Fredrickson, D., & Salt, J. (2013). Vaccination of piglets up to 1 week of age with a single-dose *Mycoplasma hyopneumoniae* vaccine induces protective immunity within 2 weeks against virulent challenge in the presence of maternally derived antibodies. *Clinical and Vaccine Immunology, 20*(5), 720-724. doi:10.1128/cvi.00078-13

Wilton, J., Jenkins, C., Cordwell, S. J., Falconer, L., Minion, F. C., Oneal, D. C., Djordjevic, M. A., Connolly, A., Barchia, I., Walker, M. J., & Djordjevic, S. P. (2009). Mhp493 (P216) is a proteolytically processed, cilium and heparin binding protein of *Mycoplasma hyopneumoniae. Molecular Microbiology, 71*, 566-582. https://doi.org/10.1111/j.1365-2958.2008.06546.x

Wilton, J. L., Scarman, A. L., Walker, M. J., & Djordjevic, S. P. (1998). Reiterated repeat region variability in the ciliary adhesin gene of *Mycoplasma hyopneumoniae. Microbiology, 144*, 1931-1943. https://doi.org/10.1099/00221287-144-7-1931

Windsor, R. S. (1977). Meningitis in pigs caused by *Streptococcus suis* type II. *Veterinary Record, 101*, 378-379.

Wise, K. S., & Kim, M. F. (1987a). Major membrane surface proteins of *Mycoplasma hyopneumoniae* selectively modified by covalently bound lipid. *Journal of Bacteriology, 169*, 5546-5555.

Wise, K. S., & Kim, M. F. (1987b). Identification of intrinsic and extrinsic membrane proteins bearing surface epitopes of *Mycoplasma hyopneumoniae. Israel Journal of Medical Sciences, 23*, 469-473.

Wise, K. S., Yogev, D., & Rosengarten, R. (1992). Antigenic variation, p473-489. In J. Maniloff, R. N. McElhaney, L. R. Finch, & J. B. Baseman (Eds.), *Mycoplasmas, molecular biology and pathogenesis* Washington, D.C.: ASM.

Woese, C. R, Maniloff, J., & Zablen, L. B. (1980). Phylogenetic analysis of the mycoplasmas. Proceedings National Academy of Science, USA, 77, 494-498.

Woolard, M. D., Hodge, L. M., Jones, H. P., Schoeb, T. R., & Simecka, J. W. (2004). The upper and lower respiratory tracts differ in their requirement of IFN-gamma and IL-4 in controlling respiratory mycoplasma infection and disease. *Journal of Immunology, 172*(11), 6875-6883.

Woolley, L. K., Fell, S. A., Djordjevic, S. P., Eamens, G. J., & Jenkins, C. (2013). Plasmin activity in the porcine airways is enhanced during experimental infection with *Mycoplasma hyopneumoniae*, is positively correlated with proinflammatory cytokine levels and is ameliorated by vaccination. *Veterinary Microbiology, 164*(1-2), 60-66. doi:10.1016/j.vetmic.2013.02.003

Woolley, L. K., Fell, S. A., Gonsalves, J. R., Raymond, B. B. A., Collins, D., Kuit, T. A., Walker, M. J., Djordjevic, S. P., Eamens, G. J., & Jenkins, C. (2014). Evaluation of recombinant *Mycoplasma hyopneumoniae* P97/P102 paralogs formulated with selected adjuvants as vaccines against mycoplasmal pneumonia in pigs. *Vaccine, 32*, 4333-4341. https://doi.org/10.1016/j.vaccine.2014.06.008

Woolley, L. K., Fell, S., Gonsalves, J. R., Walker, M. J., Djordjevic, S. P., Jenkins, C., & Eamens, G. J. (2012). Evaluation of clinical, histological and immunological changes and qPCR detection of *Mycoplasma hyopneumoniae* in tissues during the early stages of mycoplasmal pneumonia in pigs after experimental challenge with two field isolates. *Veterinary Microbiology, 161*, 186-195. https://doi.org/10.1016/j.vetmic.2012.07.025

World Health Organization. (2015). Global action plan on antimicrobial resistance. https://www.who.int/antimicrobial-resistance/publications/global-action-plan/en/

Wu, C. C., Shryock, T. R., Lin, T. L., Faderan, M., & Veenhuizen, M. F. (2000). Antimicrobial susceptibility of *Mycoplasma hyorhinis*. *Veterinary Microbiology, 76*, 25-30.

Wu, J., Yu, J., Song, C., Sun, S., & Wang, Z. (2006). Porcine eperythrozoonosis in China. *Annals of the New York Academy of Sciences, 1081*, 280-285.

Wyrsch, E., Roy Chowdhury, P., Abraham, S., Santos, J., Darling, A. E., Charles, I. G., Chapman, T. A., & Djordjevic, S. P. (2015). Comparative genomic analysis of a multiple antimicrobial resistant enterotoxigenic E. coli O157 lineage from Australian pigs. *BMC Genomics, 16*, 165. https://doi.org/10.1186/s12864-015-1382-y

Xiong, Q., Wang, J., Ji, Y., Ni, B., Zhang, B., Ma, Q., Wei, Y., Xiao, S., Feng, Z., Liu, M., & Shao, G. (2016a). The functions of the variable lipoprotein family of *Mycoplasma hyorhinis* in adherence to host cells. *Veterinary Microbiology, 186*, 82-89.

Xiong, Q., Wie, Y., Feng, W., Gan, Y., Liu, Z., Liu, M., Bai, F., & Shao, G. (2014a). Protective efficacy of a live attenuated *Mycoplasma hyopneumoniae* vaccine with an ISCOM-matrix adjuvant in pigs. *Veterinary Journal, 199*, 268-274.

Xiong, Q., Wie, Y., Xie, H., Feng, Z., Gan, Y., Wang, C., Liu, M., Bai, F., Xie, F., & Shao, G. (2014b). Effect of different adjuvant formulations on the immunogenicity andprotective effect of a live *Mycoplasma hyopneumoniae* vaccine after intramuscular inoculation. *Vaccine, 32*, 3445-3451.

Xiong, Q., Zhang, B., Wang, J., Ni, B., Ji, Y., Wei, Y., Xiao, S., Feng, Z., Liu, M., & Shao, G. (2016b). Characterization of the role in adherence of *Mycoplasma hyorhinis* variable lipoproteins containing different repeat unit copy numbers. *Veterinary Microbiology, 197*, 39-46.

Yagihashi, T., Nunoya, T., Mitui, T., & Tajima, M. (1984). Effect of *Mycoplasma hyopneumoniae* infection on the development of *Haemophilus pleuropneumoniae* pneumonia in pigs. *Japanese Journal of Veterinary Science, 46*(5), 705-713.

Yang, D., Tai, X., Qiu, Y., & Yun, S. (2000). Prevalence of *Eperythrozoon spp.* infection and congenital eperythrozoonosis in humans in Inner Mongolia, China. *Epidemiology and Infection, 125*, 421-426.

Yavlovich, A., Tarshis, M., & Rottem, S. (2004). Internalization and intracellular survival of *Mycoplasma pneumoniae* by non-phagocytic cells. *FEMS Microbiology Letters, 233*, 241-246.

Yazawa, S., Okada, M., Ono, M., Fujii, S., Okuda, Y., Shibata, I., & Kida, H. (2004). Experimental dual infection of pigs with an H1N1 swine influenza virus (A/Sw/Hok/2/81) and *Mycoplasma hyopneumoniae*. *Veterinary Microbiology, 98*(3-4), 221-228.

Yeske, P. (2002). Economics of an off-site breeding project. Proceedings IPVS, Ames, Iowa, 188.

Yeske, P. (2007). Mycoplasma eradication strategies. Proceedings AASV, Orlando, Florida, 367-370.

Yeske, P. (2008). Mycoplasma eradication from an acute outbreak using DRAXXIN® (tulathromycin). Proceedings IPVS, Durban, South Africa.

Yeske, P. (2010). Summary of *Mycoplasma hyopneumoniae* eradication experiences completed in upper Midwest USA. Proceedings IPVS, Vancouver, Canada, 141.

Yeske, P. (2017a). Intentional *Mycoplasma hyopneumoniae* exposure. James D. McKean Swine Disease Conference.

Yeske, P. (2017b). Assessment of The Likelihood of *Mycoplasma hyopneumoniae* Lateral Transmission. Proceedings Allen D. Leman Swine Conference, St. Paul, MN.

Yeske, P. (2019). What's next for disease elimination: *Mycoplasma hyopneumoniae*, PRRS, PED, or ileitis? Proceedings AASV, Orlando, Florida.

Yeske, P., Valeris-Chacin, R., Singer, R. S., & Pieters, M. (2020). Survival analysis of two *Mycoplasma hyopneumoniae* eradication methods. *Preventive Veterinary Medicine, 174*, 104811.

Yim, S. H., Hahn, T. W., & Joo, H. G. (2017). Bordetella bronchiseptica antigen enhances the production of *Mycoplasma hyopneumoniae* antigen-specific immunoglobulin G in mice. *Journal of Veterinary Science, 18*(3), 327-332.

Yogev, D., Watson-McKown, R., Rosengarten, R., Im, J., & Wise K. S. (1995). Increased structural and combinatorial diversity in an extended family of genes encoding Vlp surface proteins of *Mycoplasma hyorhinis*. *Journal of Bacteriology, 177*, 5636-5643.

You, X. X., Zeng, Y. H., & Wu, Y. M. (2006). Interactions between mycoplasma lipid-associated membrane proteins and the host cells. *Journal of Zhejiang University Science B, 7*, 342-350.

Yu, Y., Liu, M., Hua, L., Qiu, M., Zhang, W., Wei, Y., Gan, Y., Feng, Z., Shao, G., & Xiong, Q. (2018a). Fructose-1,6-bisphosphate aldolase encoded by a core gene of *Mycoplasma hyopneumoniae* contributes to host cell adhesion. *Veterinary Research, 49*, 114.

Yu, Y., Wang, H., Wang, J., Feng, Z., Wu, M., Liu, B., Xin, J., Xiong, Q., Liu, M., & Shao, G. (2018b). Elongation Factor Thermo Unstable (EF-Tu) moonlights as an adhesin on the surface of *Mycoplasma hyopneumoniae* by binding to fibronectin. *Frontiers in Microbiology, 9*, 974.

Yuan, C. L., Liang, A. B., Yao, C. B., Yang, Z. B., Zhu, J. G., Cui, L., Yu, F., Zhu, N. Y., Yang, X. W., & Hua, X. G. (2009). Prevalence of *Mycoplasma suis* (*Eperythrozoon suis*) infection in swine and swine-farm workers in Shanghai, China. *American Journal of Veterinary Research, 70*, 890-894.

Zachary, J. F., & Basgall, E. J. (1985). Erythrocyte-membrane alterations associated with the attachment and replication of *Eperythrozoon suis* - a light and electronmicroscopic study. *Veterinary Pathology, 22*, 164-170.

Zachary, J. F., & Smith, A. R. (1985). Experimental porcine eperythrozoonosis - Lymphocyte-T suppression and misdirected immune responses. *American Journal of Veterinary Research, 46*, 821-830.

Zhang, C. Y., Li, Y. F., Jiang, P., & Chen, W (2012). Use of MSG1 protein in a novel blocking ELISA for the detection of *Mycoplasma suis* infection. *Veterinary Journal, 193*, 535-538.

Zhang, H., Lunney, J. K., Baker, R. B., & Opriessnig, T. (2011). Cytokine and chemokine mRNA expression profiles in tracheobronchial lymph nodes from pigs singularly infected or coinfected with porcine circovirus type 2 (PCV2) and *Mycoplasma hyopneumoniae* (MHYO). *Veterinary Immunology and Immunopathology, 140*(1-2), 152-158.

Zhang, Q., Young, T. F., Ross, R. F (1995). Identification and characterization of a *Mycoplasma hyopneumoniae* adhesin. *Infection and Immunology, 63*, 1013-1019.

Zhang, Y. (2004). *Indoor air quality engineering*. Boca Raton, Florida: CRC Press.

Zhongyang, L., Jiansong, Z., Yijuan, S., Yuting, X., Yufeng, L., & Jiarong, X. (2017). Seroprevalence of *Mycoplasma suis* infection in pigs in eastern China as estimated by a blocking enzyme-linked immunosorbent assay. *Canadian Journal of Veterinary Research, 81*, 313-317.

Zhou, R. Q., Nie, K., Huang, H. C., Hu, S. J., Zhou, Z. Y., & Luo, H. L. (2009). Phylogenetic analysis of *Mycoplasma suis* isolates based on 16S rRNA gene sequence in China. *Veterinary Research Communication, 33*, 855-863.

Zielinski, G. C., & Ross, R. F. (1993). Adherence of *Mycoplasma hyopneumoniae* to porcine ciliated respiratory tract cells. *American Journal of Veterinary Research, 54*(8), 1262-1269.

Zielinski, G. C., & Ross, R. F. (1990). Effect of growth in cell cultures and strain on virulence of *Mycoplasma hyopneumoniae* for swine. *American Journal of Veterinary Research, 51*, 344-348.

Zimmerman, B. J., & Ross, R. F. (1982). Antibody response of swine experimentally infected with *Mycoplasma hyosynoviae*. *Veterinary Microbiology, 7*, 135-146.

Zinkernagel, R. M. (2012). Immunological memory ≠ *protective immunity. Cellular and Molecular Life Sciences, 69*(10),1635-1640.

Zinn, G. M., Jesse, G. W., & Dobson, A. W. (1983). Effect of eperythrozoonosis on sow productivity. *Journal of the American Veterinary Medical Association, 182*, 36-371.

Zou, H., Liu, X., Ma, F., Chen, P., Zhou, R., &. He, Q. (2011). Attenuated *Actinobacillus pleuropneumoniae* as a bacterial vector for expression of *Mycoplasma hyopneumoniae* P36 gene. *Journal of Gene Medicine, 13*, 221-229.

Zuo, L. L., Wu, Y. M., & You, X. X. (2009). Mycoplasma lipoproteins and Toll-like receptors. *Journal of Zhejiang University Science B, 10*, 67-76.